Ernst Cassirer and the Critical Science of Germany, 1899–1919

ADVANCE PRAISE

"The increasing interest in Cassirer makes this masterful work necessary reading. By putting Cassirer in the context of the project of the Marburg school and his relation to Leibniz, Moynahan brings out the depth and consistency in Cassirer's political and social thought, as well as its relation to his technically demanding early philosophy. It will change the way social theorists draw on Cassirer."

—*John Levi Martin, Professor of Sociology, University of Chicago*

"Moynahan's study offers deeply researched new insights on Cassirer's philosophical development. Moynahan shows that Cassirer must be understood from the beginning of his career as an independent mind who combined deep-rootedness in Marburg neo-Kantianism with the ability to theorize and reformulate innovations in the natural sciences at the beginning of the twentieth century. Cassirer's philosophical writings up to the end of his life were defined by the same combination. By taking careful account of Cassirer's early work, Moynahan's work rehabilitates Cassirer as one of most important philosophers of the last 100 years."

—*Thomas Meyer, LMU Munich, and author of "Ernst Cassirer: Eine Biographie"*

"Gregory Moynahan has written an important book on a thinker whose voice we need to bring back into the conversation of critical theory. It is in Ernst Cassirer's early works and connection with the challenging philosophy of his teacher Hermann Cohen, Moynahan demonstrates, that his relevance for contemporary debate is particularly evident. This erudite and well-argued text at once illuminates the pre–World War One reform project of the Marburg school and suggests its continued significance in Cassirer's work."

—*Drucilla Cornell, Professor of Political Science, Comparative Literature, and Women's and Gender Studies, Rutgers University*

"The last decade has witnessed a second renaissance in the scholarship on the philosopher Ernst Cassirer. With its unusual emphasis on the often-misunderstood early phase of Cassirer's development, Gregory Moynahan's book is an original and stimulating contribution to the recent literature."

—*Peter E. Gordon, Professor of History, Harvard University, and author of "Continental Divide: Heidegger, Cassirer, Davos"*

Ernst Cassirer and the Critical Science of Germany, 1899–1919

Gregory B. Moynahan

ANTHEM PRESS
LONDON • NEW YORK • DELHI

Anthem Press
An imprint of Wimbledon Publishing Company
www.anthempress.com

This edition first published in UK and USA 2014
by ANTHEM PRESS
75–76 Blackfriars Road, London SE1 8HA, UK
or PO Box 9779, London SW19 7ZG, UK
and
244 Madison Ave #116, New York, NY 10016, USA

First published in hardback by Anthem Press in 2013

Copyright © Gregory B. Moynahan 2014

The author asserts the moral right to be identified as the author of this work.

All rights reserved. Without limiting the rights under copyright reserved above,
no part of this publication may be reproduced, stored or introduced into
a retrieval system, or transmitted, in any form or by any means
(electronic, mechanical, photocopying, recording or otherwise),
without the prior written permission of both the copyright
owner and the above publisher of this book.

British Library Cataloguing-in-Publication Data
A catalogue record for this book is available from the British Library.

Library of Congress Cataloging-in-Publication Data
The Library of Congress has catalogued the hardcover edition as follows:
Moynahan, Gregory B., 1966–
Ernst Cassirer and the critical science of Germany : 1899–1919 / Gregory B. Moynahan.
pages cm
Includes bibliographical references and index.
ISBN 978-0-85728-321-4 (hardcover : alk. paper)
1. Cassirer, Ernst, 1874–1945. 2. Marburg school of philosophy. 3. Science–Philosophy. 4.
Science–Germany–History–20th century. I. Title.
B3216.C34M69 2013
193–dc23
2013016572

ISBN-13: 978 1 78308 343 5 (Pbk)
ISBN-10: 1 78308 343 3 (Pbk)

Cover illustration from Bruno Taut's *Alpine Architektur*, 1919.
Image provided courtesy of the Ryerson and Burnham Archives at the
Art Institute of Chicago.

This title is also available as an ebook.

Dedicated with love to my parents
Patricia Buchanan Moynahan and
John Francis Moynahan

CONTENTS

Acknowledgments ix

List of Abbreviations xiii

Introduction   "Reading a Mute History": Ernst Cassirer, the Marburg School and the Crises of Modern Germany   xvii

Part I   **The Marburg School and the Politics of Science in Germany**

Chapter One   The Twentieth-Century Conflict of the Faculties: The Marburg School and the Reform of the Sciences   3

Chapter Two   Cassirer and the Marburg School in the Administrative and Political Context of the *Kaiserreich*   27

Chapter Three   "The Supreme Principles of Knowledge": Cassirer's Transformation of the Tenets of Cohen's *Infinitesimal Method* (1882) and *System of Philosophy* (1902–1912)   45

Part II   **Critical Science and Modernity**

Chapter Four   Leibniz and the Foundation of Critical Science: *Leibniz's System in its Scientific Foundations* (1902)   85

Chapter Five   Science and History in Cassirer's *Substance and Function* (1910)   121

Part III   **Liberal Democracy and Law**

Chapter Six   Liberalism and the Conflict of Forms: *The Knowledge Problem* (1906–1940) and *Freedom and Form* (1916)   159

Chapter Seven   Law as Science and the "Coming-into-Being" of Natural Right in Cohen, Cassirer and Kelsen   193

Conclusion   Critical Science, the Future of Humanity and the Riddle of *An Essay on Man* (1944)   209

Index   221

ACKNOWLEDGMENTS

The earliest inspiration for this book came from my first graduate advisor, Amos Funkenstein. Having asked him which texts should be on my reading examinations in early modern European history he mentioned a few titles and then paused and added, "…and everything by Ernst Cassirer." When I noted that this entailed quite a bit of material for an exam that was only months away, he smiled and shrugged before limiting his recommendation to just the *Erkenntnisproblem* series and Cassirer's historical trilogy from the Weimar period that ended with the *Philosophy of the Enlightenment*. Sadly, soon after I left for archival research on a planned dissertation on early modern intellectual history, he died – suddenly, it seemed to his students and friends – after a short battle with lung cancer. Bereft of both an advisor and, I thought, a viable project without him, I chanced upon a further great teacher in John Michael Krois, at the Humboldt University, Berlin. Not surprisingly, the larger Cassirer reading plan was gradually revitalized. My passion for modern thought and indeed the problem of modernity, long developed in classes with Martin Jay at Berkeley, suddenly came to the fore. Thanks to Martin Jay's gracefully agreeing to take me on as an advisee and John Krois's local guidance, I was soon able to develop a workable dissertation project on the understanding of culture and "physiognomic immediacy" in Cassirer and Simmel. Martin Jay's superb guidance on the dissertation from a position somewhat skeptical of Cassirer's project was invaluable, and his particular combination of erudition and clarity remains for me unparalleled. Despite my indirect route, I thus ended upon having not just one but three of the best (and, as it turned out, most complementary) advisors any graduate student could wish for. If I can only occasionally do justice to their range of scholarship, I nonetheless always strive to bring their passion and curiosity to each moment of teaching and research.

The present book developed from core questions left unanswered in my dissertation, which I realized could only be grasped by leaving Cassirer's late work and returning to his earliest. Here too John Krois was of enormous assistance, and it was with great sadness that, soon after meeting with him in Berlin and providing an earlier version of this manuscript, I learned of his death. I am sure other friends and students will be familiar with his remarkable generosity of spirit and time, which always seemed all the more impressive given his prodigious and multifaceted academic endeavors.

I was very fortunate to have several close readers of the present work, most notably early on my colleague Roger Berkowitz at Bard College and more recently Peter E. Gordon at Harvard University. Both read through the manuscript with remarkable care and insight, and their comments fundamentally changed the direction of the text. Thomas Meyer of the Ludwig Maximilian University of Munich generously read selections of the text with his characteristic perceptiveness. The final chapter of this text benefited from discussion at the New York Area Seminar in Intellectual and Cultural History, where I would like to thank in particular Jerrold Siegel, Richard Wolin and Samuel Moyn for their comments. The same chapter appeared in the Festschrift for Martin Jay, *The Modernist Imagination: Intellectual History and Critical Theory*, and I am proud to do Martin Jay justice by now having a much more complete context in which it can be read. Of course, all shortcomings in the manuscript are my own.

A number of friends from reading groups have taken part in developing the ideas presented here. At Berkeley, an informal reading group formed around Amos Funkenstein's seminars and introduced me to a wide array of medieval and early modern thought, which was of great assistance in understanding Cassirer and the Marburg school. In retrospect these discussions with Julian Bourg, Dallas Denery, Carina Johnson, Isaac Miller, Michael Witmore and Jonathan Sheehan were a high point of my education – in my memory of them the fog is always dramatically rolling into San Francisco, the coffee perfect and Nicholas of Cusa is in the air. Also at Berkeley, Thomas Brady, Hans Sluga and Randolf Starn all provided invaluable assistance in developing earlier versions of the ideas presented here. Yehuda Elkana kindly shared his ideas on his own work on Cassirer over lunch in Berlin and noted, to my surprise, that my own projects were a logical development of my education since Amos Funkenstein "was a Cassireran" – both a concept and a category I had curiously never considered. At Bard College, a smaller writing group formed that was integral to the development of the present work, a collaboration with the ever insightful Laura Kunreuther and Julia Rosenbaum. I have further benefited from conversations with friends and colleagues at Bard including Thomas Bartscherer, Tabetha Ewing, Alice Stroup and Marina van Zuylen. Leon Botstein and David Kettler read selections from the manuscript and generously provided insight and references. Further friends who have been particularly helpful interlocutors over the years include Susan Bernofsky, Rita Chin, Lisa Cody, David Eng, Willfried Gessner, Jacqueline Goss, Haejeong Hazel Hahn, Galen Joseph, Janine Ludwig, Anne McKnight, John V. Maciuika, Elliot Neaman, Katherine Pence, Jing Tsu and Deborah Zafman.

The research contained in this work is indebted to generous fellowships from the Deutsche Akademische Austausch Dienst and the Alexander von Humboldt Stiftung Bundezkanzler-Stipendium, the latter of which was kindly given for both the earlier dissertation and a second round of research for the present book. A sabbatical leave from Bard College was instrumental to completing this work. I benefited from the help of librarians at the Beinecke Rare Book and Manuscript Library, Yale University, the Berlin State Library, the Humboldt University archives and the philosophy libraries at the Freie University and the Humboldt University. Particular thanks to Virginia Cassirer Veach, who graciously allowed me to visit her and peruse the remains of Ernst Cassirer's library and the family photographs.

I would like to thank the anonymous readers for Anthem Press as well as the earlier Martin Jay *Modernist Imagination* Festschrift. At Anthem Press, I would particularly like to thank Brian Stone, Janka Romero, Tej P. S. Sood and my editor Rob Reddick.

ACKNOWLEDGMENTS

This book is dedicated to my parents, Patricia Buchanan Moynahan and John Francis Moynahan, in admiration of their combination of humor, intelligence and humanity. My wife, Danielle Riou, has been a constant source of insight into the wider world of politics and ideas beyond the Marburg school, as well as a witty and passionate commentator on life more generally. She generously read through and perspicaciously commented on several chapters of this project. Combined with the joyful and often hilarious presence of our sons Henry and Philip, I have been blessed with a rich life outside of the demanding requirements of writing, and they have been a constant source of strength despite these demands. All of the words of innumerable books, many much more poetic, could not express my love adequately.

LIST OF ABBREVIATIONS

Works by Ernst Cassirer, arranged alphabetically by abbreviation

German texts are derived from the *Ernst Cassirer Werke* (Hamburg edition, issued by Felix Meiner Publishing since 1998, abbreviated here as ECW), and from the *Ernst Cassirer Nachgelassene Manuskripte und Texte* (Hamburg edition, issued by Felix Meiner since 1995, abbreviated here as ECN). All translations from ECW/ECN and other German texts are my own unless indicated. Known English translations of works have been used where possible, with the translation occasionally modified in relation to the German in ECW/ECN. In the interest of ready intelligibility, major works have been abbreviated by their title name rather than the corresponding edition in ECW/ECN, so that, for instance, *Freiheit und Form* is abbreviated as 'FF' instead of 'ECW7.' Frequently used shorter works are listed below both by their location in ECW/ECN and in their original source for the convenience of those without access to the collected works. Original publication dates of earlier editions are provided where applicable.

DI   *Determinism and Indeterminism in Modern Physics: Historical and Systematic Studies of the Problem of Causality.* Translated by O. Theodor Benfey (New Haven: Yale University Press, 1956 [1936]). ECW19.

DV   Cassirer, Ernst and Heidegger, Martin. "Appendix IV – Davos Disputation Between Ernst Cassirer and Martin Heidegger." In Heidegger, Martin. *Kant and the Problem of Metaphysics*, fifth edition. Translated by Richard Taft (Bloomington: University of Indiana Press, 1997), 193–207.

E   *An Essay on Man: An Introduction to a Philosophy of Human Culture* (New Haven: Yale University Press, 1965). ECW 23.

EGD   "Erkenntnistheorie nebst den Grenzfragen der Logik und Denkpsychologie." In *Jahrbücher der Philosophie, 1: Eine kritische Übersicht der Philosophie der Gegenwart* (Berlin: E. S. Mittler, 1927), 31–92. EWC19, 139ff.

EGL   "Erkenntnistheorie nebst den Grenzfragen der Logik." In *Jahrbücher der Philosophie* 3 (Berlin: E. S. Mittler, 1913), 1–59. ECW17, 13ff.

EK1   *Das Erkenntnisproblem in der Philosophie und Wissenschaft der neueren Zeit – Erster Band* (Berlin: Bruno Cassirer, 1906, 1911, 1922). ECW2.

EK2   *Das Erkenntnisproblem in der Philosophie und Wissenschaft der neueren Zeit – Zweiter Band* (Berlin: Bruno Cassirer, 1907, 1911, 1922). ECW3.

EK3	*Das Erkenntnisproblem in der Philosophie und Wissenschaft der neueren Zeit – Dritte Band: Die Nachkantischen Systeme* (Berlin: Bruno Cassirer, 1920, 1923). ECW4.
EK4	*The Problem of Knowledge: Philosophy, Science and History since Hegel Vol. 4 of Das Erkenntnisproblem.* Translated by William H. Woglom and Charles Handel (New Haven: Yale University Press, 1950, 1957). ECW5.
ETR	*Substance and Function & Einstein's Theory of Relativity.* Translated by William Curtis Swabey and Marie Collins Swabey (New York: Dover Publications, 1953 [1910]). ECW6.
FF	*Freiheit und Form: Studien zur deutschen Geistesgeschichte* (Berlin: Bruno Cassirer, 1916, 1918, 1922). ECW7.
GC	"The Concept of Group and the Theory of Perception." *Philosophy and Phenomenological Research* 5, 1 (1944): 1–35. ECW24, 209ff.
GK	"Die Grundprobleme der Kantischen Methodik und ihr Verhältnis zur Nachkantischen Spekulation." In *Die Geisteswissenschaften und Philosophie* 1 (Leipzig: Veit, 1914). ECW9, 201ff.
IC	*The Individual and the Cosmos in Renaissance Philosophy.* Translated with introduction by Mario Domandi (New York: Barnes & Noble, 1963 [1927]). ECW 14.
IG	*Idee und Gestalt: Fünf Aufsätze* (Berlin: Bruno Cassirer, 1921). ECW9, 243ff.
J	"Leibniz und Jungius." In *Beiträge zur Jungius-Forschung.* Edited by Adolf Meyer (Hamburg: Paul Hartung Verlag, 1929). ECW17, 360ff.
HC	"Hermann Cohen und die Erneuerung der Kantischen Philosophie." *Kant-Studien* 17, 3: 252–73. ECW, location N. A.
HS	Introduction and commentary to Leibniz, G. W. *Hauptschriften zur Grundlegung der Philosophie*, vols 1 and 2. Edited by Ernst Cassirer (Leipzig: Meiner Verlag, 1966 [1904/1906]). Edited by A. Buchenau, ECW9, 515ff.
K	*Kant's Life and Thought.* Translated by James Haden (New Haven: Yale University Press, 1981 [1918, 1921]). ECW8.
KI	"Der kritische Idealismus und die Philosophie des 'Gesunden Menschenverstandes.'" In *Philosophische Arbeiten*, vol. I, no. 1. Edited by Hermann Cohen and Paul Natorp (Giessen: Tempelmann, 1906). ECW9, 3–36.
KPM	"Kant and the Problem of Metaphysics: Remarks on Martin Heidegger's Interpretation of Kant." In *Kant: Disputed Questions.* Edited by M. S. Gram (Chicago: Quadrangle Books, 1967 [1931]). ECW17, 221ff.
L	*Leibniz' System in seinen wissenschaftlichen Grundlagen* (Marburg: N. G. Elwert'sche Verlagsbuchhandlung, 1902). ECW1.
LC	*The Logic of the Cultural Sciences: Five Studies.* Translated with introduction by S. G. Lofts. Foreword by Donald Phillip Verene (New Haven: Yale University Press, 2000 [1942]). ECW24, 356ff.
MS	*Myth of the State* (New Haven: Yale University Press, 1961). ECW 25.
PE	*Philosophy of the Enlightenment.* Translated by Fritz C. A. Koelln and James P. Pettegrove (New Haven: Yale University Press, 1961 [1932]). ECW15.
PR	*The Platonic Renaissance in England.* Translated by J. Pettegrove (Austin: University of Texas Press, 1953 [1932]). ECW14, 221.

PSF1	*The Philosophy of Symbolic Forms, Volume 1: Language.* Translated by Ralph Mannheim. Introduction by Charles W. Hendel (New Haven: Yale University Press, 1955 [1923]). ECW11.
PSF2	*The Philosophy of Symbolic Forms, Volume 2: Mythical Thought.* Translated by Ralph Mannheim (New Haven: Yale University Press, 1955 [1925]). ECW12.
PSF3	*The Philosophy of Symbolic Forms, Volume 3: The Phenomenology of Knowledge.* Translated by Ralph Mannheim (New Haven: Yale University Press, 1957 [1929]). ECW13.
PSF4	*The Philosophy of Symbolic Forms, Volume 4: The Metaphysics of Symbolic Forms.* Translated by John Michael Krois. Edited by J. M. Krois and D. P. Verene (New Haven: Yale University Press, 1996 [1921–1940]).
	German edition: *Zur Metaphysik der Symbolischen Formen.* Edited by John Michael Krois. Vol. 1 of *Ernst Cassirer Nachgelassene Manuskripte und Texte* (Hamburg: Felix Meiner Verlag). ECN1.
R	"Die Idee der Republikanischen Verfassung: Rede zur Verfassungsfeier am 11 August 1928" (Hamburg: Friederischsen, De Gruyter & Co., 1929). 1 v. ECW17, 290ff.
RKG	*Rousseau, Kant and Goethe: Two Essays By Ernst Cassirer.* Translated by James Gutmann, Paul Oskar Kristeller and John Hermann Randall, Jr (New York: Harper and Row, 1963).
SF	*Substance and Function & Einstein's Theory of Relativity.* Translated by William Curtis Swabey and Marie Collins Swabey (New York: Dover Publications, 1953 [1910, 1923]). ECW6.
SL	"'Spirit' and 'Life' in Contemporary Philosophy." In *The Philosophy of Ernst Cassirer.* Edited by P. A. Schilpp (Evanston, IL: The Library of Living Philosophers, 1949 [1931]). ECW17, 185ff.
SMC	*Symbol, Myth, and Culture: Essays and Lectures of Ernst Cassirer, 1935–1945.* Edited by D. P. Verene (New Haven: Yale University Press, 1979).
U	"Das Problem des Unendlichen und Renouviers 'Gesetz der Zahl.'" In *Philosophische Abhandlungen, Hermann Cohen zum 70 Geburtstag dargebracht,* 85–98 (Berlin: Bruno Cassirer, 1912). ECW9, 105ff.
WW	"Vom Wesen und Werden des Naturrechts" (1931). *Zeitschrift für Rechtsphilosophie in Lehre und Praxis* 6 (1932/34): 1–27. ECW18, 203ff.
ZWW	*Ziele und Wege der Wirklichkeitserkenntnis.* In *Nachgelassene Manuskripte und Text,* vol. 2. Edited by John Michael Krois and Klaus Christian Köhnke (Hamburg: Felix Meiner, 1999 [1937]). ECN2.

Works by Hermann Cohen, arranged alphabetically by abbreviation

Drawn principally from *Hermann Cohen Werke*, abbreviated as HCW, from the Hermann-Cohen-Archiv at the Philosophischen Seminar Zürich, general editor Helmut Holzhey, Hildesheim, G. Olms.

EL	*Einleitung mit kritischem Nachtrag zu F. A. Langes Geschichte des Materialismus*. Edited by H. Holzhey. 3rd edition (Hildesheim: G. Olms, 1984 [1883/1913]). HCW5, part II.
HCS	*Hermann Cohens Schriften zur Philosophie und Zeitgeschichte*. Edited by Ernst Cassirer and Albert Görland, vols 1 and 2 (Berlin: Akademie-Verlag, 1928).
I	*Das Prinzip der Infinitesimalmethode und seine Geschichte: Ein Kapitel zur Grundlegung der Erkenntniskritik*. Introduction by Peter Schulthess (Hildesheim: Olms Verlag, 1984 [1883]). HCW5.
KB	*Kants Begründung der Ethik nebst ihren Anwendungen auf Recht, Religion und Geschichte*. Introduction by Peter Müller and Peter A. Schmid (Hildesheim: Olms, 2011 [1910]). HCW2.
KE	*Kants Theorie der Erfahrung*. Introduction by Geert Edel (Hildesheim: George Olms, 1987 [1871/1885]). HCW1, part 3.
S	*Hermann Cohens Schriften zur Philosophie und Zeitgeschichte*, vols 1 and 2. Edited by Ernst Cassirer and Albert Görland (Berlin: Akademie-Verlag, 1928).
SP1	*System der Philosophie – Erster Teil: Logik der reinen Erkenntnis*. Introduction by Helmut Holzhey (Hildesheim: George Olms, 1977 [1914]). HCW6.
SP2	*System der Philosophie – Zweiter Teil: Ethik des reinen Willens*. Introduction by Steven S. Schwarzschild (Hildesheim: George Olms, 1981 [1907]). HCW7.
SP3	*System der Philosophie – Dritter Teil: Ästhetik des reinen Gefühls*. Introduction by Gerd Wolandt (Hildesheim: Georg Olms, 1982 [1912]). Part I, HCW8. Part II, HCW9.

Introduction

"READING A MUTE HISTORY": ERNST CASSIRER, THE MARBURG SCHOOL AND THE CRISES OF MODERN GERMANY

In July 1917, during the bleakest days of the First World War, the writer Hermann Bahr (1863–1934) published an essay on the current situation in Germany and Austria-Hungary for the popular monthly *Die Neue Rundschau*.[1] Largely forgotten now, Bahr was at this time one of the more influential writers in Central Europe and something of an impresario of the Viennese modernist movement.[2] A cofounder of the Viennese newspaper *Die Zeit*, the author of over thirty books and plays and a founding member of the "Young Vienna" group of poets, he was also a correspondent with many of the principal thinkers of his native Austria, including Hugo von Hofmannsthal, Robert Musil, and Arthur Schnitzler. The 54-year-old Bahr was just finishing a promotional year of travel in Germany for his *Expressionism* of 1916, a text that proved one of the more enduring statements on this Central European art movement. In his article, however, Bahr turns from art to contemporary politics and society. Does contemporary Germany, he asks, have any figure that could grasp the nation's contemporary crisis and lead Germany to a new role in the future?

Surveying prominent writers as intellectually and politically diverse as the theologian Ernst Troeltsch, the philosopher Max Scheler, the arch-nationalist Houston Stewart Chamberlain and the historian Friederich Meinecke, Bahr searches for what he calls a "contemporary Goethe" who could encompass and recast the frightful forces of the modern era. What was needed was a figure who could harness the desire for change suggested by philosophers such as Nietzsche, yet who "prepares the way for the future in clear words."[3] Ultimately, Bahr concludes, the only contemporary figure with this capability was the philosopher and historian Ernst Cassirer (1874–1945). In the manner of an expressionist painter, Bahr writes, Cassirer's work sketches the underlying forces at work in Germany and reveals their basic tendencies and import, even as it redefines their potential. In both style and substance, Cassirer's is the voice that could redefine the purpose of modern Germany in its hour of crisis.

Looking back on Bahr's piece and its context, readers at all familiar with the works of Cassirer or the situation of Germany in 1917 can only marvel at Bahr's conclusion, and indeed the entire heroic tone of his article. Cassirer is known today as one of the most important philosophers of science and culture of the first half of the century, but by no means the political and cultural redeemer Bahr depicts. Even the date of Bahr's work is surprising, since few of Cassirer's works published before 1917 would appear on the surface even to be "political" in our sense of the term. By the summer of 1917, Cassirer's writing was rooted almost entirely in the philosophy of science and in a remarkable series of works entitled *The Knowledge Problem in the Philosophy and Science of the Modern Period*, which linked changing ideas of science and philosophy with changing definitions of subjectivity and objectivity.[4] Yet somehow Bahr, working at the twilight of the Wilhelmine era, held that Cassirer's critique of the "structure of knowledge" could not only influence the institutions and structures of society, but could also be converted into political insight and indeed national harmony.

Until recently, most commentators would probably not even concur with the philosophical importance Bahr ascribes to Cassirer. At the time it was written, though, Bahr's text would in this regard be quite intelligible. For Cassirer was by this point, at the age of 43, widely considered one of the leading philosophers of his generation, and his reputation would increase in the coming years. Twelve years later, for instance, the 1929 debate in Davos, Switzerland between Cassirer and Heidegger was considered the pivotal intellectual event of the era. Where this "Davos debate" is now usually remembered as marking Heidegger's appearance as a "philosopher of the future," at the time it was equally telling for situating Cassirer as one of the pre-eminent philosophers of the present. Cassirer's German reputation was founded foremost in his sophisticated prewar works from his youth and middle age, works largely unacknowledged in his North American reception in favor of his often more accessible thought from the Weimar period, which was popularized and translated following his emigration to the United States in 1941.

Bahr's own political affiliations and interests increase the mystery posed by his appreciation of Cassirer. Bahr had earlier drifted in and out of a nationalistic pan-Germanism, occasionally spiked with anti-Semitism, that make his support for the cosmopolitan writings of the Jewish German Cassirer surprising.[5] Although Bahr's career is remarkable in part for his later turn to philo-Semitism – his 1894 series of interviews *Antisemitismus* was largely an attack on the movement and a renunciation of his earlier work with the Far Right politician Georg von Schönerer, and he redoubled this position in the 1920s – he is a surprising advocate for Cassirer's work.[6] Whereas Cassirer's only ostensibly political essay, *Freedom and Form: Studies in German Intellectual History* of 1916, was an explicitly cosmopolitan text that painted a broad history of German political thought in its European context, Bahr had written a jingoistic and militant pro-German screed as recently as 1915.[7] Bahr's article also strikes a surprising note in the manner it discusses Cassirer in the same breath as such virulent nationalists as Chamberlain. For Cassirer's role in Germany in the coming years was defined by his place as one of the most prominent liberal philosophers and perhaps the most prominent Jewish German academic of his generation. By the time Cassirer became one of the first prominent Jewish German rectors of a German university at Hamburg University in 1929, the shift in political climate would have made Bahr's comparison impossible.[8]

Bahr's reception of Cassirer highlights one prominent, albeit unusual, case where Cassirer's early work, now often considered entirely apolitical, was popularly received as directly political and rhetorically sophisticated. With the eyes of a contemporary from the Wilhelmine period, Bahr allows us to perceive what was occluded by Cassirer's

Weimar reception: the sophistication of his attempt to redefine the science, politics and self-definition of imperial Germany, and the salience of his attempt to recast mainstream German discourse. For Bahr, Cassirer's work advanced a clear political agenda for Germany, but did so in a manner that was oblique to the particular arguments of each of his works even as its form was clear across their broader structure. Cassirer's ability to recast key moments of German history – "as in a symphony," Bahr writes – creates new moments of coherence and political possibility through their transformation of traditional narratives.[9] Nor was Bahr's sense of Cassirer's project far off from Cassirer and his teacher Hermann Cohen's (1842–1918) vision of their collective work in the so-called Marburg school. For within the form of a philosophy of science and a synoptic history of ideas, Cassirer was indeed trying to present the conditions for the possibility of a new understanding of Central European society and a new foundation for liberal democratic politics. The political distance bridged between Cassirer and Bahr suggests the potential range of reception for Cassirer's project, not despite but because of its often understated presentation. It suggests as well the possible efficacy of this project at the dusk of Wilhelmine politics, even as the generational timing of Bahr's reception also suggests why Cassirer's early philosophy ultimately failed to gain a popular audience and would indeed be increasingly viewed as dated in the Weimar period.

Precisely due to his prominence, Cassirer was one of the first Jewish Germans to go into exile in 1933. First in England and Sweden, and then, in 1942, in the United States, Cassirer learned new languages and philosophical idioms, while a number of his German texts went unfinished and unrecognized.[10] Delayed by immigration and war, several prominent works were left unpublished until the late 1990s, such as his crucial 1937 systematic overview of *The Knowledge Problem* series, entitled *Objectives and Paths of Reality-Knowledge*.[11] A number of Cassirer's most important books, notably the first three volumes of *The Knowledge Problem* itself, have still not been translated into English. Only recently, with the publication of Cassirer's complete works in German and new editions in French and Italian, has a more complete picture of Cassirer's thought begun to emerge. Yet both in Europe and particularly in North America it is largely Cassirer's post–First World War writings that have been recognized, while those of the first half of his life, on which Bahr's review and Cassirer's contemporary fame rested, have until recent years been largely ignored outside of specialized academic disciplines.

Bahr recognized – as did another commentator, the young George Lukács – that far from simply recapitulating German history, Cassirer's symphonic presentation in works such as *Freedom and Form* uses the methodology of his earlier functionalist philosophy of natural science for the social and human sciences, and does so as part of a distinct political and ethical program.[12] For Bahr, the goal of this program was to illuminate new connections, forms of possible action and modes of understanding the "new truth," as Bahr describes it, of a world based entirely on relations.[13] It operates both as a "history of the construction of the modern sciences," in all of their forms, and also as an ethical presentation of the central problems of the era, which, rather than posing obvious solutions, "leaves us standing before the central dilemmas."[14]

Bahr's reception of Cassirer, particularly when placed in conjunction with figures from the Left such as Lukács and centrists such as Ernst Troeltsch, also suggests how both friends and enemies could readily interpret Cassirer's early work as part of a sweeping reform project when interpreted in conjunction with the late works of his teacher Hermann Cohen and the Marburg school, as it almost always was either tacitly or directly.[15] Now largely read from the perspective of his later philosophy, this early work is often misunderstood as either part of a "scientistic" phase of Cassirer's career or a straightforward expansion of Kantian philosophy. It was, however, directed against the blind belief in science, the *Wissenschaftsgläubigkeit* of the age, and formed part of a fundamental redefinition of German philosophy, particularly of neo-Kantianism, developed in Cohen's late works, notably in his three-volume *System of Philosophy* (1902–1912).[16] Initiated through Cassirer's historical study *Leibniz's System in its Scientific Foundations* (1902) and summarized in his *Substance and Function: Investigations Concerning the Fundamental Questions of the Critique of Knowledge* (1910) and *Freedom and Form: Studies in German Intellectual History* (1916), Cassirer developed a powerful yet comprehensible translation of Cohen's demanding late philosophy that aimed to recast the self-definition of Wilhelmine Germany through its natural, social and human sciences.[17] As such, Cassirer's work aimed to establish the conditions for the possibility of a progressive and liberal Germany in its focus on the processes, institutions and epistemologies that tacitly underlay politics.[18]

Against a crisis of meaning in both traditional cultural authority and the concept of truth in the sciences, Cassirer followed Cohen in seeking a third way in which the sciences could constructively facilitate a new political society and modern culture, particularly a liberal democratic society based on a constant critique of science, knowledge and institutions.[19] In this regard Cassirer's work before the First World War was the capstone of the decades-long Marburg school project founded by Friederich Albert Lange (1828–75) and Cohen, which hoped to use this "critique of knowledge" and science to refashion German society and thought on the most fundamental level.[20] Cassirer's early work attempted to further what he took to be the school's tremendous importance for the future of Germany, and sought to facilitate the mainstream reception of Cohen's thought.[21] The basis of Cassirer's commitment to Cohen was not simply personal loyalty or academic progress – from a powerful family of industrialists, Cassirer's options beyond academia were vast and he only reluctantly entered the field – but his perception of the importance of Cohen's demanding late philosophy for the future of Central Europe.

The ambitions of the Marburg school are difficult to overstate, even as Cassirer's work underplayed nearly all of them in its attempts to broaden the appeal of the school and of Cohen's late philosophy. Thomas Nipperdey's standard history of the period depicts these ambitions appropriately when he writes that the Marburg school set out to establish nothing less than a model for a "political-social theory of reform, idealistic-socialistic democracy, political reconciliation and general human development."[22] Even though Cassirer was careful to avoid the distinct terminology of the school and to minimize direct reference to its members, his role as the acknowledged successor to Cohen made the interpolation of a political message in his works, however rarified, characteristic of his reception by figures such as Bahr or Troeltsch. Indeed, precisely in his personal and rhetorical ability to recast the Marburg project in general terms and histories, and to reveal the key themes of the school woven throughout German intellectual history, he was taken as its ideal proponent. Thus in a letter from 1911, his colleague Paul Natorp describes Cassirer's advantage for the school as precisely his "decisive impression of quiet objectivity," an objectivity that was combined with a brilliant and occasionally countervailing rhetorical ability to develop key themes of the school in precisely the "symphonic" style that enamored Bahr.[23]

The plausibility of Cassirer's early project, however, and its ability to generate enthusiasm comes into focus once the influence of the Marburg school is considered as a whole. Even as it was to fade rapidly after the First World War in direct influence, the school was immensely influential, and is now considered to be pivotal – either through attraction or repulsion – for a truly immense array of German and European thought in the first third of the twentieth century. To list only a few of the fields and figures that have been closely linked with the school, we could cite: the formation of the ideology of German social democracy (Bernstein, Vorländer, Eisner); the development of modern existentialism and ontological philosophy (von Hartmann, Heidegger, Lévinas, Rosenzweig); the development of modern history and social theory (the Bakhtin Circle, Elias, Panofsky, Kantorowicz); the redevelopment of the history and philosophy of science (Koyré, Meyerson, Wind); and the reformulation of law that would lead to Hans Kelsen's theory of "pure law" (*Reine Rechtslehre*) and with it, in part, both the Austrian Constitution of 1920 and the United Nations Charter.

The key to understanding this influence was that the Marburg critique was, despite Cassirer's attempts at "preparing the way" for a wider popular reception of its tenets, largely aimed at transforming the academic disciplines and their vast administrative application in Germany. Often mistakenly understood as focused entirely on the natural sciences, the central goal of the Marburg school under Cohen was the transformation and critique of all of the sciences – including particularly the social and human sciences, or humanities – starting with the institutional matrix of the university. The historical model was the influence of Leibniz and then Kant on German society through the university system in its pervasive role in state and confessional administration, social organization and cameralism in the eighteenth and nineteenth centuries. In this regard, the Marburg school project epitomized what Fritz Ringer described as a "Mandarin" attempt at political change from above, or indeed what Kevin Repp and others have defined as an "antipolitics" that sought change through administrative activity outside of party politics (even as Lange and Cohen, but not Cassirer, were quite active in the politics of the Social Democratic party as well).[24]

Since Leibniz, "science" (*Wissenschaft*) had been defined in Central Europe as the critical reflection on almost any system of organized endeavor, and entailed the active role of the social, cultural and natural sciences for transforming society and developing its confessional and political structure. For many in Wilhelmine Germany, it was axiomatic that the Germany of the Second Industrial Revolution was defined by the attempt to apply, as one contemporary North American observer of Germany put it, "scientific thought to every process and every social and industrial problem."[25] Although unreflexively embraced by many elites as the key to Germany's rapid rise as a world power, for many others this scientific turn was deeply problematic, and the "politics of cultural despair," as Fritz Stern described it, opposed this form of modernism with an equally strident call for a return to *Kultur* and neo-Romanticism.[26]

The Marburg school's third way beyond these positions was through an immanent critique of the meaning and limits of science itself. In doing so, they sought to delimit scientific power and the administrative control of the technocratic state with which it was connected, while retaining an open-ended definition of "humanity" that developed in large part through the sciences, yet superseded them both ethically and as an immanent reality. The proximate means to this end began with the transformation of the idea of science and its application in the university system, but the goal was ultimately – as Cohen in particular emphasized – the transformation of the institutions and culture of imperial Germany.

Once Cassirer's early work is understood in this context, we can begin to see both why it would so fascinate Bahr and also why this aspect of Cassirer's thought would remain largely unrecognized after the Wilhelmine era. Written in the idiom of the philosophy of science or history, Cassirer's early work may not immediately appear to us as political. Yet for a wide array of readers it was apparent that the range and ambition of this project was vast, and that Cassirer sought to redirect the earlier Marburg project, to redefine the relation of science and society and to recast German history into, as he called it in *Freedom and Form* following Goethe, a more "productive" formulation of its "foundational ideas."[27]

Ironically, however, Bahr's 1917 appraisal came right as Cassirer himself was beginning to recognize the limitations of Cohen's "third way" – and, indeed, of the language of science – for transforming Germany's situation or for reaching a wider audience. With the founding of the constitution of the Weimar Republic in 1919 (the end point of this study), Cassirer largely redirected his efforts towards a straightforward but less nuanced support of the existing republic on the one hand, and to an investigation of aspects of myth, language and perception that he found newly political on the other. His earlier arguments were enthusiastically received and developed in a number of specific academic fields, notably law, history, art history, literature and the philosophy of science. The complex formulations of his early work were gradually refined in many of his Weimar writings by a democratic style that would become highly regarded for its accessibility and clarity.

Read carefully, however, Cassirer's later work is consistently informed and often explicitly based on the premises of his early philosophy. At nearly every key juncture in his later work, he directs readers back to his earliest texts, particularly *Substance and Function*, as a touchstone for understanding his philosophy as a whole. He clearly intended his work to be read as one continuous whole, albeit as a whole open to fundamental permutations as it developed, a model of change developed from his early understanding of group theory in its relation to history.

The narrative arc of this book follows nearly the opposite path of many descriptions of Cassirer's work and life by placing the focus on the "early Cassirer," and claiming that key aspects of his later work are either conceptually dependent on this initial project or are a popularization – and in some cases a simplification – of it. The text follows the trajectory of other revisionist readings of the "early" phase of a philosopher – say, those of Dieter Henrich on Hegel or the Frankfurt school's reading of Marx – by re-establishing lost elements of the author's historical context, notably here through the meaning of science to the Wilhelmine period, providing the basis for a new reading of the entirety of his or her work and from this basis establishing a new approach for understanding the author's texts in relation to contemporary intellectual trends.[28] Such a reading has the advantage of following Cassirer's own technique for interpreting authors – notably Leibniz, Kant and Goethe – by starting with their initial questions, and then tracing their work out as a dialogic whole of changing, but related, forms in functional continuity. In this reading, the initial problematic, but not necessarily initial intent, of an author is of crucial importance and usually takes on a surprising, but decisive, form in relation to later stages of their work.[29]

Cassirer's post–First World War reception was markedly different than his actual work during the Wilhelmine period. Following the First World War, Cassirer was often defined simply as a "good" liberal or defender of the Enlightenment tradition. Such a reading, which gained particular prevalence in North America in the 1950s, effaces much of the complexity – and durability – of his thought. Whereas Cassirer's work with Hermann Cohen before the First World War is often described as a necessary but expendable way station to his mature views on the philosophy of culture, Cassirer was deeply concerned with transforming into contemporary idiom what he perceived – and reaffirmed throughout his life – as the genius of Cohen's key works. These works were, as he told students in 1935 at Oxford University – using terms he almost always repeated when mentioning Cohen publically – of unequalled value but haunted by the "grave flaw" that they were written in a style that was nearly incomprehensible: "Wrapt in such a mystery of philosophical style and technical language," as he quoted his earlier teacher Georg Simmel as saying, "that nobody as yet could understand them perfectly."[30] Cassirer's work compensated for this flaw, indeed perhaps overcompensated, by recasting Cohen's basic themes into a rhetorical and argumentative pattern that was exquisitely clear in each of its facets. Read in combination, however, with the sort of "symphonic" overview Bahr recommended, they provide a compelling synthesis of key aspects of Cohen's complex work even as over time they recast and critiqued it.

By the Weimar period, Cohen and Cassirer's work from the decade before the First World War was already often incomprehensible to younger audiences. The dominant tone of the era was perhaps summarized by Siegfried Kracauer's statement – himself a generation younger than Cassirer – that the Weimar period was marked by a "'hatred of science' rampant among the best of today's academic youth."[31] In this context, both the intent and content of the Marburg critique that Cassirer was promoting became both unintelligible and suspect, and Cassirer himself markedly shifted his rhetorical and philosophical focus away from the approach that would have engaged members of an older generation such as Bahr.

Reading a Mute History: Historicism and Cassirer's Reception

Writing in 1966, nearly fifty years after Hermann Bahr's review, the philosopher Michel Foucault reviewed Cassirer's *The Philosophy of the Enlightenment* (1932), placing it in the context of the early volumes of the philosopher's *Problem of Knowledge* series (vols 1–3, 1906–1920; vol. 4, 1940/1950). Although as idiosyncratic in a manner as Bahr's review, Foucault's essay serves as a complement to it by offering an introduction to the often forgotten influence of Cassirer's early work on the specialized academic disciplines in the twentieth century, an influence that ranged from history and the history of science to linguistic theory and law. Like Bahr, Foucault presents a view of Cassirer at odds with the present anglophone consensus on the philosopher, one based on a close reading that has occasionally been eschewed by Cassirer's audience in the assumption that his work is self-explanatory or commonsensical. If Bahr alerts us to the manner in which Cassirer's work was political in a sense different from our own use of the term, Foucault highlights the fact that the methodology of this politics was nonetheless received and adapted in surprisingly influential ways.

In a tone initially surprising to those who have read Cassirer's classic *The Philosophy of the Enlightenment*, Foucault contends that Cassirer's work reads "like a manifesto" and "founds the possibility for a new history of thought. It is indispensable that it be known, for it is from here that we others need to depart."[32] Although written thirty years earlier, Cassirer's 1931 *The Philosophy of the Enlightenment* "belongs to our own time."[33] It quickly becomes clear, however, that by a "new history of thought" Foucault is referring not just to *The Philosophy of the Enlightenment*, but to what he describes as Cassirer's "great philosophical works (especially his *Knowledge Problem*)," as they developed from the earliest phase of his career.[34] Indeed, although Foucault highlights the particular importance of *The Philosophy of the Enlightenment* as a response to the rise of Nazism, Cassirer's "manifesto" summarizes a fairly consistent lifelong approach to history anchored in the *Problem of Knowledge*. Programmatically established in the introduction to the first volume of that series in 1906 and developed through its following four volumes, the same approach formed the conceptual basis of Cassirer's notable – and more widely read – chronological trilogy of texts from the late 1920s and early 1930s, each of which focused on key figures from the *Knowledge Problem* series, *Individual and Cosmos* (1927; Nicholas of Cusa), *The Platonic Renaissance in England and the Cambridge School* (1932; More, Cudworth, Shaftesbury) and finally *The Philosophy of the Enlightenment* (1932; Leibniz).

Cassirer's historical style, Foucault writes, is based on an intentional and radical attempt to "isolate from other histories (that of the individual, or of groups)" the "autonomous space of theory: and with its eyes to discover a history that has remained mute."[35] The overall purpose of Cassirer's work is to provide a space for completely reassessing the historical context of an era by focusing solely on its surface discourse and reassembling it into its basic structural forms. These reassessments reveal the "fatalities of reflection and of knowledge" in the constitution of modern thought and society.[36] It is in recovering this "mute history" of an era that Cassirer is truly heir to the Enlightenment tradition of "putting into question the forms and limits of our knowledge," which has continued "from Marx to Lévi-Strauss." Against the French emphasis on psychologism, Cassirer's work presents a historical analysis of how subjectivity and objectivity themselves develop from the forms of thought of a given era.[37] Cassirer's central innovation is to demonstrate that "thought and discourse, or rather their inextricable bond, far from offering a pure and simple manifestation of what we know, rather constitutes the birthplace of all understanding."[38]

Although Cassirer is often referred to as a "neo-Kantian," Foucault insists that this should not in this case be understood as a movement or school of thought. Rather, it is "the impossible position Western thought finds itself in," in attempting to overcome the "cut" (*coupure*) established by Kant in the modern ideas of subjectivity and objectivity: "Neo-Kantianism (in this sense, we are all neo-Kantians) is the relentless injunction to revive this cut, both to recover its necessity and draw from it the utmost."[39] The "fundamental abstraction" at the heart of Cassirer's text, Foucault writes, moves his work from the usual diplomatic, political or social history based on the intention of actors to a "history of problems" (*Problemgeschichte*) based on ideas and practices. It allows, as Foucault notes in terms again resonant with Bahr's earlier review, "a paradoxical *dècoupage*," a new break and juxtaposition of key elements, which "ruptures the most familiar lineages" of history. Cassirer's obvious limitation of biographical, economic or social background, far from being an oversight, is similar to the "iconoclastic gesture by which all of the great disciplines have been founded."[40] It is identical to how political economy was established through its isolation of production from the concrete means of producing wealth, or linguistics developed in its separation of language (*langue*) from speech (*parole*).

Foucault's reading of Cassirer thus situates the German philosopher in terms similar to Foucault's own *The Order of Things* (1966), which provides a broad reading of different configurations of knowledge – "*epistemes*" – that constitute the European "order of things" since the early modern period. In a manner largely consistent with Cassirer's theory of historical transitions as laid out in *The Knowledge Problem* and *Substance and Function* (1910), each episteme presents for Foucault an entirely new configuration of knowledge, which shares some of the content but none of the internal functional meanings of the previous epistemes.[41] The rhetorical payload of this approach for Foucault's work, and the sign of its apparent distance from Cassirer's developmental histories, becomes clear in the final chapter of Foucault's text, where it is revealed that the reader of 1966 is caught between epistemes. Our world is on the brink of dissolution: the present image of "man" will be erased "like a face drawn in the sand at the edge of the sea."[42]

Although at first glance Foucault's review appears suspiciously close to his own theory of history, his vision of the radicalness of Cassirer's approach is confirmed both in *The Philosophy of the Enlightenment* (particularly in its central reading of Leibniz as it develops from Cassirer's earlier *Leibniz's System*) and in the methodological foundation of this work found in Cassirer's *Knowledge Problem*.[43] Cassirer's introduction to *The Knowledge Problem* does in fact claim to develop a new field on the basis of a "fundamental abstraction" through historical forms of knowledge, one which is, moreover, elaborated in *Substance and Function*, as well as in *Freedom and Form*.[44] Cassirer's reading of Leibniz and Kant in his early work, and his development of Cohen's philosophy of origin, did argue for a "cut," and a break with any substantive definition of subject and object, almost as disorienting as Foucault suggests, albeit one that ultimately resolves into a very different modernist narrative of progress.

At the beginning of his *Knowledge Problem*, Cassirer writes that he has set out to study how knowledge posits the relation of "subjectivization and objectivization" in different historical eras, a project developed from Herman Cohen's *Erkenntniskritik*, or "critique of knowledge."[45] The centerpiece of Cassirer's historical project was that "subject and object," like nearly all the classical dichotomies, "are not given and obvious possessions of thought," but rather products of the "driving [*treibendes*] knowledge ideal" that must be "reconstructed from the entire intellectual movement of an age."[46] Ultimately, there is indeed no given separation of subject and object at all, but only a common form of experience from which these poles can be variously defined. "The problem of 'experience' itself," Cassirer writes, "is not just given but is one of the most difficult problems and must be derived from the practices and theories of each age, not only in what it says, but also in its activities."[47] Although the notion of a complete "break" between epistemes does not exist for Cassirer in the same manner as for Foucault – for Cassirer there will always be at least one form of functional continuity between eras – his understanding of transformation based on the model of group theory of mathematics could readily be construed as constituting a change of epistemes as radical as that which Foucault suggests.[48] A slight change in any one element of a group has the potential to change the meaning of the whole, and conversely a new definition of the whole completely changes the meaning of any element, including by analogy any particular definition of objectivity, subjectivity or their division.

Cassirer's early philosophy marked the impossibility of objective knowledge of the transcendental subject (to put it in Kantian terms) other than as experience itself, the related flexibility of (empirical) subject and object positions in history, and an emphasis on the necessary limitations of any particular science before experience. Foucault's description of Cassirer's own historical model as revealing "a history that has remained mute" reflects an attempt to outline underlying structural relations that first condition the possibility of individual experiences of subjectivity, objectivity and, indeed, reality itself.[49] In a 1912 essay on Cohen, Cassirer outlined the basic tenet of this form of historical thought, which holds that "we only understand a historical event when we grasp it in the context that would remain hidden from it itself."[50] In recasting the deep structural foundations of different eras, what Cassirer calls "secret-yet-fully-open"[51] (*geheimnisvoll-offenbaren*) foundations of an era, through studies of the changing definitions of subjectivity, objectivity, temporality and truth, Cassirer hoped to recast the possibilities for present and future action.[52]

Foucault's review of *The Philosophy of the Enlightenment* ultimately concludes that Cassirer "retreated from the possibilities he uncovered" by returning to a narrative focus on the individual over structure, and on ideas over institutions. Although Cassirer claims to focus on practice as much as philosophy, "he grants to both philosophy and thought a primacy that he never defers from questioning: as if the thought of a particular era had its place [...] in a theory of the world rather than in positive science, in aesthetics rather than the work of art, in philosophy rather than the institution."[53] Yet for Foucault, as for many key thinkers of the twentieth century, Cassirer's work as it developed from the *Problem of Knowledge* constituted a touchstone of twentieth-century thought. Indeed, it is telling that in a 1984 retrospective of his own work written under pseudonym, Foucault would even define his project as a neo-Kantian "critical history of thought" in a manner that seems close to Cassirer and Cohen's "knowledge critique": "To the extent that Foucault fits into the philosophical tradition, it is the critical tradition of Kant, and his project could be called a Critical History of Thought. [...] If what is meant by thought is the act that posits a subject and an object, along with their possible relations, a critical history of thought would be an analysis of the conditions under which certain relations of subject to object are formed or modified, insofar as those relations constitute a possible knowledge."[54]

Despite the fame Cassirer's early volumes of *The Knowledge Problem* and *Substance and Function* gained before the First World War, Foucault's recognition of their influence is typical of their role largely within specific academic fields – in this case the history of science in which *The Knowledge Problem* had a prominent role and the history of ideas in *The Philosophy of Enlightenment* – and often not as part of a broader philosophical reception. Perhaps for this reason, one of the leading philosophers of late twentieth-century Germany, Hans Blumenberg (1920–1996), could describe Cassirer's *Substance and Function* as singular in being one of the most influential, yet least recognized, works of the first half of the century.[55] If we look at Cassirer's influences within the particular historical fields, we can see that others often used Cassirer's theoretical plan and model of historical methodology in more concrete applications. These moments of influence suggest Cassirer's initial innovation of using a critical functional science as a means towards social and political transformation, a strategy born out of the weakness of "Mandarin" academic elites, but one that would – perhaps tellingly – come to have a broad influence throughout academia and arguably society. Thus, recent studies have highlighted the pivotal role Cassirer's early work had on the German tradition of the history of science that was notably further developed by Ludwik Fleck and, in the United States, by Thomas Kuhn. As Michael Friedman has noted, *Substance and Function* – along with *The Knowledge Problem* – exerted a decisive influence on the philosophy of science in figures such as Émile Meyerson, Léon Brunschvicg, Hélène Metzer, Anneliese Maier and Alexander Koyré, and can be described as perhaps the key influence on Thomas Kuhn's synthesis of these broadly neo-Kantian views.[56] Certainly for Foucault's teachers in the philosophy of science, and particularly for Georges Canguilhem, Cassirer's influence would have been notable.

The sociologist Pierre Bourdieu suggests a further point of influence from Cassirer when he writes, "The philosophical glosses that, for a time, surrounded structuralism have neglected and concealed what really constituted its essential novelty – the introduction of the structural method or, more simply, of the relational mode of thought that, by breaking with the substantialist mode of thought, leads one to characterize each element by the relationships that unite it with all the others in a system and from which it derives its meaning and function."[57] Both here and elsewhere in his work, Bourdieu makes it clear that his use of the terms substance and function are derived from Cassirer's *Substance and Function*. Thus, describing his own idea of the "cultural field," Bourdieu writes, "Constructing an object such as the literary and cultural field requires and enables us to make a break with the substantialist mode of thought (as Ernst Cassirer calls it) that tends to foreground the individual, or the visible interactions between individuals, at the expense of structural relations – invisible or visible only in their effects – between social positions that are both occupied and manipulated by social agents that may be isolated individuals, groups or institutions."[58]

Similarly, even the sociologist Norbert Elias, a thinker who at first appears remote from Cassirer's interests, was apparently influenced by Cassirer's early works, which he came to know through his teacher Richard Hönigswald. Cassirer's particular definition of functionalism has been read as directly influencing Elias's criticism of philosophy and his move towards "thinking structurally and in terms of relationships."[59] In sociology, a further continuity with Cassirer's early work is the "functionalism" of Talcott Parsons and later the systems theory of Niklas Luhmann, which prove closely related to Cassirer's early philosophy. For Luhmann, Cassirer's early works were a "provocation" against the "common-sense" background assumed by much of twentieth-century thought – a summary again nearly completely at odds with Cassirer's North American reception.[60] One recent overview of Luhmann's work concludes that "the work of Ernst Cassirer in his 1910 *Substance and Function* was of extraordinary importance for Talcott Parsons and Niklas Luhmann in the development of systems-functionalism (Parsons), or functional systems theory (Luhmann). Only with the background of Cassirer's work is Luhmann's social theory to be understood at all."[61]

Finally, if we look outside the social sciences proper we will find an equally powerful and delimited reception of Cassirer's early functionalist philosophy in law, or "legal science." When the philosopher Siegfried Marck wished to write an overview of the state of German legal philosophy in 1925, he summarized the basic disputed issues by citing Cassirer's earlier work through a repetition of its German title: *The Substance Idea and Function Idea in the Philosophy of Law*.[62] Marck described the development of law in the previous thirty years as characterized by a movement from "the idea of substance to the idea of function."[63] The terms, he noted in a preface, are derived from Cassirer's "fundamental book *Substance and Function*."[64] "Considering the present setting of problems in the current philosophy of law," Marck writes, "making use of this terminology appears to be the most pregnant means of expressing its present oppositions and questions of contention."[65] This functionalist approach allowed Marck to survey the wide expanse of problems linking representation with the problem of personality, a subject to which Marck turns in his penultimate chapter on "the problem of personality in law," which serves as a prelude to a "general theory of the state."[66] It was indeed this problem of the "personality" of state, peoples and legal individuals that was the starting point of *Freedom and Form* (1916), and to which Cassirer applied the same functionalist critique in the realm of the human and social sciences as he had earlier, in his *Substance and Function*, to the concept of the object in the natural sciences.

Cassirer's Philosophy and the "Metapolitics" of the Late Marburg School

In placing Cassirer's philosophy in its Wilhelmine context, two related approaches present themselves. In the first place, it can be seen that Cassirer judged and interacted with his own period using much the same method of reconstructing a "mute history," as Foucault described it, of subterranean structural issues and concerns for the present as he used to define the past. A second approach to Cassirer's "mute" or structural studies of events, however, serves as something of a tragic counterpoint to this first sense. For here we can see how Cassirer's methodology led to a self-censorship from his own writings of a number of motives and issues – economics, contemporary politics, religion, family – that may have been particularly important to him. Nonetheless, far from turning from an apolitical interest in science in his early years to a new defense of liberalism in the Weimar period, Cassirer's published texts long had a political intent consistent with his later personal involvement in the Weimar Republic. Indeed, Cassirer's innovative development of themes such as Humboldtian liberalism, a defense of rights, and a redemptive reading of future humanity occur with particular clarity in his early works, and are then often assumed as background to his more often straightforward defense of Weimar liberalism.

Where many accounts of Cohen's philosophy see it culminating in a reading of the exact sciences, which is indeed the focus of the first volume of *System of Philosophy*, Cassirer's early work departed from the later volumes of Cohen's *System*, which covered ethics, aesthetics and – in a planned but unpublished final work – psychology and pedagogy. Cassirer's work in this regard provides a window into understanding why a school initially famous for its focus on the natural sciences was ultimately equally, if not more, influential for the historical social and human sciences.[67] Cassirer translated Cohen's complex "philosophy of origin" into what would now be described as a "process philosophy."[68] Emphasizing process or function over substance, and becoming over being, this process philosophy served for Cassirer as the distinctly "modern" foundation for society and politics. The result was similar to later forms of structuralism in that it explained the particular event solely through its structural context, but it went beyond most forms of structuralism by seeking within "structural relations" underlying generative "laws of process" of various forms, whether on the level of the narrative processes of history or the causal processes of the exact sciences.[69] Placed in its historical context, Cassirer's work thus constituted one of the earliest and most developed forms of process philosophy in the twentieth century, different but analogous to the project of Alfred North Whitehead.

That the apparently neutral functionalist critique contained in works such as *Substance and Function* had broad implications for the social sciences and politics is evident when it is read in the series beginning with *Leibniz's System* and culminating in Cassirer's functionalist reading of the social sciences, law and state in *Freedom and Form: Studies in the German Intellectual History*. Concurrently with these works, Cassirer developed the historical reconstruction of "knowledge" in different epochs in *The Knowledge Problem*, which provided a history of changing definitions of experience or reality in the modern period, covering the historical permutations of such apparently immutable topics as the definitions of subjectivity and objectivity, spatiality and temporality.

Although Cohen's complex philosophy of origin and its foundation in the problem of the "infinitesimal" were often ridiculed for their incomprehensibility and for their occasionally opaque use of science, Cassirer saw that their importance lay in providing a foundation for – in the particularly terminology of the Marburg school – a "functional" (that is, fully relational and process-based) understanding of society and reality. The term "functional" for the Marburg school had a characteristically broad application as a mathematical metaphor used to define any moment of experience as pure "relational" determination.[70] Where later linguistic structuralists would use language as the basis of a relational definition of meaning, the Marburg school found mathematics a more productive and less fraught schema. Cassirer's early work sought to "correct" Cohen's occasionally bombastic philosophy of science while at the same time translating this central aspect of Cohen's philosophy into a more durable and comprehensible philosophical and political form.

The core challenge of presenting this philosophy, most clearly evident in *Substance and Function*, is that it is a philosophy based on difference in which experience is defined, as Cassirer writes, not through "the identity of ultimate substantial things, but the identity of functional orders and correlations. These functional correlations do not exclude the element of diversity and change, but are determined only in it and by it."[71] Describing a philosophy in which identity is determined only by "diversity and change" was, as Cassirer was well aware, not easy; as he notes in the first volume of his *Knowledge Problem*, a direct depiction of such a functional philosophy "is as crucial and unavoidable as it is difficult."[72] Moreover, the initial mode of understanding how unity exists only within difference is, in Cohen's work, through the challenging model of calculus. The unity in difference of functional relations is depicted as a relation similar to that of the integral and differential, and thus as a derivative directionality or movement within difference. Cohen's *System of Philosophy* was a sustained effort to depict such a philosophy of difference (or more accurately, philosophy of the differential) in epistemology, ethics and aesthetics. Cassirer considered *Substance and Function* an analogous argument, expanded to use new aspects of a "mathematical theory of the manifold" such as group theory, and using contemporary case studies from the exact sciences and psychology.[73] In neither case was the prospect of a wide reception of the foundation of this philosophy particularly promising outside of narrow academic circles.

A major early discovery of Cassirer's, however, was that he could outline this same challenging philosophy in a far more intuitive and clear manner through historical presentation. As developed already in the first volume of *The Knowledge Problem*, the historical presentation of changing definitions of subjectivity, objectivity and forms of knowledge allowed for a narrative and methodology that in turn allows, as Cassirer writes, "the phantom of the absolute to disappear from the first," and would illuminate a philosophy of difference "practically without effort and in full clarity."[74] In Cassirer's reading this philosophical project went beyond epistemology, and certainly beyond the epistemology of natural science to which it was frequently linked. Its focus was rather on the interrelation of knowledge, ethics and teleology in defining "reality," a project that anticipated and formed the foundation of his later critique of the ontological philosophies of Nicolai Hartmann and Martin Heidegger. On the practical level, this argument culminated in the related narrative of *Freedom and Form* ten years later in a surprising definition of liberalism based not on the "enclave" of the insular subject, but on understanding the conditions for the possibility of the creation of different forms of subjectivity and objectivity, and indeed reality, by the sciences and institutions of society themselves.

The intrinsic relation of science, in the broad sense encompassing the social and human sciences, with politics was the keystone of Cassirer's early work, as clearly outlined in his first major text, *Leibniz's System* (1902). The study culminated in a reading of four areas of the "human sciences" (*Geisteswissenschaften*) – history, law, aesthetics and a secularized reading of Leibniz's "theodicy" – that negatively defined the conditions for the possibility of the individual (in the broadest sense) and humanity.[75] In this reading, any particular form of subjectivity or association is defined through its contextual setting, and the task of the human sciences is to critique and cultivate this process. Although radically different in form, the underlying logic for this inflection of individuals is, in Cassirer's reading, the same as that which underlay Leibniz's calculus. In both cases the "individual" – a term Cassirer frequently uses in place of Leibniz's terminology of the monad – is real and empirically given, yet defined solely through its dynamic context. On this basis Leibniz provided, despite this apparently challenging foundation, the basis for a new definition of liberalism, rights and democracy that was, as Cassirer puts it later in *Freedom and Form*, "more sharply argued and more consequently executed [...] than by Locke, Montesquieu or Rousseau."[76] The basic impetus of Cassirer's early work was in applying this quintessentially modern "Leibnizian" approach to all of the modern sciences – from "political science" and "literary science" (*Literaturwissenschaften*) to the exact sciences – as an extension (and, as we will see, a critique) of the Marburg project he developed from Cohen.

The contemporary value of this endeavor was noted by Cassirer in a 1902 self-commentary on *Leibniz's System* published in the journal *Kant-Studien*, where he notes that the book highlights Leibniz's enduring importance, not simply in his influence for the natural sciences or mathematics (both widely evident by the turn of the century) or on the later philosophy of Kant. Rather, he concludes, the "most important" consequence of Leibniz's philosophy, specifically his monadology, for the present was in its "fruitfulness for the human and social sciences [*Geisteswissenschaften*] – in history, ethics and aesthetics – where it outlived the particular form of the system and has remained consistently productive in the progress of German idealism."[77] It was in extending this productive aspect of Leibniz's work, as modernized through Cohen, that Cassirer's early work would most powerfully develop, and in which it would find its particular political importance.

Hermann Cohen had earlier initiated a development of this aspect of Leibniz's work under the rubric of a "metapolitics." Parallel, Cohen writes, to how the Leibnizian "metageometry" or "hyper-Euclidean" geometry recontextualizes key aspects of geometry, and has been developed recently by the "new geometers" in redefining basic themes such as point and line, so a metapolitics would seek out foundational transformations to the basic definitions of politics, such as subject or state.[78] It thus entailed the "constructive" role of the sciences in defining new modes of subjectivity and objectivity from which any active politics first develops, and the demand for a constant critical reflection on the meaning and possible fungibility of different sciences.[79]

Cassirer's work eschewed the striking terminology of a "metapolitics," but developed many of its key themes in his studies of the sciences from *Leibniz's System* through *Freedom and Form*. Indeed, by creatively using the theme of group theory to explain the possible radical permutation of any structure through its elements, Cassirer developed a foundational approach to the theory of knowledge throughout his work that substantially improved on Cohen's early analogy to the "modern geometers."[80] For Cassirer subjectivity as well as objectivity are never completed projects but are defined by the developing functional ensemble in which they occur, and through which they can thus be thoroughly redefined. "The concept of the I," Cassirer writes, "is foreign to [Leibniz's system]," and the same proves true on similar grounds of Cassirer's own early work.⁸¹ Andrea Poma, one of the pre-eminent scholars of Cohen, concluded that Cohen's late work both anticipates and pre-emptively subverts many postmodern critiques and deconstructions of the subject, since in Cohen's philosophy "there would be nothing to deconstruct, the subject being the target of a construction process that had never been carried out, an infinite task."⁸² The same could be said of Cassirer's further development of Cohen's work, and it is indeed not least the model of this "infinite task" that has made Cassirer's Wilhelmine work both so challenging and so influential.

The foundation of Cassirer's early project can be suggested by noting its close historical and philosophical relation to the "negative theology" of earlier modern writers such as Moses Maimonides, Nicholas of Cusa and Pico della Mirandola. This relation was highlighted by both Cohen and Cassirer: Cohen's *System of Philosophy* began by situating Nicholas of Cusa as the modern founder of his philosophy of origin, and Cassirer's *Knowledge Problem* similarly began with Cusa as the historical foundation of his "critique of knowledge."⁸³ Just as in negative theology the goal was to avoid assumptions about the nature of divinity or absolutes by only defining aspects negatively ("God is not blind"), so for Cassirer and Cohen humanity and reality come to be defined as part of a single open-ended and unfathomable project, and yet one anchored in real historical forms and circumstances. It was not the medieval transcendental assumptions of negative theology that were pivotal for this approach, however, but rather the method through which this theology had defined its objects. The distinctly modern form of this method was first developed by Leibniz, particularly through the "new" and immanent form of determination epitomized by the logic of calculus, through which an object was defined negatively by its context yet was assumed to be "real" as a form of movement and development. In this way the world became part of an immanent "system of forces," or as Cassirer describes it, "an intensive manifold of form-giving functions [*gestaltender Funktionen*], which link with one another into a complete system of activities."⁸⁴ Here each individual entity (monad) is only known dynamically through its relation with its context and any provisional "form" of experience is open to infinite empirical exploration and creative transformation.

The refinement of Leibniz's philosophical method occurred in a development from Kant and Salomon Maimon (1752–1800) through Cohen, and was centered on a logical form of infinite (or limitative) judgment that was at the center of Cohen's late philosophy of origin. Already Jakob Gordin (writing in consultation with Cassirer, Albert Görland and Julius Guttmann) had noted in a book on the topic in 1929 that this apparently arcane problem – he concludes by suggesting the more approachable term "question judgment" – as it derived from negative theology could be considered the core of the late Marburg project as it is epitomized by Cohen's work, and the basis of Cohen's different understanding of dialectic and history from Hegel.⁸⁵ Particularly in conjunction with a related problem – the judgment of reality, also first suggested by Maimon – this theme defined a distinct turn in Cohen's late thought away from any latent form of subjective idealism and towards a new form of the immanent critique of science.⁸⁶ It functions, as Gordin summarizes it, as the "methodological fulcrum of philosophy," which reveals how the "transcendental method, the method of philosophy, is also the method of science, and not only the mathematical natural sciences, but also the social-historical sciences and not in any lesser degree that sphere of cultural thought in which is crystallized the facta of religion and the related religious science [*Religionswissenschaften*]."[87]

Cassirer developed Cohen's approach to these core themes into the foundation of his philosophical and historical method, most notably through a reading of what we will term the problem of reality, while often substituting historical or thematic presentation for Cohen's often highly technical terminology.[88] As developed in *Substance and Function* and *Freedom and Form*, this approach formed the basis of a process philosophy that allowed him to avoid any substantial definition of objects or of subjects, but rather assume that these were co-defined as a unitary continuum by the system of knowledge and relations in which they were found. The "being" of humanity and reality, as earlier of God, is defined negatively and relationally, even as it is taken as an immanent but continuously changing presence.[89] Cassirer's frequent touchstone in this regard was the Renaissance definition of humanity in which, in Pico della Mirandola's words, humanity is a "work of indeterminate form […] to which nothing of its very own could be given," so that it is instead progressively shaped by its own history and actions.[90]

Far from leading to an abstract view of humanity, Cassirer's approach focuses on the underlying logic of practices in history and the sciences and the application of this study to society. Cassirer summarizes the basis of this Leibnizian view in stating that "being, particularly the being of spirit [*Geist*], is only revealed and unfolded in activity."[91] Leibniz's work forms a comprehensive theory of the interrelation of the individual, knowledge, society and reality that Cassirer in turn suggested could still serve as a model for the sciences in the twentieth century. For Cassirer, Leibniz's philosophy is the basis for both the Romantic tradition of aesthetics and the Enlightenment tradition of science, and returning to the common philosophical foundation of this historical development of experience (particularly through the problem of "determination" (*Bestimmung*) of the particular) is central for understanding modernity.[92] Despite the rationalism of Cassirer and Cohen, a key aspect of the late Marburg project holds that both individually and collectively humanity is in a state of productive, structured and emergent creation – and that this creation occurs as much in what Leibniz called the "confused" synthesis of the aesthetic perception of reality as it does in discrete rational calculation. The development of the social, ethical and natural sciences hinges on a constant awareness of this creative synthesis of reality as a means of skepticism, a search for new axes of continuity within processes, and an ethical orientation towards the future.

Although the clearest version of such continuities resemble the Kantian *a priori* in what Cohen described as "pure thought," even these forms are theoretically capable of functional permutation and reinterpretation – a process that is first clearly defined by Cassirer's so-called "invariant theory of truth" as it develops from the group theory of Felix Klein.[93] Here Cassirer used group theory to define a series of "invariants" between any permutations of sets, and these invariants form by analogy the only definition of a "truth" within a particular experiential system or perception. Moreover, in a second marked departure from Kant, *a priori* truths for Cassirer are themselves considered as the limit case of a continuum of invariants of form that stretches from laws and principles on one side to mores or indeed any consistent phenomenon on the other.

Invariants of experience are dialectically revealed in a manner that leads to richer forms of human knowledge and sociation, but these invariants are not reducible to Kantian categories nor, as was sometimes assumed in the reception of Cohen and Cassirer, a development similar to the Hegelian Idea.[94] Instead, history and the sciences have a dialogic and open-ended quality, even as they can also reveal real continuities in the world.[95] The power of this model can be suggested from an example drawn from the science of law at the end of *Freedom and Form* but of universal application, where Cassirer endorses Fichte's dictum that "the constitutional moment is not the historical past of a country [...] but its future."[96] In the same manner, by finding analogous underlying "constitutional" logics of any science or practice, to broaden Fichte's statement, Cassirer held that new developmental transformations of the sciences and developmental futures for humanity could be promoted.

The narrative presented here largely follows a chronological survey of Cassirer's key works from his earliest dissertation on Descartes in 1899 (published in 1902 with *Leibniz's System* and "necessarily woven together with it" as part of an "originary plan") through 1919, the year of the Weimar Republic's constitution, with occasional excursus where this work is illuminated by Cassirer's late philosophy.[97] In this reading, Cassirer's defense of the new republic effectively superceded his attempt to develop a particular form of Wilhelmine liberalism and his related critique of the sciences. In 1918, the end of the First World War, Hermann Cohen's death and the founding of the Weimar Republic effectively brought to an end the middle phase of Cassirer's life. Only with the establishment of the Weimar Constitution in 1919 following the Spartacist Uprising, however, can we truly imagine Cassirer starting out on a new philosophical and political direction.[98] These most clearly came to light in 1921, with the beginning of Cassirer's published work on a philosophy of symbolic forms.[99]

Cassirer clearly wanted his early work to be accessible to any educated (albeit dedicated) reader of his day, even though he also often availed himself of what now appears to be a daunting range of examples and ideas from a number of disparate sciences. In presenting one historical and philosophical path through his early work, I have attempted to renew this accessibility to the best of my ability, even as I am well aware that in a number of places my expertise falls short of the subtleties of Cassirer's arguments. My hope is that in presenting the broad contours and contexts of Cassirer's early work as a synoptic narrative whole, I have partially compensated for any specific deficiencies in presentation.

Bibliographic Context

In a manner characteristic of the split of academic study into "two cultures" of natural science and the humanities following the First World War, most English language accounts of Cassirer have assumed that his early work on science was largely removed from his later philosophy of culture. This early work is in any case taken as of value primarily as a chapter in the philosophy of natural science, rather than as part of a wider critique of

Wilhelmine society or the development of the social and human sciences. An important current of continental thought has provided a counterpoint to this approach by linking Cassirer's work on natural science with his broader philosophy, and Cassirer's early work to Cohen's late philosophy, but the focus has not principally been on the political and social dimensions of the Marburg critique of the sciences. By tracing the development of Cohen's central ideas as the leitmotif to Cassirer's approach to a reform of science and society, the present work hopes to provide a single vantage for understanding what is original and historically important in Cassirer's early work.

A brief overview of recent work on Cassirer, with a particular focus on work in English, suggests the importance of reconstructing the early phase of his career, and the ease with which his late work can be misread without this foundation. The persistence of a "dualistic" reading of Cassirer, split between his scientific and cultural work as well as his early and Weimar periods, is epitomized by the most recent philosophical biography of Cassirer's Weimar work by Edward Skidelsky. Skidelsky's superbly written survey, *Ernst Cassirer: The Last Philosopher of Culture*, focuses largely on Cassirer's studies of culture from the Weimar period. As such the text understandably does not develop the historical meaning of Cassirer's Wilhelmine work, and indeed defines Cassirer's maturity precisely through his wartime break from Cohen's philosophy and purportedly domineering personality.[100] Skidelsky follows a number of earlier commentators in characterizing Cohen's central philosophy of the infinitesimal as a "bizarre doctrine" that anchored a "mathematical dogmatism" and philosophy of science of little worth.[101]

This dismissal of Cohen's philosophy was certainly not shared by Cassirer, however, who wrote several spirited and entirely cogent defenses of the universal philosophical importance of Cohen's use of the infinitesimal and theory of science. Moreover, Cassirer's pivotal first book on Leibniz – a text not mentioned by Skidelsky – provides an extensive and sensible contextual reading for understanding the place of the infinitesimal in Leibniz's work, and by extension in Cohen's as well. Looking back on his career in 1943, Cassirer considered his relation with Cohen the most vexed and central aspect of his development. "My attachments with him," he said, "and my disentanglement from him, both are vital."[102] By avoiding these "attachments," Skidelsky's interpretation strays from Cassirer's written statements. The pivotal transformation of Cohen's ideas in Cassirer's *Substance and Function*, for instance, is elided by claiming that the text is heavily indebted to the logic of Bertrand Russell, which is – despite a relative paucity of references to the work – said to save Cassirer from Cohen's mathematical functionalism. *Substance and Function* culminates, however, in a claim that Cassirer is advancing the opposite view from Russell's on "the general tendencies dominating present epistemology" in his understanding of the "constructive" value of logic.[103] Cassirer argues that his own version of this "constructive" logic developed from his reading of the "mathematical theory of the manifold," which is clearly a continuation of both Cohen's philosophy (as well as Leibniz's) and, as he states, that of Cassirer's teacher Benno Erdmann's theory of groups – much less so of Frege or Russell, despite his recognition of their importance.[104]

For Cassirer, this general "constructive" logic is of the broadest possible import, developing the theme of the infinitesimal and limit analogously to that of Leibniz and Cohen in a manner that forms, as Cohen states, "a law of the production of content in aesthetics particularly, as well as in logic and ethics."[105] By failing to outline the intrinsic relation of Cohen's theory of the infinitesimal to his logic, ethics and aesthetics, Skidelsky incorrectly assumes that Cohen has "no forms of objectification over and beyond natural science."[106] On this basis, Skidelsky's text in turn fails to address Cohen's actual critique of the sciences in Wilhelmine Germany as well as the originality of a project of reform in the human and social sciences that arguably led Cassirer to become his principal advocate.

By sidestepping the admittedly thorny world of Cohen's writings, Skidelsky mistakes key continuities and ideas in the basic structure of Cassirer's philosophy as it develops and critiques Cohen, most importantly the role of the infinite in his philosophy, his threefold definition of the sign (sign, signified and the infinite horizon of context connecting the two) and the connection of this theory with phenomenology and the critique of the ontological philosophies.[107] Ultimately, this leads his interpretation of the later Cassirer, particularly in relation to Heidegger and ontology, to misinterpret or ignore key programmatic statements.[108] The famous debate with Heidegger at Davos, for instance, is usually considered to hinge on their respective positions on finitude and infinitude, but Cassirer's central reading of the latter problem is fully comprehensible only through his extensive prewar texts on this issue, such as *Substance and Function*, and their relation to Leibniz and Cohen's theory of infinitude.

Although Cassirer strongly criticized aspects of Cohen's philosophy, and was indeed skeptical of aspects of his philosophy of science, his gradual "disentanglement" from this philosophy already in the first half of his career forms the basis of his own work, both in his early texts on the sciences and in his later critiques of the life and ontological philosophies. In this regard, Peter E. Gordon's recent *Continental Divide: Heidegger, Cassirer, Davos* has markedly raised the level of discussion of both Cassirer's work and his famed Davos debate with Martin Heidegger.[109] Yet although the text provides a lively reconstruction of Cassirer's later statements on this debate, and the surprising commonalities of his position to that of Heidegger through Cohen, it does not fully recognize Cassirer's earlier work as a basis of key themes in the discussion, again notably in the starting point from *Leibniz's System*. In this regard the present book can be considered as something of a prelude to Gordon's later account.

Despite the range of Cassirer's and Cohen's philosophies, their works were nonetheless first a critique of contemporary science, broadly defined, and it is in this regard notable that some of the most important recent works on Cassirer have been in the field of the philosophy of the natural sciences. The most illuminating insights for the present work have been those of Michael Friedman, Norbert Ihmig, Thomas Ryckman and Christiane Schmitz-Rigal.[110] Ryckman's *The Reign of Relativity: Philosophy in Physics, 1915–1925*, to take as an example, overturns dated readings of Cassirer as a traditional neo-Kantian to revisit his conflict with the logical positivists in the philosophy of science, arguing for the importance of Cassirer's emphasis on a "relativized *a priori*," that is, a rejection of final knowledge of the *a priori* in favor of one in which the Kantian *a priori* becomes, somewhat paradoxically, historically fungible. The *a priori* both reveals invariables in experience *and* is capable of development and transformation. It has, Ryckman writes, "*both* a 'constitutive' and an ideal 'regulative' *a priori* significance."[111] Ryckman in turn tentatively links this position to what has been described as a "structural realism," in the sense that reality can only be revealed within a structural ensemble, an argument further developed recently by Georges Ibongu and others.[112] Such a position, developed by Cassirer in parallel with Weyl and Eddington, explains in Ryckman's words "how and why *a priori* constraints of reasonableness can be imposed on nature without proudly (but naïvely) presuming them to be inherent in nature itself."[113] Ryckman argues that none of these authors ever presented a "fully worked out epistemology of science" in relation to relativity or this position.[114] The present work attempts to provide a prehistory of this concept in Cassirer's work. In this regard it also forms something of an introduction to the equally important recent work on Cassirer's philosophy of science by Christiane Schmitz-Rigal. Beginning with *Die Kunst offenen Wissens: Ernst Cassirers Epistemologie und Deutung der modernen Physik* (2002), her work has depicted Cassirer as developing an "art of open-knowledge" and process philosophy in a manner that the present essay aims to trace in its earlier development from Cohen.[115]

Recent work has emphasized, and occasionally overemphasized, the parallels of Cassirer's philosophy with "postmodern" thinkers such as Deleuze, Derrida, Foucault and Lacan, but relatively little attention has been directed to a historical basis for these continuities in Cassirer's early transformation of Cohen's functionalism and philosophy of difference. Thus, S. G. Loft's *Ernst Cassirer: A Repetition of Modernity* (2000), for instance, presents an introduction and synopsis of Cassirer's work in its relation to contemporary theory, but the focus is largely in terms of philosophical family resemblance, not historic development. Similarly, Drucilla Cornell's excellent *Moral Images of Freedom: A Future for Critical Theory* (2008) and *Symbolic Forms for a New Humanity* (with Kenneth Panfilio, 2010) attempt to reinvigorate critical theory by using aspects of Cassirer's later symbolic forms project. A vital historical and philosophical source for the reading of symbolic forms in this regard could be found in Cassirer's early development of Cohen's philosophy of the infinitesimal and origin. As Andrea Poma, the eminent scholar of Cohen, has noted, the contemporary relevance of Cohen's work is best approached by understanding it as a philosophy of difference that is distinctly modern even as it anticipates – and allows for a critique of – key themes of postmodernity (his particular example is Cohen's influence on Deleuze's *Identity and Repetition*).[116] A similar approach can be applied to Cassirer's work. Rather than valorizing, as some later postmodern philosophies will, the totalization of difference or change itself, Cassirer's underlying goal is to explain the paradoxical appearance of unity and meaning in difference (as argued in *Substance and Function*), and with this both the logic of change between differing forms of experience and the basis for progress or freedom, however defined, in modern society (as further developed in *Freedom and Form*).

The recent continental revival of neo-Kantianism as a critical contemporary philosophical field is in this regard crucial to the present study.[117] The works of philosophers such as of Fabien Capeillères, Massimo Ferrari, Amos Funkenstein, Wolfgang Marx, Andrea Poma, Enno Rudolph and Ulrich Sieg, among others, have all greatly illuminated my understanding of Cassirer's early works. The importance of Cassirer's philosophy to the development of the sciences and the Marburg critique of Wilhelmine society has been partly developed by Sieg and Köhnke in the history of the Marburg school, and Meyer and Ferrari through Cassirer's philosophical biography.[118] The dialog of Cassirer's work with that of Cohen has been developed by Massimo Ferrari, Andrea Poma and Ursula

Renz in a manner that informs my reading here, although without the emphasis on the role of science, metapolitics and "the problem of reality" I have tried to develop.[119]

Although concurring with most of the content of the recent reinvigoration of neo-Kantianism, and hoping to spur a renewed appreciation of the creativity and breadth of this movement, the present work avoids the terminology of "neo-Kantian" in relation to Cassirer's early works, as it is partially misleading. Cassirer's critique is certainly developed from the neo-Kantian focus on the transcendental question, that is, an attempt to understand the transcendental rules underlying any "facta" or given, but in important ways Cassirer follows the late Cohen beyond this tradition. Indeed, the following essay assumes that despite its evident value, the term "Neo-Kantian" is a foil to understanding both Cassirer and his relation with Cohen, particularly for an anglophone audience. As John Michael Krois noted early on in the recent revival of Cassirer scholarship, this is certainly true if the term is taken to equate with "a kind of subjective idealism" or reducible to an epistemology of the natural sciences.[120] Yet it can also be true in more sophisticated readings of neo-Kantianism. Whereas the Marburg school is undoubtedly "neo-Kantian" in its origins and often in its methodology, Cassirer's philosophical starting point was in the *late* work of Cohen, where Cohen notably departed from key elements of Kant's philosophy in his so-called philosophy of origin (*Ursprung*).[121] Here Cohen opened a path for a new philosophy, which by 1914 he described as presenting a "drastic challenge [...] to the key features" of Kant's philosophy, even as he argued that it was nonetheless in continuity with its innermost logic.[122] The themes of the infinitesimal judgment and the judgment of reality, for instance, develop within Kant's reading of the schematism as the basis of unifying understanding and sense but ultimately supersede it in a new manner, and this has far reaching influences for a definition of society, reality and science.[123]

Due to this starting point, which was of central importance not only to Cassirer but to his later philosophical antagonist Martin Heidegger, describing Cassirer's work as "neo-Kantian" often inadvertently obscures key moments of its originality. As Cassirer suggests in a 1914 essay on post-Kantian thought, the post-Kantian reception gradually recast each of the Kantian oppositions (noumenal/phenomenal, subject/object, being/ethics) as part of a continuum, and Cohen's work is implicitly the end point of this development.[124] The term "neo-Kantianism," on the other hand, is often mistaken, at least in anglophone thought, as re-establishing the "obvious" Kantian divisions in Cassirer's thought in these oppositions, such as the split of the consciousness from experience, or ethics from epistemology, and is thus misleading from the outset.[125] Indeed, the mistaken assumption that Cassirer's work begins with the traditional Kantian divisions – such as those we might find in Heinrich Rickert's contemporaneous *The Limits of Concept Formation in Natural Science* (1902) – make its key tenets difficult to grasp, notably Cassirer's emphasis on difference, immanence and reality as they are first developed in *Leibniz's System*, and as they form the basis of his continued dialog with phenomenology and ontology.[126] Cassirer's work, moreover, primarily develops from the later volumes of Cohen's *System of Philosophy*, which are focused on ethics, aesthetics and (in a never-completed third volume) psychology and pedagogy, and entail problems that go beyond epistemology.

Like a gestalt form or group, such as the multistable image of a "Rubin's Vase" that can also be seen as two faces regarding each other, the term "neo-Kantian" reveals one set of relations but may conceal others. It functions similarly to what might be an equally accurate description of Heidegger as a "neo-Kierkegaardian" or Marx as a "neo-Hegelian": these statements may be true, but the form of statement itself conceals a key aspect of what is new in their work. The same is true, albeit to a lesser degree, for categorizing Cassirer as neo-Kantian. Later moments in Cassirer's philosophy where he directly claims to supersede Kantian or Husserlian approaches, such as in his definition of "symbolic *Prägnanz*," can be traced back to the earlier moments in his work where the Kantian philosophy had already been radically transformed, as we will argue here in his development of Cohen's logic of the limit, infinite judgment and infinitesimal.[127]

For this reason, the present study uses Leibniz as an alternate approach for understanding the Marburg school's originality, and the means through which it transformed the Kantian project.[128] Cassirer's reception of Leibniz, in particular, functioned both as a "critique of Kant," as Enno Rudolf has it, and as the outline of a new philosophy. In this regard, I agree with Massimo Ferrari's recent conclusion that "Cassirer [is] a neo-Kantian methodologically speaking," and indeed is perhaps the truest of the Marburg school to Cohen's vision of the priority of understanding "*a priori* forms […] as processes."[129] Precisely as a process philosophy that also attempts to understand the foundation of various facets of "reality," however, the originality of Cassirer's work can also profitably be read from the perspective of Leibniz. Leibniz's philosophy was applied by Cassirer in particular to illuminate the leading edge of contemporary science and conceptualization, notably in fields such as group theory and logic, and to create a distinctly modern philosophy. Yet Cassirer also used Leibniz as the basis of his interpretation of nearly the entirety of German classicism: "Neither Lessing, Herder, nor Goethe," he states, "are followers of Leibniz's system, but they are all dependent on the forms that were created from within it."[130] Leibniz's logic also illuminates an earlier history developing from the negative theology of figures such as Nicholas of Cusa and Maimonides that allowed for a radical recasting of the problem of appearance and reality. Most importantly, Cassirer saw Leibniz's work as the basis for a political definition of rights and democracy. Speaking at the annual "Constitution Day" of the Weimar Republic at Hamburg University in 1928, Cassirer singled out the importance of Leibniz's work in this regard: "work that is ever more astonishing in the range and depth of its theoretical accomplishments."[131] Although known for its importance in many other fields, Cassirer writes, it is the "brilliance" of Leibniz's political writing and ideas that is central for understanding the German roots of the modern idea of rights and the Weimar Constitution.[132] When we search for Cassirer's argument for this Leibnizian politics and its relation to Leibniz's wider work, however, and particularly his understanding of the sciences, we find it not in the Weimar period so much as in his earliest works and process philosophy.

Notes

1 Bahr, Hermann. "Über Ernst Cassirer," in *Die Neue Rundschau*, vol. 28 (Berlin: S. Fischer, 1917), 1484.
2 The term appeared most prominently for a Bahr symposium and associated publication, *Hermann Bahr – Mittler der europäischen Moderne: Hermann-Bahr-Symposion, Linz 1998*, ed. Kurt

Ifkovits and Lukas Mayerhofer (Vienna: Adalbert-Stifter-Institut des Landes Oberösterreich, 2001).
3 Bahr, "Über Ernst Cassirer," 1484.
4 Cassirer, Ernst. *Das Erkenntnisproblem in der Philosophie und Wissenschaft der neueren Zeit*, vols 1–4 (Darmstadt: Wissenschaftliche Buchgesellschaft, 1994 [1906–1940]) (ECW 2–5). I have departed from the English title translation of the fourth volume, the only volume printed in English, *The Problem of Knowledge: Philosophy, Science and History since Hegel*, trans. William H. Woglom and Charles W. Hendel (New Haven: Yale University Press, 1978). Henceforth it shall be referred to as *The Knowledge Problem*, more suggestive of Cassirer's dual meaning as both a study of problems within knowledge and knowledge as itself a problem.
5 On Bahr see Berlage, Andreas. *Empfindung, Ich und Sprache um 1900: Ernst Mach, Hermann Bahr und Fritz Mauthner im Zusammenhang* (Frankfurt am Main: Lang, 1994); Daviau, D. G. "Hermann Bahr: The Catalyst of Modernity in the Arts in Austria During the Fin de Siècle," in Daviau, D. G. *Understanding Hermann Bahr* (St Ingbert: Röhrig Verlag, 2002), 95–112; Daviau, D. G. *Hermann Bahr*, Issue 744 of World Authors Series (Woodbridge, CT: Twayne, 1985); Dittrich, Rainer. *Die literarische Moderne der Jahrhundertwende im Urteil der österreichischen Kritik: Untersuchungen zu Karl Kraus, Hermann Bahr und Hugo von Hofmannsthal* (Frankfurt am Main: Lang, 1988); Farkas, Reinhard. *Hermann Bahr: Dynamik und Dilemma der Moderne* (Vienna: Bohlau, 1989).
6 Bahr, Hermann. *Der Antisemitismus: Ein internationales Interview* (Weimar: VDG Weimar, 2005 [1894]).
7 Cassirer, Ernst. *Freiheit und Form: Studien zur deutschen Geistesgeschichte*, henceforth *Freedom and Form*. Cassirer had, moreover, taken the unusual step for him of publically speaking on the most political section of the text, its final chapter on the relation of the "Freedom Idea and the Idea of State" (FF 303), in his 1917 lecture "Der deutsche Idealismus und das Staatsproblem," in *Zu Philosophie und Politik, Nachgelassene Manuskripte und Texte*, vol. 9, ed. John Michael Krois and Christian Möckel (Hamburg: Felix Meiner, 2008), 3–27, commentary 277ff. On Bahr's wartime writing, see Ifkovits, Kurt. "'Nur noch Deutsche!' oder 'slawisches West-Reich' – Hermann Bahrs Kriegspublizistik in den Jahren 1914/15," in *Das Gewebe der Kultur: Kulturwissenschaftliche Analysen zur Geschichte und Identität Österreichs in der Moderne*, ed. Johannes Feichtinger and Peter Stachel (Innsbruck: Studienverlag, 2001), 209–231.
8 The claim that Cassirer was the first Jewish German rector of a German university was made at the time and since, and was assumed to be a continuation of Cohen's unusual academic success as a nonconverted Jew. Cassirer was not in fact the first, however, and the number of earlier rectors who may be considered Jewish or partially Jewish has been noted to be as high as four, as claimed for instance in an aside by Peter E. Gordon in his review of Thomas Meyer's *Ernst Cassirer* (Hamburg: Ellert & Richter, 2006) in the *Association for Jewish Studies Review* 32 (2008): 436.
9 Bahr, "Über Ernst Cassirer," 1488.
10 See Hansson, Jonas and Svante Nordin. *Ernst Cassirer: The Swedish Years* (New York: Peter Lang, 2006).
11 ZWW, henceforth *Objectives and Paths*.
12 Mannheim, Károly. "Rezension von Ernst Cassirer, Freiheit und Form: Studien zur deutschen Geistesgeschichte," *Atheneum* 3 (1917): 409–13.
13 Bahr, "Über Ernst Cassirer," 1489, 1497–8; Bahr had earlier, and more famously, linked the philosophy of Ernst Mach with the movement of impressionism, declaring that Mach was the "philosopher of the era" who had "most clearly spoke to the life-movement of our generation, of our feeling for the world." Bahr had specifically declared in 1903 that Mach's "Analysis of Sensation" (1881) was the philosophical and scientific counterpoint of artistic impressionism. Arens, Katherine. *Functionalism and Fin de Siécle: Fritz Mauthner's Critique of Language* (Bern: Peter Lang, 1984), 253; Bahr, "Impressionism," in *Dialog vom Tragischen* (Berlin: S. Fischer, 1904), 113. Cassirer's work was apparently taken by Bahr as the next stage of this philosophical functionalism, and was assumed to be linked to wider cultural forces. The concept that a philosophy of science

could define an era and be closely linked to contemporary artistic, political and cultural currents was entirely intelligible to Bahr's *fin de siècle* world in a manner that was rapidly becoming all but indecipherable.

The surprising futurity of Cassirer's extension of the Marburg Project – its combination of modernism with a focus on transformative potential – is what appealed to Bahr and proves so difficult to recover today. It is in this regard that the cover image of this book –taken from the penultimate image of architect Bruno Taut's utopian *Alpine Architecture* (1917/18) – is intended to suggest the futurity and inventiveness of form Bahr found in Cassirer's work. It also highlights Cassirer's understated biographical relations to his time and society. Ernst Cassirer was close to his more politically and artistically progressive cousin Paul Cassirer, who ran the leading impressionist gallery in Berlin and worked with Taut on the *Arbeitsrat für Kunst*, an artists' collective that emerged in 1918. (Taut, Bruno. *Bruno Taut – Alpine Architecture: A Utopia*. Edited with an Introduction by Matthias Schirren. New York: Prestel Publishing, November 2004.) The original image can be found in *Alpine Architektur des Architekten Bruno Taut*, (Hagen Folkwang-Verlag, 1919), available online at the Historischer Buch- und Zeitschriftenbestand der Weimarer Kunst- und Bauhochschulen: http://goobipr2.uni-weimar.de/viewer/image/PPN679693351/77/ (accessed 12 December 2012).

14 Bahr, "Über Ernst Cassirer," 1498, 1492.
15 Troeltsch defines Cassirer largely through his relation to the Marburg school, "the philosophical school to which Cassirer is first and foremost a member" and considers his work a somewhat one-dimensional defense of the Enlightenment, in which "[the] concept of humanity, which is entirely limited [here] to the formal and rational aspects, goes along and whose approach it is to answer all questions in relation to forms of reason, and from this form to gain the content" (Troeltsch, Ernst. Review of *Freiheit und Form: Studien zur deutschen Geistesgeschichte* (Berlin: Bruno Cassirer, 1917), reprinted in *Aufsätze zur Geistesgeschichte und Religionssoziologie*, ed. Hans Baron (Tübingen: J. C. B. Mohr, 1925 [1917]), 698). Troeltsch ultimately does not appear to follow Cassirer's functional argument as a whole, however, recognizing for instance that the discussion about Goethe's definition of the form is the center of the book and that the text concludes by applying this definition of form "to ethics and state philosophy […] where Hegel is something of the Goethe of Social Philosophy" (697). He does not address, however, how this would fundamentally change the nature of Hegelian dialectics and their application to the social sciences and state theory, particularly liberal Humboldtian theory, which is the center point of the book and of Cassirer's actual relation to the Marburg philosophy. Troeltsch's real complaint is that Cassirer does not truly address either medieval or Romantic ideas of freedom, and particularly what he sees as the tragic relation of the "irrational fantasy drive of the gothic" era as it comes into contact with "foreign form principles." Instead, Cassirer is thus said, inaccurately, to present freedom as a "bounded world of individual-rational autonomy" or "aesthetic rational autonomy" (Troeltsch, Ernst. *Deutscher Geist und Westeuropa: Gesammelte Kulturphilosphische Aufsätze und Reden* (Tübingen: Scientia Verlag, 1966 [1925]), 232).
16 In this regard I agree with Edward Skidelsky: "The Marburg school's philosophy of science was primarily an effort to *overcome* naturalism and positivism, thereby demonstrating the affinity of science with ethics, aesthetics and theology" (Skidelsky, Edward. *Ernst Cassirer: The Last Philosopher of Culture* (Princeton: Princeton University Press, 2008), 244n57). I would stress, however, that although there is a science of ethics (law), aesthetics (art history, among others) or theology, Cassirer and Cohen also stress that in different ways the phenomena of ethics, aesthetics and theology all stand beyond the ambit of any particular science or of the concept of science more generally.
17 With the exception of the Leibniz text, these are the key books that Bahr claims represent the "curriculum vitae of the new truth" (Bahr, "Über Ernst Cassirer," 1498). The term "human sciences" or "social and human sciences" will be used to translate *Geisteswissenschaften*, a notoriously untranslatable term (literally, "sciences of the spirit"), which as we will see had a particularly broad meaning for the Marburg school.

18 Bahr, "Über Ernst Cassirer," 1493.
19 In this regard, Cassirer's understanding of the "crisis" of his era was fairly consistent throughout his career; it was a "crisis in man's knowledge of himself," as he put it in his 1944 *Essay*, in which the "homogeneity of human nature" was in doubt as a cultural assumption, a scientific tenet and a philosophical theme. This crisis and the related "crisis of historicism" contained in his view the possible basis of their own resolution if understood from the perspective of a functionalist philosophy of difference (E 19). On the vast literature of "crisis," see the overview in *Heidegger, Dilthey, and the Crisis of Historicism* (Ithaca, NY: Cornell University Press, 1995), 3–19; and the recent overview (with an emphasis on the later postwar period, but relevant here) of Gordon, Peter E. *Continental Divide: Heidegger, Cassirer, Davos* (Cambridge, MA: Harvard University Press, 2010), 43–86.
20 Timothy Raymond Keck's valuable dissertation, *Kant and Socialism: The Marburg School in Wilhelmine Germany* (Madison: University of Wisconsin, 1975) provides a sense of this transition in arguing at once that Cassirer was the "genuine heir to Cohen and Natorp," even as the next step of politics from their engaged socialism does not occur, in his view, until Cassirer's *Myth of the State* in 1945 (414). An alternate reading, however, is that although Cohen's move to Berlin in 1912 could be read as the "disintegration as a unified movement" of Marburg neo-Kantianism, it was also the culmination of Cohen and Cassirer's common work on a critical science of society, one that Cohen developed through specific focus on *Religionswissenschaft*. The strong current of Marburg socialism that Keck ably depicts was with this turn, however, effectively over.
21 Cassirer, Toni. *Mein Leben mit Ernst Cassirer* (Hamburg: Felix Meiner Verlag, 2003), 40.
22 Nipperdey, Thomas. *Deutsche Geschichte, 1866–1918, Erste Band: Arbeitswelt und Bürgergeist* (Munich: Verlag C. H. Beck, 1998), 681.
23 Paul Natorp to Albert Görland, 10 July 1911, in Staats- und Universitätsbibliothek Hamburg, Nachlaß Görland, n. 195, cited and further discussed in Sieg, Ulrich. *Aufstieg und Niedergang des Marburger Neukantianismus: Die Geschichte einer philosophischen Schulgemeinschaft* (Würzburg: Königshausen & Neumann, 1994), 344.
24 Repp, Kevin. *Reformers, Critics and the Paths of German Modernity: Anti-Politics and the Search for Alternatives, 1890–1914* (Cambridge, MA: Harvard University Press, 2000); Pieter M. Judson, *Exclusive Revolutionaries: Liberal Politics, Social Experience and National Identity in the Austrian Empire, 1848–1914* (Ann Arbor: University of Michigan Press, 1996); Ringer, Fritz K., *The Decline of the German Mandarins: The German Academic Community, 1890–1933* (Hanover, NH: Wesleyan University Press, 1990).
25 Repp, *Reformers*, 1, quoting Howe, Frederic C. *Socialized Germany* (New York: Scribner, 1915), 4. The Second Industrial Revolution is usually defined by the application of research science to industrial processes, notably in the chemical, electrical and automotive industries, and had its most marked success in Germany and the United States. The term gained popularity with David S. Landes (*The Unbound Prometheus: Technological Change and Industrial Development in Western Europe from 1750 to the Present* (New York: Cambridge University Press, 2003 [1969]), 4, 231ff.) and is nicely developed through Joel Mokyr (*The Gifts of Athena: Historical Origins of the Knowledge Economy* (Princeton, NY: Princeton University Press, 2004), 64ff.). It was first used by Patrick Geddes (*Cities in Evolution: An Introduction to the Town Planning Movement and the Study of Civics* (London: Williams & Norgate, 1915 [1910]) 46, 59) and described at length for "oil fuel, electrical industries, etc." following the "the very foremost of all Prometheans of electricity in Lord Kelvin" (60ff.). The term serves as a useful reminder of the era in which Cassirer was writing and the primacy of the concept of science that he and Cohen saw in all aspects of society.
26 Stern, Fritz. *The Politics of Cultural Despair: A Study in the Rise of the Germanic Ideology* (Berkeley, CA: University of California Press, 1974).
27 FF 192. Cassirer concludes *Freedom and Form* by writing that "German intellectual history" will have the value of having revealed the "deeper foundations" for the development of the

problem of freedom and the state, and with it a transformation of the nostalgic "longing" into something productive (386). The popular reception of works such as *Freedom and Form* is suggested by the large number of reviews recommending the work for its accessibility as an early university or advanced highschool (*Gymnasium*) primer on the topic, see for instance the overview of reviews by Walter Eggers and Sigrid Mayer (*Ernst Cassirer: An Annotated Bibliography* (New York: Garland Publishing), 243–71), in particular examples such as C. L. A. Pretzel's review in *Die Schule* 21 (1917): 619.

28 Specifically, a parallel could be drawn to the relation of Hegel's early theology (*Einheitstheologie*) to his philosophy, as suggested by Dieter Henrich's classic *Hegel im Kontext* (Frankfurt am Main: Suhrkamp, 1981 [1967]), or in turn to an "early" Marx more closely linked to Hegel following the various receptions of the 1844 manuscripts.

29 The strongest argument for this methodology is found simply in Cassirer's later continued reference back to his early works as the foundation for reading his later works. Indeed, frequently footnotes citing his earliest works, if followed carefully, intrinsically redefine the meaning of later statements.

30 Cassirer, Ernst. "Cohen's Philosophy of Religion," *Internationale Zeitschrift für Philosophie* 1 (1996): 90. Reprint of discussion from June 1935 to the Oxford Jewish Society.

31 Kracauer, Siegfried. *The Mass Ornament*, trans. Thomas Y. Levin (Cambridge, MA: Harvard University Press, 1995), 213–14.

32 Foucault, Michel. "Une histoire restée muette," *La Quinzaine Littéraire* 8 (1966): 3–4.

33 Ibid.

34 Foucault, "Une histoire," 2.

35 Ibid., 4.

36 Ibid.

37 Ibid.

38 Ibid.

39 Ibid.

40 Ibid.

41 Following the model of group theory, any change in a historical era for Cassirer can completely change the functional meaning of each aspect – the only connecting element will be the questions that first caused this transition. In this regard, Cassirer's model of epistemic change, as we'll see below, is ultimately closer to that of Hans Blumenberg than to Foucault. "Here it is the 'functional form' itself that changes into another; but this transition never means that the fundamental form absolutely disappears, and another absolutely new form arises in its place. The new form must contain the answer to the questions, proposed within the older form; this one feature establishes a logical connection between them, and points to a common form of judgment, to which both are subjected" (SF 268). Foucault does, however, use a similar model of "transformable groups" for historical change in his later *Archaeology of Knowledge and the Discourse on Language*, trans. Rupert Sawyer (New York: Vintage, 1982), 127.

42 Foucault, Michel. *The Order of Things* (New York: Vintage, 1973), 387.

43 Nor is it surprising that Foucault would be well aware of them, both in his extensive exchanges on the history of science with Georges Canguilhem and in his own work. Foucault had been closely connected with the University of Hamburg in 1959–60 during his directorship of the French Institute in Hamburg, for instance, where Cassirer's influence on the philosophy department and indeed the university as a whole remained strong.

44 EK1 v ff.; ECW2 ix ff.

45 I 6ff.

46 EK1 8, 10; ECW2 7, 9.

47 EK1 10; ECW2 9. An even stronger commonality of Foucault's work with that of Cassirer is found in Foucault's further development of his theory in *The Archaeology of Knowledge*. Here Foucault puts forth the demand for a "historical *a priori*" that is "not a condition for the validity

of judgments, but a condition of reality for statements." As we will see, a similar development of the problem of reality and the "Real" from Cohen was foundational for Cassirer's work, and similar to Cassirer's understanding of the ultimate synthetic goal of the "driving knowledge ideal of a period" that defines each era's "experience." Cassirer was presumably aware that his colleague Georg Simmel called his own attempt to find more foundational rules the "historical *a priori*." The two planned to teach a course together in 1908 on the knowledge theory, and Cassirer's use of forms of knowledge to address the problem of reality would presumably be recognized by both.

48 Florence, Maurice (Michel Foucault). "Michel Foucault," in *Dictionnaire des Philosophes*, ed. Denis Huisman (Paris: Presses Universitaires de France, 1984), 941ff., reprinted as "Foucault" in *Michel Foucault: Aesthetics, Method, and Epistemology*, ed. James Faubion, trans. Robert Hurley (New York: New Press, 1998), 459–63. The text was provided, and partially written, by François Ewald, Foucault's assistant at the Collège de France, but is considered to be written almost entirely by Foucault (941n1).

49 Ibid., 4.

50 HC 272; ECW9 137. This essay has been translated by Lydia Patton as "Herman Cohen and the Renewal of Kantian Philosophy," in *Angelaki* 10, 1 (2005): 95–108, but I unfortunately only discovered this translation as this book went to press, so was not able to use it exclusively here.

51 This usage follows Lydia Patton's translation.

52 Ibid.

53 Foucault, "Une histoire," 3.

54 Of course, particularly as a text written under a pseudonym, this self-assessment needs to be taken with a grain of salt as one of many possible frames of reference and influence. Although no doubt also influenced by the neo-Kantian tradition at the *College de France*, Foucault's description arguably finds its closest analog to Cassirer's work and the Marburg tradition of "Knowledge Critique" (*Erkenntniskritik*) epitomized by Cassirer's *Knowledge Problem*, particularly if understood through the broad influence on the history of science that Michael Friedman and others have noted.

55 Blumenberg, Hans. *Wirklichkeiten, in denen wir Leben* (Stuttgart: Philipp Reclam, 1981), 164–5.

56 Friedman, Michael. "Ernst Cassirer and Thomas Kuhn: The Neo-Kantian Tradition in the History and Philosophy of Science," in *Neo-Kantianism in Contemporary Philosophy*, ed. Rudolf A. Makkreel and Sebastian Luft (Bloomington: University of Indiana Press, 2010), 180. Friedman establishes the opposition of Cassirer's emphasis on a move to "pure" functional relations with Meyerson's more dialectical emphasis on the necessity of "substantive" turns in thinking. In my view, Cassirer's full definition of reality is a necessary counterpoint to his functionalism: science does not purely "disembody" reality into mathematical functionalism, but rather provides a way new way of seeing what Cassirer had earlier referred to as "relative being," which changes the nature of this problem.

57 Bourdieu, Pierre. *The Logic of Practice*, trans. Richard Nice (Stanford, CA: Stanford University Press, 1990), 4 (translation modified). Bourdieu goes on in this paragraph to mention Cassirer's formulation of this problem.

58 Bourdieu, Pierre. *The Field of Cultural Production: Essays on Art and Literature*, ed. Randal Johnson (New York: Columbia University Press, 1993), 3 (translation modified). In a survey of Bourdieu's work, Frederic Vandenberghe writes, "More than any other sociologist, Bourdieu, who published Cassirer in the collection which he directed at the Editions de Minuit, is influenced by Cassirer." In "'The Real is Relational': An Epistemological Analysis of Pierre Bourdieu's Generative Structuralism," *Sociological Theory* 17, 1 (1999), 33n3.

59 Maso, Benjo. "Elias and the Neo-Kantians: Intellectual Backgrounds of the *Civilizing Process*," *Theory, Culture and Society* 13, 3 (1995): 57, 66. Maso similarly develops Elias's complex relationship with Cassirer's work, which he had known since his days as a student, at length in "The Different Theoretical Layers of *The Civilizing Process*: A Response to Goudsblom and Kilminster & Wouters," *Theory, Culture and Society* 12, 3 (1995): 127–45. Essentially, he

claims that even as Elias was hostile to Cassirer's lack of focus on social processes, he found his philosophical grounding of sociology the best available at the time, and he took to it in part in opposition to the neo-Kantianism of his teacher Richard Hönigswald. Maso's second article is a reply to other contributors in this volume, who are skeptical of his position.
60 Luhmann, Niklas. *Soziologische Aufklärung IV* (Opladen: Westdeutscher Verlag, 1994).
61 Horster, Detlef. "Niklas Luhmann" in *Philosophen der Gegenwart*, ed. Jochen Hennigfeld and Heinz Jansohn (Darmstadt: Wissenschaftliche Buchgesellschaft, 2004).
62 Marck, Siegfried. Foreword to *Substanz- und Funktionsbegriff in der Rechtsphilosophie* (Tübingen: J. C. B. Mohr, 1925); on legal personification, 23–5, 83–147; on von Gierke, 92–104.
63 Marck, foreword to *Substanz- und Funktionsbegriff*, i.
64 Ibid.
65 Ibid.
66 Marck, *Substanz- und Funktionsbegriff*, 83–147, 148ff.
67 In this regard, this book offer an alternate history of the "crisis of historicism" to that developed around Heidegger's conflict with the Baden school of Rickert and Windelband outlined in Charles R. Bambach's excellent study, *Heidegger, Dilthey, and the Crisis of Historicism* (Ithaca, NY: Cornell University Press, 1995). Bambach notes at the beginning of his book the "great paradox" that although the Baden school of neo-Kantianism, notably Windelband and Rickert, was "preoccupied with the methods of history, it was precisely the natural-scientifically focused Marburg neo-Kantians that significant historical interpretations of the philosophical tradition developed," notably in Cassirer, Natorp and Cohen, whose "works radically changed the historical understanding of philosophy." Bambach goes on to "seize on this paradox as a way of reading the whole historicist tradition by finding in this separation between historical method and historical experience within the Baden school the *aporia* of the crisis in German thinking" (59). In this, Bambach succeeds admirably, but occluded in this reading is the more knotted but equally important dialog of the Marburg school's reading of history and historicism as it developed in Cohen and Cassirer's work and formed an alternate reading of historicism to that of Heidegger. Here we find a rich historical tradition developing not in spite of the initial focus on natural science, but precisely through its reflections on historicism and science as they define how reality comes to be experienced differently in different epochs.
68 Donald Verene thus writes, "Cassirer can rightly be described as a process philosopher in metaphysics" and compares his work to the late Whitehead (Verene, Donald. "Cassirer's Metaphysics," in *The Symbolic Construction of Reality: The Legacy of Ernst Cassirer*, ed. Jeffrey Andrew Barash (Chicago: University of Chicago Press, 2008), 100). Verene, however, claims that Cassirer's focus on "man" means this process would only apply to "non-human reality," a claim I would argue that Cassirer's early work – and the manner in it connects with the "fourth" volume of *The Philosophy of Symbolic Forms* – belies. Thus in *Substance and Function*, Cassirer argues that both the process of immediate experience and the processes revealed by scientific experience have to be understood as different modes of symbolic function, and that these in turn can only be understood as part of a larger philosophy of process (SF 201ff.). Similarly, the "process of perception" and the "process of judgment" are always correlated, and in turn also fit into a wider theory of process (SF 345–246). By the time of Cassirer's *Einstein's Theory of Relativity* in 1923, this emphasis on a process philosophy is made even more explicit, with Cassirer concluding that neither the claim for understanding the objective process of Newtonian physics nor the "subjective" process of Bergon's *dureé réelle* represent an absolute reality, but rather "a partial experience is hypostatized as the whole" (SF 455).
69 SF 220.
70 On this use of the term function and functionalism in the Marburg school, see "Funktion" in *Historisches Wörterbuch der Philosophie*, ed. Joachim Ritter (Stuttgart: Schwabe, 1972), 1140–41.
71 SF 324.
72 EK1 vi; ECW2 x.

73 SF 3.
74 EK1 vi; ECW2 x.
75 ECW1 478.
76 FF 331.
77 Cassirer, Ernst. "Selbstanzeige," *Kant-Studien* 7 (1902): 376; ECW9 439.
78 EL 116. In its application to "metapolitics" we find the point where the Marburg Project is most directly relevant to the development of the social sciences, particularly anthropology, economics, history and sociology.
79 EL 4, 116.
80 In a late essay from 1945, Cassirer claims he is essentially the first philosopher to grasp the philosophical importance of group theory, a process he begins in *Substance and Function* and developed in his writing from the 1930s, although he only emphasizes here its epistemological value, not its wider structural value for the theory of knowledge. A fuller discussion of the same topic is presented in "The Concept of the Group and the Theory of Perception," trans. Aron Gurwitsch, *Philosophy and Phenomenological Research* 5 (1944): 1–35. In a discussion with the Columbia Philosophy Club in 1945, he says, "What I would like to emphasize in this paper is the fact that in spite of this abstract character – or perhaps just *because of* this abstract character – the concept of the group is able to elucidate some problems that are of general interest and great import for a theory of empirical knowledge. Yet in this field of inquiry the concept of the group has, to my mind, been unduly neglected. That the concept has not only a mathematical and physical but a universal *epistemological* interest is a fact that, as yet, has not been fully recognized" (SMC 273).
81 FF 37. In short, in Kantian terminology the "transcendental ego" is for Cassirer, as for Leibniz, ineffable and thus not describable, while the "empirical ego" is for Cassirer an object like any other in the world and thus only known through form.
82 Poma, Andrea. *Yearning for Form and Other Essays on Hermann Cohen's Thought (Studies in German Idealism)* (New York: Spring Verlag, 2010), 367.
83 Cohen, SP1, i.
84 FF 21.
85 Gordin, Jakob. *Untersuchungen zur Theorie des unendlichen Urteils* (Berlin: Academie-Verlag, 1929), v; on the development from medieval negative theology, 4. He defines the "question judgment" as a third class in comparison with affirmative and negative judgments (129). Jakob wrote the book with the support of the *Hermann Cohen-Stiftung bei der Akademie für die Wissenschaft des Judentums*, with the advice of Cassirer, Albert Görland and Julius Guttmann (vi). On the comparison with Hegel, see appendix two, "Die Dialektik des Ursprungs (Cohen) and die Dialektik des Systems (Hegel)" (142–67).
86 Ibid., 32.
87 Ibid., 133, 131.
88 In particular, we will argue that Cassirer's key source for understanding the problem of reality will be the Leibniz of *Leibniz's System*, not Kant. In this manner he avoided much of the complexity of both Maimon and Cohen's evolving attempts to define this problem in terms of the terminology and architecture of Kantian philosophy, even as he benefited from their conclusions. On the complexity of Maimon's often "contradictory formulations" of this problem and of Cohen's gradual clarification of it, see Gordin, *Untersuchungen*, 32ff. Gordin nonetheless finds the "key" to both philosophies in Maimon's "correct" formulation based on his "conception of the qualitative infinitesimal elements of reality [Realität]" that form the basis of the "rule for the development of objects" (32). Precisely since he uses Leibniz's broader formulation of this problem, Cassirer often uses the terms *Realität* and *Wirklichkeit* interchangeably in a manner that does not correlate with technical Kantian usage, where the former is a category of quality opposed to negation and the latter is, as Marco Giovanelli summarizes it, "a modal category opposed to mere possibility." Both will usually be translated as "reality," unless the context suggests clarification between them is necessary. For an exploration

of this problem in Cohen's work, see Giovanelli, Marco. *Reality and Negation – Kant's Principle of Anticipations of Perception: An Investigation of its Impact on the Post-Kantian Debate* (New York: Spring Verlag, 2011), 19ff.; particularly on the "problem of reality," 21, and n. 63.

89 In this regard, the present work takes Cohen's late work as the historical starting point for John Michael Krois's argument that for Cassirer in some cases Kant's transcendental method is "ruled out by definition," notably in application to what Goethe called *Urphänomen*, which Cassirer describes as "life, reality, being, existence […] understood as names of a process" (Krois, John Michael. "Cassirer, Neo-Kantianism and Metaphysics," *Revue de Metaphysique et de Morale* 96, 4 (1992): 444 [437–53]). Cassirer's work leads to both a critical science based on the transcendental method and a recognition that the immanent process of "reality" in history, ethics and aesthetics supersedes any particular science, form of knowledge or individual. As Cassirer quotes Hobbes at the beginning of ZWW, "Appearance itself is the most curious and wonderful fact," and this amazement is based on what Cassirer describes as an "ineluctable duality of determination," first outlined by Leibniz (ZWW 3–4). Its foundation is Leibniz's definition of "unity in diversity" between the establishment of the particular, through which the transcendental rules and scientific setting of any object can be established, and the general differential of this particular in the inexhaustible plenum of immanent meaning revealed as life or reality (ZWW 3–4). Even as this is a duality, it is precisely a unity, and as such arguably forms the basis of what Cassirer later considered the unity of "spirit and life" (see SL 866ff.). For a survey of the relation of Cassirer to Cohen in this regard, see Paetzold, Heinz. "Die Frage nach Ernst Cassirers Neukantianismus mit Blick auf Cohen und Natorp," in *Sinn, Geltung, Wert: Neukantianische Motive in der modernen Kulturphilosophie*, ed. Christian Krijnen and Ernst Wolfgang Orth (Würzburg: Königshausen & Neumann, 1998), 219ff. The specific point of connection I see with Cohen and Leibniz's earlier philosophies lies in Paetzold's sixth thesis on the importance of symbolic prägnanz, which I see having a stronger link to earlier ideas of determination that, as linked with Gordin's ideas of infinite judgment, go beyond traditional ideas of neo-Kantianism even as they are informed by it precisely on this point (234). See also Renz, Ursula. *Die Rationalität der Kultur: Zur Kulturphilosophie und ihrer transcendentalen Begründung bei Cohen, Natorp and Cassirer, Cassirer Forschung 8* (Hamburg: Felix Meiner Verlag, 2002), 266ff.

90 EK1 19–22; EK1 120. Della Mirandola, Giovanni Pico. *On the Dignity of Man*, trans. Charles Glenn Wallis (Indiana, IN: Hackett, 1998), 4.

91 L x.

92 Leibniz is the "author" of the primacy of "feeling" in German thought. His work defined for the first time "in the dark underground [Untergrunde] of consciousness" the concept of "pure delight" (*reinen Lust*) in which the distinction is removed between "delight in seeing" and the material transmitted by sense (ECW1 414). The role of Leibniz's theory of aesthetics as the basis for a new theory of experience that spans Romantic and Enlightenment traditions (notably through the theory of determination), is put forth most clearly in FF 43ff., 293, and in PE 275ff. Notably, the latter culminates in the claim that in Alexander Gottlieb Baumgarten (1714–62) this problem is most fully developed, although in *Leibniz's System* Cassirer notes that Baumgarten's work is only a partial development of Leibniz foundational perception (L 462). With Lessing, Cassirer holds that the "problem of the beautiful leads not only to the foundation of systematic aesthetics but to the foundation of a new 'philosophical anthropology'" (PE 353). This new definition of humanity in Baumgarten and Lessing, in which "man should not transcend the finite, but explore it in all directions" (354) will be similar to Cassirer's own approach to Leibniz as the key to an immanent reading of aesthetics, ethics and history. For an analogous poststructuralist reading of Baumgarten (with reference to Cassirer) on the problem of determination in linking problems of aesthetics, history and experience, see Gasché, Rudolf. "Of Aesthetic and Historical Determination," in *Post-structuralism and the Question of History*, ed. D. Attridge, G. Bennington and R. Young (New York: Cambridge University Press, 1991), 139–61; on the concept of determination, 145.

93 Ihmig, Karl-Norbert. *Cassirers Invariantentheorie der Erfahrung und seine Rezeption des "Erlanger Programs"* (Hamburg: Felix Meiner, 1997).
94 An early and valuable approach to this problem is found in the material on Cassirer and Cohen in Levy, Heinrich. *Die Hegel-Renaissance in der deutschen Philosophie: mit besonderer Berücksichtigung des Neukantianismus – Philosophische Vorträge*, vol. 30 (Charlottenburg-Berlin: Pan-Verlag, 1927); on Cohen, 34ff.; on Cassirer, 45ff. An excellent overview of the Hegelian aspects of Cassirer's thought is found in the seminal article by Verene, Donald Phillip. "Kant, Hegel, and Cassirer: The Origins of the Philosophy of Symbolic Forms," *Journal of the History of Ideas* 30, 1 (1969), 33–46. The emphasis on Cohen's late work as presenting a path different from both Kant and Hegel in the present work is distinct from Verene's own nuanced attempt to define Cassirer's work in terms of a synthesis of the two in the recent *The Origins of the Philosophy of Symbolic Forms: Kant, Hegel, and Cassirer* (Evanston, IL: Northwestern University Press, 2011).
95 Levy appropriately notes that, unlike Hegel, Cohen's logic and dialectic develops an "open-system" based on a "labile and variable" play of relations (ibid., 35). The use of the term "dialogic" seems appropriate for this open system, particularly since it was in part influential in the Bakhtin circle in developing their later use of term. For futher information on Bakhtin's relation to Cassirer and the Marburg school see: Brandist, Craig. "Bakhtin, Cassirer and Symbolic Forms," *Radical Philosophy* 85 (Sept/Oct 1997): 21–7; Brandist, Craig. *The Bakhtin Circle: Philosophy, Culture and Politics* (Sterling, VA: Pluto Press, 2002), 105–108; Poole, Brian. "The Philosophical Origins of Bakhtin's Carnival Messianism," in *Michael Bakthin*, vol. 1, ed. Michael Gardner (London, ON: University of Western Ontario, 2002); Steinby, Liisa. "Hermann Cohen and Bakhtin's Early Aesthetics," *Studies in East European Thought* 63, 3 (2011): 227–49.
96 FF 342.
97 L x.
98 As I will argue, the appearance of the constitution as a form of law was of particular importance to Cassirer. It is telling in this regard that immediately on hearing of the Enabling Acts (*Ermächtigungsgesetz*) in 1933, Cassirer decided that Germany was doomed, and the Cassirers left the country almost immediately. For Cassirer as for Cohen, the form of law and that of the ethics, mores and politics of a society are inseparable. For this reason the date of 1919 seems an apt end point for this book, even as I recognize that the "middle" phase of Cassirer's career is often placed in 1921 with the textual initiation of the symbolic forms project (Cassirer, Toni. *Mein Leben*, 195). On the war as an "epochal event" in Cassirer's life, see Ferrari, Massimo. "Zur politischen Philosophie im Frühwerk Ernst Cassirers," in *Cassirers Weg zur Philosophie der Politik*, ed. Enno Rudolph, Cassirer-Forschungen, vol. 5 (Hamburg: Felix Meiner Verlag, 1999), 43.
99 The fairly standard dating of 1921 as the beginning of Cassirer's explicit consideration of symbolic forms is suggested most recently by Skidelsky (*Ernst Cassirer*, 82ff.), based on helpfully locating this theme in the essay "Goethe and Mathematical Physics: A Consideration in Knowledge Theory" (ECW9 268). The inspiration of Cassirer's idea of symbolic forms, however, occurred earlier in 1917, as Dimitry Gawronsky describes it: "Cassirer once told how in 1917, just as he entered a street car to ride home, the concept of symbolic forms flashed upon him; a few minutes later, when he reached home, the whole plan of his voluminous work was ready in his mind. [...] It is not true that only the human reason opens the door which leads to the understanding of reality, it is rather the whole human mind [...] which determines and moulds our conception of reality" ("Cassirer: His Life and His Work," in *The Philosophy of Ernst Cassirer*, ed. P. A. Schilpp (Evanston, IL: The Library of Living Philosophers, 1949), 25). I would further add that from early on – namely in the Leibniz book in the section on organism – Cassirer is quite clear that this is not only mind, but mind as a unity with body (L 364).
100 Skidelsky, *Ernst Cassirer*, 65.
101 ES 47–51, 65. The claim made by Skidelsky, based on comments by Frege and Russell, that Cohen's actual interpretation of calculus is false is called into question by Lydia Patton in

Hermann Cohen's History and Philosophy of Science (PhD diss., McGill University, 2004), 112ff. Using the notation and terminology of Wilhelmine Germany, Cohen's basic argument is both coherent and consistent with calculus of his time.
102 Cassirer, Toni. *Mein Leben*, 94.
103 Skidelsky, *Ernst Cassirer*, 50. Cassirer does make wide use of Russell's work in his study of the philosophy of number, but where Russell claims that "the 'objectivity' of pure concepts and truths is accordingly put on a plane with that of physical things," Cassirer's concept of group theory will demand that "this whole cannot be presented like a quiescent object of perception, but, in order to be truly surveyed must be grasped and determined in the law of its construction" (SF 316–17). With this will hang, in turn, Cassirer's central concept of the productive role of imagination in the relation of part and whole as it develops from Leibniz, and ultimately his concept of both symbol and function. Tellingly, Cassirer also provides a critique of Russell's book on Leibniz for making a similar mistake, namely reducing Leibniz's generative logic and philosophy down to a few simple precepts (L 532–41).
104 SF 3; for further reading of the critical reaction of Cassirer to Frege and Russell, see Heis, Jeremy. "'Critical philosophy begins at the very point where logistic leaves off': Cassirer's Response to Frege and Russell," *Perspectives on Science* 18, 4 (Winter 2010): 383–408.
105 HCS2 321–2.
106 Skidelsky, *Ernst Cassirer*, 50. Even Cassirer's brief *Encyclopædia Britannica* entry on neo-Kantianism vividly contradicts this statement, noting that the logic at the base of Cohen's theory of the infinitesimal is used to anchor the claim that "reality is never 'given' in any sense," and that from this starting point it traces the "various ways and directions in which thought moves in this 'production of this object'" through the entirety of human activity, a project developed through his later volumes of the *Logic of Pure Knowledge* on ethics and aesthetics ("Neo-Kantianism," in *Encyclopædia Britannica*, 14th edition, XVI: 214–15). Ironically, as with many other commentators, Skidelsky's assumptions about the shortcomings of Cohen's reading of "natural science" blinds him to the fact that Cohen's work was a critique of the assumptions of science as a totality or given unity. In particular, by not recognizing the meaning of Cohen's reading of limitative judgments and continuity, Skidelsky fails to recognize that Cohen's work was from the very beginning able to recognize "the possibility of categorical yet nonetheless nonmathematical relations – and thus of nonmathematical objectivization," which he imagines only occurs through Cassirer's use of Russell and Frege (*Ernst Cassirer*, 65).
107 As John Krois summarizes: "Mistaking the triadic character of Cassirer's theory of signs and symbolism for a dyadic one results in great confusion because such a perspective overlooks the pragmatic dimension of Cassirer's thought" (*Cassirer*, 52). Although Krois's goal here is to link Cassirer's work with Peirce, the deeper basis of this triadic definition is precisely in the logic suggested first by the infinitesimal, as Cassirer already highlights in his studies of Leibniz and Jungius (J 361ff.). As Cassirer notes in *Substance and Function*, this theory precisely goes beyond the work of Russell and Frege in suggesting "a movement of thought" as the basis of any judgment and theory of signs (SF 317).
108 Thus, for instance, he never explains or even notes Cassirer's key statements on the meaning of infinite and "infinitude" in the debate, and never engages with Cassirer's theory of reality.
109 Gordon, *Continental Divide: Heidegger, Cassirer, Davos* (Cambridge, MA: Harvard University Press, 2010).
110 In addition to the works of Ryckman and Schmitz-Rigal discussed below, particularly valuable have been Ihmig, *Cassirers Invariantentheorie der Erfahrung* and *Grundzüge einer Philosophie der Wissenschaften bei Ernst Cassirer* (Darmstadt: Wissenschaftliche Buchgesellschaft, 2001) and Friedman, Michael. *A Parting of Ways: Carnap, Cassirer and Heidegger* (Chicago: Open Court, 2000).
111 Ryckman, Thomas. *The Reign of Relativity: Philosophy in Physics, 1915–1925* (New York: Oxford University Press, 2005), 15.

112 Ryckman, *Reign of Relativity*, 242–3; Ibongu, Georges. *Cassirer's Structural Realism* (Berlin: Logos Verlag, 2011); Gower, Barry. "Cassirer, Schlick and Structural Realism: The Philosophy of the Exact Sciences in the Background to Early Logical Empiricism," *British Journal for the History of Philosophy* 8, 1 (2000): 71–106.
113 Ryckman, *Reign of Relativity*, 12.
114 Ibid.
115 Schmitz-Rigal, Christiane. *Die Kunst offenen Wissens: Ernst Cassirers Epistemologie und Deutung der modernen Physik* (Hamburg: Meiner, 2002); also her "Ernst Cassirer: Open Constitution by Functional A Priori and Symbolical Structuring," in *Constituting Objectivity: Transcendental Perspectives on Modern Physics*, ed. Michel Bitbol, Pierre Kerszberg and Jean Petitiot, *The Western Ontario Series in Philosophy of Science*, vol. 74 (New York: Springer Verlag, 2009); and "Science and Art: Physics as Symbolic Formation," *Synthese* (2009): 21–41; "Die Kunst der Wissenschaft," *Philosophia Naturalis* 40, 2 (2003): 255–91. Schmitz-Rigal writes in her 2009 article, and I concur, that it is "astonishing – and regrettable" that the central contribution of Cassirer's work outlined in *Die Kunst offenen Wissens* has still "not yet been fully perceived and acknowledged by most researchers," namely that "by favouring a *functional* perspective, already nascent in Kant's work, he reaches an unlimited adaptability for *all* components, *a priori* included. This generative conception is based on the creative constitution of 'orders of meaning' as [a] transcendental condition for all objectivation: a perspective that goes well beyond the Kantian framework" ("Science and Art," 22).
116 To demonstrate his argument, Poma examines the influence of Cohen's work on Deleuze's *Difference and Repetition* ("Yearning for Form," 313–67).
117 Some of the key points of this scholarship have been nicely developed for the collection *Neo-Kantianism in Contemporary Philosophy*, ed. Rudolf A. Makkreel and Sebastian Luft (Bloomington: Indiana University Press, 2010).
118 Sieg, Ulrich. *Aufstieg und Niedergang des Marburger Neukantianismus: Die Geschichte einer philosophischen Schulgemeinschaft* (Würzburg: Königshausen & Neumann, 1994); Köhnke, Klaus Christian. *The Rise of Neo-Kantianism: German Academic Philosophy Between Idealism and Positivism*, trans. R. J. Hollingdale (New York: Cambridge University Press, 1991); Meyer, Thomas. *Ernst Cassirer* (Hamburg: Ellert and Richter, 2007), 36; Ferrari, Massimo. *Ernst Cassirer, Stationen einer philosophischen Biographie: Von der Marburger Schule zur Kulturphilosophie* (Hamburg: Meiner Felix Verlag GmbH, 2003).
119 Poma, Andrea. *The Critical Philosophy of Hermann Cohen*, trans. John Denton (Albany, NY: State University of New York Press, 1997); Poma, Andrea. *Yearning for Form and Other Essays on Hermann Cohen's Thought (Studies in German Idealism)* (New York: Spring Verlag, 2010), 367.
120 Krois, John Michael. *Cassirer: Symbolic Forms and History* (New Haven: Yale University Press, 1987), 7.
121 Historians of neo-Kantianism have developed a very different reading of Cohen's work by focusing on the middle period of his productivity, a reading similar to one that Cassirer was at pains to correct in his early defenses of Cohen. Authors who have defined the Marburg school through Cohen's work from this period have emphasized its radically constructive character and emphasis on the mathematical sciences in a manner that was, as Cassirer emphasized, not true at least of his later works. Thus Klaus Christian Köhnke's otherwise invaluable standard work on neo-Kantianism summarizes Cohen's philosophy as somewhat ridiculous in both its focus on pure construction of objects and its emphasis on the mathematical sciences beyond all other topics, culminating in the "'construction or production of the object' which appeals to the proposition […] that we know a priori of things only what we ourselves have put in them" (*The Rise of Neo-Kantianism*, 186). Whether true or not for Cohen's middle phase, Cassirer would strongly object to nearly every aspect of this characterization for his late work, starting with the subjective idealism implied by "we ourselves" and the spatial assumptions of "put in them." To use Cassirer's terminology as a short hand, invariants such as *a priori* are discovered in the world even as they are capable of permutation, and these are how we first

know both subjects and objects. In a similar manner, Hans Blumenberg takes Cohen as the epitome of a theoretical arrogance in which only science defines the world, so that (to take the example of the heavens): "The pure heavens that are worthy of theory do not lie behind the visible firmament, but in front of it, in man's possession, in the construction of hypothesis" (*The Genesis of the Copernican World*, trans. Robert M. Wallace (Cambridge, MA: MIT Press, 1987), 110).

122 Cohen, Hermann. *System der Philosophie, Erster Teil: Logik der reinen Erkenntnis*, 2nd edition (Berlin: Bruno Cassirer, 1914), xi. The introduction to the 1902 edition makes no mention of this distance from Kant, although the content of the text is not particularly different. This passage is cited and further discussed in Poma, *Critical Philosophy*, 79.

123 On the role of the judgment of reality in resolving the "gap between totally heterogeneous principles of understanding and sense" in a manner not possible for the categories or schema alone, see Gordin, *Untersuchungen*, 32.

124 Briefly stated, describing Cassirer as a neo-Kantian becomes confusing since nearly all of the dualisms that might be applied to Kant's work are overcome in Cassirer's early philosophy, notably in *Leibniz's System* and *Substance and Function*. In the first place, Cassirer's early work is radically immanent, there is no opposition of transcendental ego and the manifold, intellect and sense or knowledge and ethics. The notion of static *a priori* knowledge is replaced with a dynamic and historical *a priori*, as are the static definitions of time and space. Most centrally, his work uses Leibniz and then Cohen's theory of judgment to develop a position of "open constitution" that, as Christiane Schmitz-Rigal argues, appears to move decisively beyond Kant. For a complete exegesis of this break from Kantian dualisms, see Schmitz-Rigal, *Die Kunst offenen Wissens*, 24–85; Rudolf, Enno. "Substanz als Funktion: Leibnizrezeption als Kantkritik bei Cassirer," in *Ernst Cassirer in Kontext* (Tübingen: Mohr-Siebeck, 2003), 31–40. For a brief overview of this theme, see Cassirer's reading of the role of "post-Kantian speculation" in transforming Kant through a more general philosophy of origin or unity in GK. A nuanced study of this problem is found in Massimo Ferrari's essay, "Is Cassirer a Neo-Kantian Methodologically Speaking?" in *Neo-Kantianism in Contemporary Philosophy*, 293–314. By the terms of Ferrari's definition, however, we would have to conclude that Cassirer's opening statement at Davos is correct: his work can only be considered "methodologically" neo-Kantian as part of a historical reception, and is surprisingly no less or more dependent on this tradition than his opponent, Martin Heidegger.

125 For a basic outline of some of the assumptions of the term "neo-Kantianism," see Krois, *Cassirer*, 6ff.; for and account of Cassirer's gradual development from neo-Kantianism, see Krois, "Cassirer, Neo-Kantianism and Metaphysics," 437–53. I would thus stand by Krois's claim that "the Marburg school itself was broader than the label 'neo-Kantianism' implies" (453).

126 Rickert, Heinrich. "The Limits of Concept Formation," in *Natural Science: A Logical Introduction to the Historical Sciences*, ed. and trans. Guy Oakes (New York: Cambridge University Press, 1986 [1902]). Cassirer himself noted the inaccuracy of applying the terminology of neo-Kantianism to his work. At the beginning of his Davos debates with Heidegger in 1929, he objected to the use of the term "neo-Kantian" to define his philosophy, going so far as to claim "there is an absence of existing neo-Kantians" – presumably including himself – and then claiming that, if anything, in relation to a functional development *from* neo-Kantianism he "found here in Heidegger a neo-Kantian" (Cassirer, Ernst and Martin Heidegger, "Appendix IV: Davos Disputation Between Ernst Cassirer and Martin Heidegger," in *Kant and the Problem of Metaphysics*, 5th edition, trans. Richard Taft (Bloomington: Indiana University Press, 1997), 193–4). As with much of the Davos debate, Cassirer's point is not fully comprehensible unless Cohen's work is understood as a key point of departure for both his opponent and himself. Heidegger had, in his earlier lectures towards *Being and Time* (now published in English as *The History of the Concept of Time*), clearly distinguished the value of Cohen's work – along with the historicism of Dilthey, Cassirer's other principle teacher – from other trends in

neo-Kantianism, notably those of the so-called south-western school of "Windelband and Rickert," who he said "leveled and trivialized [Cohen and Dilthey's initiatives] and twisted their problems beyond recognition," particularly in relation to history (*History of the Concept of Time: Prolegomena*, trans. Theodore Kisiel (Bloomington: University of Indiana Press, 1992), 17). At Davos, when asked to define neo-Kantianism, however, he intentionally groups Cohen with precisely this "trivialization" using simply a list of names: "Cohen, Windelband, Rickert, Erdmann, Riehl" ("Davos Disputation," 193). In defining Cassirer's work as "neo-Kantian," we risk the same misrecognition, and with it overlooking how Cassirer followed not only Cohen's initial insights into the Kantian system, but his later originality in moving beyond what Heidegger termed the "science of consciousness" of Cohen's early work (*History of the Concept of Time*, 16). For more on the important connection of Cassirer and Dilthey, see Schmitz, Heiko. *Von der "Kritik der historischen Vernunft" zur "Kritik der Kultur": Über die Nähe der Projekte von Wilhelm Dilthey und Ernst Cassirer* (Würzburg: Königshausen & Neumann, 2006).

127 As explained below at length, we will retain the German for the second term "*symbolische Prägnanz*" since it is largely untranslatable, and easily misunderstood in its rendering as "pregnance."

128 This use of Leibniz as an approach to Cassirer's work is not in itself new, although it is relatively rare in the secondary literature. Two recent helpful guides have been: Rudolf, Enno. "Substanz als Funktion: Leibnizrezeption als Kantkritik bei Cassirer," in *Ernst Cassirer im Kontext: Kulturphilosophie zwischen Metaphysik und Historismus* (Tübingen: J. C. B. Mohr, 2003), 31–40; and Ferrari, Massimo. "Symbol und Ausdruck: Die Leibnizschen Quellen der Philosophie der Symbolischen Formen," in *Stationen einer Philosophischen Biographie, Cassirer-Forschungen*, vol. 11, trans. Marion Lauschke (Hamburg: Felix Meiner, 2012), 39, 46ff., 163ff. Schmitz-Rigal also develops the Leibnizian aspects of Cassirer's work throughout her study, particularly in regard to its process qualities (*Die Kunst Offenen Wissens*, 88–9).

The role of Cassirer's *Leibniz's System* as an historical introduction and companion to Cohen's *System der Philosophie* is more widely recognized in the literature on Cohen, but has not been extensively developed in readings of Cassirer. See, for instance, even the introductory encyclopedia entry on Leibniz by Helmut Holzhey, the doyen of contemporary studies of Cohen, and Vilem Mudrock, where they write, "It was only with Ernst Cassirer's book *Leibniz's System* [...] that a comprehensive interpretation of Leibniz was presented by one of the Neokantians [...]. Cassirer defended the view that it was the logical principles that were the roots of Leibniz's thought. [...] Cassirer understood 'logic' in the Neokantian sense of a theory of the principles of scientific cognition. Cassirer's *Leibniz* can therefore be seen as providing the historical underpinnings of Cohen's *Logik der reinen Erkenntnis* (1902)" (*The A to Z of Kant and Kantianism* (New York: Scarecrow Press, 2010), 167). On Cassirer's general reception of Leibniz, see: Seidengart, Jean. "Cassirer, Reader, Publisher, and Interpreter of Leibniz's Philosophy," in *New Essays on Leibniz Reception* (Basel: Springer, 2012), 129–42.

129 Ferrari, "Is Cassirer a Neo-Kantian," 307.

130 FF 43.

131 ECW17 395. He is speaking in particular about the gradual publication of material from the Hannover collections, but this statement holds more generally for his view of Leibniz.

132 The speech does claim, however, that the German idealist tradition culminated in Kant's work but it is Leibniz's work, particularly in relation to Blackstone and the Western tradition, that receives surprising emphasis (ECW17 291–2).

Part I

THE MARBURG SCHOOL AND THE POLITICS OF SCIENCE IN GERMANY

Chapter One

THE TWENTIETH-CENTURY CONFLICT OF THE FACULTIES: THE MARBURG SCHOOL AND THE REFORM OF THE SCIENCES

Ever since its first publication in 1883, Hermann Cohen's text *The Principle of the Infinitesimal Method and Its History: A Chapter on the Foundations of the Critique of Knowledge* has formed a puzzle for all but close followers of the Marburg school.[1] The text is widely considered to mark a shift in Cohen's thought and to establish the most influential direction of the school, and with it began the school's enormous influence in German philosophical, legal and scientific thought in the three decades before the First World War. Intended, however improbably, to form something of a "popularization" of Cohen's ideas, the text formed the germ of Cohen's own philosophy as initiated in the second edition of his *Kant's Theory of Experience* two years later and developed in his late *System of Philosophy* (1902–1912). It is true that the puzzle for many readers of *Infinitesimal Method* was simply its contents, for despite its purportedly approachable form, it was widely considered extremely difficult and occasionally accused of being needlessly opaque. The philosopher Friedrich Kuntze noted that "Cohen's book is generally considered to be one of the most difficult books of German philosophy, and one can hear from considerate people that the only thing they understood in the work were the citations from foreign authors."[2] Kuntze claimed that this difficulty stemmed from the challenging conceptual nature of the infinitesimal as much as it did from Cohen's style. Other readers, notably Gottlob Frege, argued that this incoherence was based on the book's flawed reading of the history of calculus itself. "Cohen's style of writing," Frege wrote, "[…] is by no means distinguished for its clarity, and […] is sometimes even illogical."[3]

Yet the central puzzle for most readers since 1883 has simply been how and why Cohen (and the Marburg school) derived from such a recondite history and philosophy of calculus a school of philosophy that would ultimately claim to encompass aesthetics, law, religion and ethics. In worlds of reception seemingly totally removed from the history of mathematics, Cohen's work was of decisive significance for figures ranging from the Bakhtin circle and the social democrat Eduard Bernstein to philosophers such as Franz Rosenzweig, Martin Heidegger and Emmanuel Lévinas. It moreover exerted a vast institutional influence within a wide array of academic disciplines ranging from law to theology. By focusing on Cohen's development of the infinitesimal problem, and Cassirer's transformation of this key theme, we gain a singular vantage on the late Marburg school project. Although this approach will necessarily avoid other important figures and aspects of the school – particularly the work of

Paul Natorp and Nicholai Hartmann – it provides a single theme through which to reconstruct a key element of both the school and Wilhelmine thought in general.

The most common answer to the puzzle of Cohen's use of the infinitesimal has been simply that the Marburg school was at its core a philosophy of science, and thus a reading of the history of calculus could epitomize the meaning of this philosophy. For proponents, connecting the Marburg school with the philosophy of science initially appeared as its greatest triumph and source of legitimacy, at once linking it to the key problems of modernity and reconnecting it with the central problems of Kant's philosophy. As Cohen argued in a lengthy 1912 introduction to F. A. Lange's *History of Materialism*, his reading of the centrality of the infinitesimal method as a philosophical turning point in modern thought was consistent with the work of Hertz, Planck and Einstein.[4] Indeed, Cohen noted, Planck's *The Principle of the Conservation of Energy* (1887) had described its own theoretical foundation through the use of the term "infinitesimal-theorie" in a manner Cohen found similar to his own.[5]

For opponents, however, the dominance of the philosophy of science was the Achilles heel of the school. In large part due to its reception as a philosophy of science, by the 1920s the Marburg school was widely considered to be concerned with only a subsection of the philosophy of science, to be anchored in an outdated narrative of progress, and to no longer address the principle issues of the day. By the time of the Davos disputation between Martin Heidegger and Cassirer in 1928, the Marburg school was closely linked with a failed "technical" or ontic project in opposition to Heidegger's boldly relevant new ontological philosophy.

Although recent work has emphasized the often surprising proximity of Cassirer and Heidegger, the contemporary judgment is indicative of how heavily the Marburg school was linked with a reading of natural science, and how far the spectrum of natural science was by then isolated from social, political and existential concerns at least for many younger auditors. The philosophy of science was increasingly understood as, at best, an isolated speciality within philosophy and, at worst, in the interpretation of the "life" philosophies of Henri Bergson or Ludwig Klages, based on an inherently alienated and static reading of the world. By contrast, for the Wilhelmine period the future of human endeavor and thought had often been intimated in the world of contemporary science, and this fact illuminated not only academic but educated discourse more generally.

As Cassirer's reviews and comments on Cohen repeatedly stressed, Cohen's focus on the infinitesimal was the basis of his entire late philosophy, but it hinged on a logical and even phenomenological set of issues of far broader importance than mathematics.[6] The philosophical themes stemming from Cohen's reading of Kant's schematism and the anticipation of perception, the key philosophical issues at hand, could however in Cassirer's view initially have been presented in a number of ways, and indeed the emphasis on mathematical calculus was only loosely retained by members of the Marburg school itself as it developed over the years. Cassirer's own *Leibniz's System* (1902) demonstrated that in fact many of the key philosophical themes Cohen derived solely from calculus could be elucidated from an array of other examples. Throughout his career Cassirer wrote essays emphasizing the importance of Cohen's philosophy – notably in 1912, 1918, 1931 and 1942 – but nearly every time he would note that it suffered from the defect of being written in an often impenetrable style. The clearest example of this defect was perhaps Cohen's literal use of the "infinitesimal method" as the basis of his philosophy itself. Even as Cassirer agreed with Cohen's conclusions, he devoted considerable effort to recasting this theme philosophically and rhetorically.

Although the Marburg school was later reviled as part of a broad cultural trend subordinating philosophy to the prestige of the natural sciences, Cassirer emphasized that it in fact held nearly the opposite agenda in the first decade of the twentieth century: to subject the sciences to a critique and redefinition by philosophical means, which would in turn reveal the true creativity of both the sciences and humanity. By changing the modern definition of organized knowledge in all branches of the sciences – the natural, social and "human" sciences (*Geisteswissenschaften*) – the Marburg school sought to transform society, and indeed the definition of humanity itself. Cohen's relentless emphasis on the problem of the infinitesimal and the philosophy of the exact sciences often occluded this program. The underlying logic that connected the infinitesimal problem to Cohen's wider philosophy, however, was indeed in Cassirer's view the basis for a new understanding of the world and the human project of unparalleled importance. This philosophy culminated in a relational and open-ended definition of experience based on what Cohen called a "critique of knowledge" (*Erkenntniskritik*), emphasizing both science and the limits of science before ethics, aesthetics and, ultimately, the meaning of humanity.

Cassirer's work would on the one hand attempt to emphasize this broader significance while on the other seek to clarify – and in some cases redefine – Cohen's occasionally heavy-handed use of the philosophy of science. This project defined Cassirer's work through the First World War and provided a bridge to Cohen's work for the uninitiated, while also, viewed retrospectively, defining Cassirer's gradual departure from his teacher. The present chapter will address the meaning and stakes of Cohen's larger project of reform as it developed from his functional philosophy, which will then be socially and politically contextualized in Chapter Two, before we return, in Chapter Three, to Cohen and Cassirer's detailed reading of the infinitesimal theme as the central point of their philosophies of difference.

The Modern Conflict of the Faculties and Cohen's Program of Reform

Despite the apparently abstruse nature of the infinitesimal problem, Cohen's late work used it as the keystone for what could be described as a radical resolution to the "conflict of the faculties," in the sense first outlined by Kant in his famous essay, across all of the sciences and forms of knowledge.[7] Kant saw the "lower" faculty of philosophy as working to critique the "upper," that is, state, faculties of law, theology and medicine. In Cohen's vision, however, philosophy would redefine the meaning of science in all of the faculties and would transform not only the traditional higher faculties but all the new disciplines of science, particularly the "sciences of the state" and humanities, as well. In the present Anglo-American context, this project might at first appear solely as a scheme for university reform. In Wilhelmine Germany, however, Cohen intended this reform as a global critique of German society, administration, politics and indeed all definitions of humanity. The goal was to "avoid all dogmatism" in assumptions about reality, and to provide a new understanding of a humanity open to radical transformation in the future. Cohen saw this as a nationalist project that would advance and even catalyze the scientific development of Germany in every dimension, even as it was a cosmopolitan project that would open the path to a new set of potentials for humanity.

Couched in Cohen's arcane terminology and challenging style, this project provoked venomous reactions throughout Germany. As the first major nonconverted Jewish German tenured philosopher, Cohen's work became the target of unprecedented anti-Semitic attacks, not least for attacking the "common sense" understanding of German nationalism and developing a "desiccating" philosophy. It is in this context that Cassirer became convinced of the actual importance of Cohen's work beneath his often counterproductive bombast, and focused his considerable philosophical and rhetorical skills on advancing Cohen's project. As the historian Thomas Meyer accurately puts it, the two became a "team" advancing the same root arguments using radically different techniques and styles.[8] Cohen, who was 33 years older than Cassirer, suggests some of the nature of this relationship when he joked to Cassirer's wife: "You see my dear Mrs Cassirer, this is what is lovely about our Marburger School – that all sorts of temperaments are represented, from impetuous youth to mellow old age. In which, of course, I represent the youth and your husband the old age."[9] Particularly in relation to resituating Cohen's philosophy of science on far more durable foundations, Cassirer's work would prove pivotal to Cohen's project precisely for his restraint. Conversely, however, the ambitions and range of Cohen's original work forms the necessary frame in which Cassirer's early writings need to be interpreted.

Science for both philosophers was redefined as a functional process that unfolded truths, but truths that were conditional since they could be further reinscribed in new forms. Both did not hold that science was a conclusive endeavor that unveiled final substantive truths separate from their context. The goal was a more skeptical and fungible definition of each science, as well as a critique of the limits of sciences themselves in relation to ethics, aesthetics and reality.[10]

The key to this argument was indeed summarized in Cohen's definition of the infinitesimal, but Cassirer's early work clarified that its importance could be understood apart from this initial demonstration in the philosophy of mathematics and was in fact of global significance. Writing in a polemic against the philosopher Leonard Nelson's claim that Cohen promoted a "mysticism" through his use of the infinitesimal, Cassirer pithily summarized the full implications of Cohen's use of calculus as part of a larger project of a process philosophy or, in the specific use of the term of the day, a functionalist philosophy that would avoid the pitfalls of materialism and idealism, empiricism and rationalism.[11] Cassirer described Cohen's philosophy through its widest category of transcendental logic (simply "logic" in the text below), which will in turn ground a philosophy of science as well as a functional definition of humanity and human history:

> The basic idea of Cohen's work can be stated quite briefly: if we want to achieve a true scientific grounding of logic, we should not begin from any sort of completed existence [*fertiger Existenz*]. What naïve intuition [*Anschauung*] takes as its obvious and secure possession, this is for logic the real problem; what it assumes as directly "given,"

this is what must be critically analyzed and taken apart in its crucial conditions for thought [*gedanklichen Bedingungen*]. We should not begin with any sort of objective Being [*gegenständlichem Sein*], no matter of what sort and no matter what relation we place ourselves to it: for every "Being" is in the first place a product and a result which the operation of thought and its systematic unity has as a presupposition. A foundational conceptual setting of this sort, an intellectual condition in which we can first speak of "reality" in the scientific sense, is found by Cohen in the idea of the infinitesimal as it is detailed and fixed in modern mathematics.[12]

The infinitesimal and calculus, then, will ultimately form for Cohen the model for a nonessentializing process, function or "operation of thought" of determining particular moments of experience. Neither the "subjects" nor "objects" in the world, nor any putative distinction between them, can be taken as a "completed existence." Rather they are all defined through process, and the infinitesimal is simply the best (and for Cohen first) pure example of such a process. In this regard, Cassirer can further respond to Nelson: "The infinitesimal is not a thing [*Ding*] but a condition [*Bedingung*], not any sort of reality at hand, but an instrument of thought for the discovery and construction of true being [*wahrhaften Seins*]."[13] For Cohen, the concept of the "infinitely small" is given neither in intuition nor in sensation, as Cassirer writes in a later summary, but is creatively extended from the "pure thought" – in this case simply the concept of the infinitely small itself – to shape, or better reveal, our perception of the world, and this process can be extended to any phenomenon.[14] "Pure thought" could reveal many aspects of the world, and encompassed the vast development of different fields of calculus and the evolution of scientific thought itself.

Cassirer presents a simplified but consistent interpretation of Cohen's basic insight in which the crux of this "functional" definition in calculus can by analogy be extended to any phenomenon, so that "the particular appears as a differential that is not fully determined and intelligible without reference to the integral."[15] Technically, the model is the first law of calculus, which explains the relation of an integral of a function and a differential of a function, and suggests that this relation allows, to put it simply, the particular conditions of the function to describe its global conditions and vice versa.[16] Putting aside the details of this interpretation, Cassirer's basic point is that "all that is given *means* something that is not directly found in itself; but [...] there is no element in this 'representation' which leads beyond experience as a total system." Only through such a model, Cassirer argued, can the constellation of rules or laws that explain any particular "fact" or event – whether in natural science, aesthetics, law or in other fields of human endeavor – be properly illuminated in relation to the whole without essentialism.

The ultimate goal of this project, Cohen writes, will be a "critique of knowledge" (*Erkenntniskritik*) that will explain "the synthetic principles or foundations of knowledge, from which science builds itself and from which its validity hangs."[17] The functional relation of part to whole in thought will for Cohen be correlative to their relation in experience. It is this project that will define Cohen's philosophy of origin or *Ursprung*, in which the appearance of any phenomena is explained in terms of the next order level of assumptions in which it takes part. The relation of the whole of the horizon of these assumptions to the appearance of the particular raises questions of existence and "reality" that would be central in one direction to the ontological philosophies of Heidegger and Nicolai Hartmann, in another to Cohen's theology and in quite another to Cassirer's own philosophy of reality.

By claiming that his thought was founded on the infinitesimal problem, then, Cohen in Cassirer's reading did not principally mean a mathematical technique or its history, but a form of logic that Cohen claimed was "a basic rule of consciousness."[18] It is in this sense that Cohen could make the otherwise unintelligible claim that the infinitesimal formed the base of not only a philosophy of science but "a law of the production of content in aesthetics particularly, as well as in logic and ethics."[19] In each of these fields, the play of experience understood functionally would replace assumptions of "objective Being," and this project far from presenting a turn towards mathematical physics would be an attempt to reveal the historical transformation of reality in all domains. In Cassirer's presentation of Cohen, the natural sciences, particularly calculus in mathematics and the problem of force in physics, were simply where the power and subtlety of this epistemological turn was most immediately evident – not the literal application of the calculus itself.[20] Cohen, on the other hand, occasionally understood the evolution of "pure thought" to provide a single developing matrix of meaning, unifying all of the sciences, and saw calculus as literally the point around which a definition and actualization of "pure thought" first crystallized into the modern sciences.

Undoubtedly, part of the initial irritation with Cohen's style and philosophy was that, precisely due to his broad approach to the infinitesimal, its initial reception largely occurred in the humanities and social sciences, not the natural sciences themselves. By the turn of the century, Cohen's theory of the infinitesimal was received as a model or even metaphor of a purely relational philosophy in such a manner that, against his own wishes, many readers appear to have eschewed a precise understanding of its meaning. This popular reception was undoubtedly a factor in the countervailing frustration with the opacity of his writing, and the not entirely unjustified attacks on its technical foundations by figures such as Gottlieb Frege, particularly in Cohen's literal approach to the relation of the various fields of calculus to pure thought. Thus, as the critic Ian Hunter notes, Cohen's work was "widely absorbed by humanist intellectuals" and followed a reception pattern similarly to later ideas such Noam Chomsky's generative grammar: even as its technical details were often glossed, its popularity derived from a "cultural message," which in this case was "that all apparently objective things are the products of formal differential relation."[21] Thus, within the Bakhtin circle in Russia, for instance, Cohen's work was adopted not for its focus on natural science, but precisely for its open-ended and creative role in understanding process. M. I. Kagan, a student of Cohen and member of the circle, summarized the core idea of this reception in a eulogy to Cohen in 1920, arguing that Cohen demonstrated that "the concept of appearance [*Erscheinung*] is not something absolute through which the significance of being might be exhausted. [...] In the sciences we are constantly dealing with the process, with the progressive path, of the cognition of being. Being is always problematic, is never divined. The *Ding an sich* ("thing in itself") is a border, a limit in the endless path of the idea of cognition."[22] Applied to the humanities and social sciences, the result of this approach was a critique and phenomenology of the given in terms of context that proved widely influential. For Cohen, however, it also represented a means to recast the social sciences and politics of Germany as a whole.

Cohen's Critique of "Naïve" Science and its Distortion of Modern Philosophy

In 1912, Cassirer and Cohen both presented public overviews of Cohen's philosophy with characteristically different, but ultimately complementary, tenors. A brief comparison of the two can provide a clearer sense of both the content of Cohen's philosophy and its wider goals. In a 1912 retrospective of his teacher's philosophy, "Hermann Cohen and the Rejuvenation of Kantian Philosophy" in the journal *Kant-Studien*, Cassirer highlighted the aims of Cohen's work within its mainstream "neo-Kantian" reception while also signaling its innovations. Cohen's work is situated as part of a universal reform of the sciences and philosophy, with his philosophy of the natural sciences important primarily as a critique of normative ideas of science, which in turn allows a critique of the sciences in relation to ethics, aesthetics and psychology.

Both the title and the forum for the article at first appear to anchor Cohen in the familiar rubric of neo-Kantianism, but the piece already highlights the more innovative claims of Cohen's *System of Philosophy* (1902–1912) that move beyond several of Kant's basic premises.[23] Cassirer's awareness of the differences between Cohen's early and late work in this regard is suggested by an article on neo-Kantianism from the same year of 1912 for the *Encyclopædia Britannica*. Here Cassirer describes Cohen's early books on Kant as "the high point of neo-Kantianism," but depicts Cohen's *System of Philosophy* as an entirely separate development based on his concept of the infinitesimal – clearly he was aware that his teacher's innovations were a potentially radical shift from Kant, even as he downplayed this to the audience of *Kant-Studien*.[24] Cohen's 1912 introduction to the third edition of Lange's *Materialism*, to which we will turn to next, highlight precisely this difference and the more radical ends to which he hoped his reform of the sciences and university would lead.

Although Cassirer's essay is more of a synoptic overview of Cohen's conclusions than a demonstration of his argument, the piece provides an excellent corrective to the popular misreadings of Cohen's philosophy and an important overview of its structure. Rather than the study of natural science being an end in itself, Cassirer emphasizes that Cohen saw a critique of contemporary conceptions of natural science as the starting point for a broader reformulation of the social and human sciences. Cohen's central argument, Cassirer stresses, is with the mainstream understanding of "the relation of philosophy and science" in late nineteenth-century German society, an understanding that had permeated popular culture. "One must see anew the entire epoch in which Cohen's studies of Kant began," Cassirer writes, and against which his criticism was focused, in regard to the overarching and misplaced faith in natural science that dominated the period.[25] This epoch, Cassirer writes, was one in which the problems of humanity and philosophy were ever more determined solely by the specific sciences, yet the understanding of these sciences and their relation to reality was itself "naïve."[26]

It is entirely wrong, Cassirer notes, to claim that Cohen's critique of knowledge is based on a "one-sided focus on mathematical theories of nature."[27] Not only are his theories of ethics and aesthetics equally important, but even within the natural sciences his redefinition of the biological sciences is as important as the exact sciences. Cassirer thus highlights that Cohen's work is foremost a critique of the pervasive role of naturalism and false readings of science in the late nineteenth century, even in thinkers who appeared to be aloof from any reading of science whatsoever. These tacit assumptions block the possibility of an understanding not only of science, but also of humanity itself. The exact sciences of the era tacitly present a definition of "reality" that itself needs to be open to critique, and it is for this reason that any critique of society in the late nineteenth century had to begin with a critique and redefinition of science.

Indeed, from Cassirer and Cohen's perspective every modern philosophy contained a tacit philosophy of science, and the less this was recognized, usually the more egregious and absolute its assumptions were. To the degree any philosophy assumed a substantive "reality" – whether positively as a world of "things" or negatively, as in Romanticism, as a totality of that which is unknown – they misread the relational and developmental nature of experience as understood "functionally." Conversely, every use of science and the philosophy of science had both a philosophical and political importance.[28]

Cohen's work in short was not the high point of nineteenth-century scientism, as some commentators have proposed, but an attempt at its philosophical critique.[29] Cassirer's article on Cohen notes that the absolute domination of the naturalism of the period was seen even in those who appeared to oppose it, such as Schopenhauer, who retains, in Cassirer's words, "the language of the natural sciences in a completely naïve manner utterly lacking in the capability for criticism, particularly in regard to physiology."[30] Cohen's approach, by implication, was precisely the path to which a critical relation to science and physiology could be developed. Even writers who appeared to be cautious in their use of science were in Cassirer's view distorting its implications. The physicist and popularizer Hermann von Helmholtz uses the language of science with much greater precision, but he "applies it in a manner that far exceeds the limits of its real applicability."[31] Cohen's work, Cassirer suggests, helped define these limits.[32] Similarly, even Friederich Albert Lange, the cofounder of the Marburg school, remained bound to the false premise that the "psychophysical organization" of the organism of man determines the essence of humanity. This approach, Cassirer says, "undoubtedly raises the question of human knowledge more than it solves it."[33] Conversely, a reading of science and philosophy "capable of criticism" was the center the project begun by Cohen.

The path out of false readings of science lay in redefining scientific reality so that it is understood as a creative but mutable understanding of relations in the world. This insight, Cassirer continues, hinges on the "method of thought" Cohen developed first in his *Infinitesimal Method* and in the second volume of his *Kant's Theory of Experience*, which focused on the critique of "the given" that historically began with the invention of the infinitesimal method. Its second key moment is to use these insights to transform the center of the Kantian project from a static description of the categories to a dynamic reading of the schematism and principles, with the focus on the Kantian "anticipation of perception."[34] Here the philosophy of relations finds in Cohen's philosophy its fullest philosophical justification. Rather than making assumptions about "existing reality and its causal interrelations [*Wechselwirkung*]" and particularly "the existence of things," this approach will define science as a more flexible system of "ideal relations between truths" – essentially laws, axioms, and category-like patterns of consistency – open to permutation.[35] The goal for Cohen is not a splintering of experience through methods of mathematical physics, as his terminology occasionally suggests, but a definition of the richness of experience through a functional understanding of fungible "relations among [these truths] of interrelational dependency."[36]

Having situated Cohen's relation to natural science as the first stage of critique, Cassirer continues his philosophical overview of Cohen's work by examining the central claims of his philosophy and some of their consequences. The first of these claims, and for the broader social and human sciences the most important, is that "the opposition of subject and object" is false, and is indeed not even a "correct disjunction."[37] Kant's original use of the terminology of subject and object, Cassirer writes, is a mistaken vestige of eighteenth-century psychologism and has been "outgrown."[38] "The transcendental 'subjective,'" Cassirer writes, "is precisely that which is revealed as a crucial and general factor in any knowledge [*Erkenntnis*]; but precisely this is what for us is the highest achievable definition of 'objectivity.'"[39] Once this relation is recognized correctly, it can be seen that the assumption of given "natural scientific realities" can no longer be considered the starting place for a critique of knowledge, since these themselves are revealed on closer analyses to be partly "ideal structures" that spontaneously reveal experience within a historical and developmental process. Definitions such as "material and movement, force and mass" are revealed as "instruments of knowledge" rather than things in themselves: they allow perceptions of new aspects of reality and new modes of inquiring into reality, rather than acting as end points to knowledge.[40] Similarly, the social sciences are not concerned with the study of given subjects, but with the historical and structural determination of these subjects and their interrelation with differing definitions of "objects" in their world.

Cohen's key innovation was to describe how such "instruments of knowledge" had a historical component that was inextricably tied with their transcendental foundation. Although real consistencies and truths were revealed in the world, these were never absolutely known and were always capable of rearticulation within the horizon of a common – and never fully knowable – set of transcendental functions. Only a comprehensive survey of the forms of knowledge of different periods could reveal these structures as they have developed to the present, the process that Cohen described as a "critique of knowledge."

Cohen's philosophy, Cassirer continues, also overcame the traditional opposition of "empiricism" and "rationalism," since now it is revealed that "reason [...] presents itself in the system of experience" as a set of immanent realities.[41] "The structure of the 'spirit' [*Geist*] which idealism seeks," Cassirer writes, "can be read nowhere other than in the structural interrelation [*Strukturzusammenhang*] of natural sciences, just as is also true in ethics and aesthetics" – that is the first, second and third volumes of Cohen's *System of Philosophy*.[42] There are transcendental invariants, such as those proposed in Kant's categories, but also numerous other forms of invariants progressively revealed within fields such as economics, chemistry or even the arts. Transcendental invariants are not known in themselves but only in relation to a particular pattern of knowledge. In all cases, the "empirical" and "rational" are completely unified in Cohen's reading: rationality develops in and through the world of the empirical, and the empirical itself first "appears" particularly in configurations and continuities through the rational as an "instrument of knowledge."

Such forms of knowledge are theoretically open to radical transformation in the future, as Cohen thought had already occurred to the Kantian categories in the development of nineteenth-century non-Euclidean mathematics and in contemporary physics. Cohen, for instance, noted in his own essay from 1912 the recent example of Einstein's special theory of relativity, where the concept of space and time are revealed as relational to one another (ultimately as what was later defined as Minkowski's space/time).[43] Kant's original transcendental intuitions of time and space had to be recognized not as absolute, but as only a partial definition of more complex forms, a process that would be true of the categories as well. Cohen's goal was a description of the "pure" form of these invariants, essentially the relational definitions at the base of transcendental claims that could explain all such variation. A precise model for understanding how invariants could be both stable and "true," yet also variably open to permutation, only appears with Cassirer's application of the model of group theory to the problem, but is assumed in Cohen's discussion.

Cohen's Critique of the Sciences and Development of a "Metapolitics"

Cohen's lengthy 1912 introduction to F. A. Lange's *History of Materialism* (henceforth, "Introduction") is perhaps the clearest statement of his ambitions for the Marburg school, particularly its role in a modern "conflict of the faculties." If Cassirer's overview of Cohen's project presents a relatively conservative estimation for the scholarly audience, Cohen's own statement is considerably bolder. Cohen's plan was to transform the organization of knowledge in Wilhelmine Germany, which is to say science in all of its forms, towards a more open-ended understanding of humanity, one which recognizes – as Cohen puts it simply but profoundly in his "Introduction" – that the "spirit of humanity is not absolute; it produces itself in its productions [*erzeugt sich in seinen Erzeugnissen*]."[44] Lange's *History of Materialism*, a founding text of the Marburg school and still a best-seller nearly five decades after its publication, provided Cohen with a forum for introducing and defining the Marburg school project at a high point of its influence.

In the "Introduction," Cohen writes that the Marburg school seeks to recast Wilhelmine society by creating a common philosophical base of the "human" and "natural" sciences, which would redefine not only the organization of knowledge but "ethics and politics" as well.[45] Indeed, a key goal of the Marburg school, Cohen writes, is to develop a metapolitics that would, following a template first developed by the reception of Stoicism in the Renaissance, combine ideals with practical knowledge, the written law of jurisprudence with the unwritten laws of custom and practice.[46] The necessary relation of philosophy – presumably the Marburg school's philosophy – and the reformation of the university epitomized this theme. Since the time of Luther, Cohen writes, the public high school system and affordable university has been the "central symptom of [German] political culture" and the "critical means for national progress." Although it is not always evident, "the inner power" of their practical import and cultural force for the modern world was increasing.[47]

In Cohen's concept of university reform, nothing less than a reframing of the "future of the German university system" was necessary. This would entail the philosophical transformation of the teaching of the natural sciences (*Naturwissenschaften*), the redefinition of the human sciences (*Geisteswissenschaften*) and the systematic reformulation of all of the social sciences (*Gesellschaftswissenschaften*).[48] The Marburg program would even, Cohen proposes, reformulate the meaning of religion so as to reduce confessional strife.[49] Finally, this philosophy would culminate in a new form of nonmaterialist socialism (i.e., non-Marxist) that was harmonious with government planning and patriotism.[50]

Both present philosophy and contemporary common sense were in Cohen's view dominated by vestigial ideas about the nature of the world that made a philosophical return to immediate experience and first principles impossible without a fundamental critique of knowledge. As in Cassirer's review essay, the key issue Cohen emphasizes is the false understanding of natural science that underlay *other* philosophies of the late nineteenth century, and which effectively defined both common sense (in the subject/object distinction, the concept of causality, etc.) and movements that claimed to completely reject the tenets of natural science, such as neo-Romanticism. Although Cohen's criticism of neo-Romanticism and mysticism have been widely recognized, the primacy of his more fundamental critique of the normative science of the late nineteenth century has often been overlooked. The epicenter of this conflict in contemporary Germany was Kant's own view on the meaning of science, which would determine for Cohen nothing less than the "world-historical fight over the meaning of philosophy itself."[51]

As an object of criticism, the stakes for reinterpreting science for Cohen were nothing less than confronting what he took to be the twin evils of his era: on one hand, a fully determined mathematization of the human spirit, and on the other, a mysticism and voluntarism that claimed immediate intellectual intuition of truth. Thus, in his "Introduction," Cohen describes empirical psychology as closely related to the "reactionary tendencies of the time" since it bypasses the critical use of the humanities and social sciences (*Geisteswissenschaften*). Empirical psychology was "corrupting" (*Verderben*) the critical meaning of both philosophy and culture.[52] The enormous ambitions of empirical psychology were in fact a direct institutional and philosophical threat to the Marburg school. The scale of this threat can be suggested by the so-called "*Psychologismusstreit*," or "battle over psychology," that raged in the decades around the turn of the century and inspired Cohen's disdain of the concept of an independent psychological faculty. At the Third International Congress for Psychology in Munich (1896), Theodor Lipps declared that all that had previously been the domain of philosophy – aesthetics, ethics and logic – would in the future become the domain of an empirical psychology.[53] In debates over attention, determination and gestalt, new claims to the scientific reading of the human spirit were advanced.

The converse extreme in Cohen's view was "romantic metaphysics – which one always wants to wake from its death-sleep, and which is encouraged through general politics, where it is developed again into the calling of a simplistic national party-politics – and against the true classical national philosophy."[54] The outline of this "classical national philosophy" of course was largely taken to follow the Marburg critical interpretation of Leibniz, Kant, Fichte and Hegel. The best summary of this tradition, Cohen notes, is found in Cassirer's synoptic works, particularly *Leibniz's System* and *The Knowledge Problem*, which highlight its foundation in the relation of the finite and the infinite, as it developed from Cusa and Leibniz through Kant and Salomon Maimon.[55]

The relational definition of the sciences Cohen is promoting, however, has its greatest impact in the advancement of a new understanding of the social and human sciences. The social sciences, Cohen writes, lead to a new form of "metapolitics," or means of recontextualizing and defining traditional politics within a broader understanding of society and nature. Cohen's definition of metapolitics is based on the "important expansion" of the concept of politics first developed by the liberal theorist August Ludvig von Schlözer (1735–1809).[56] Metapolitics broadens the field of politics to include nearly any activity of society, as perhaps suggested by the new academic disciplines of the day such as sociology, and was meant to call into question the tacit assumptions of legal, social and state theory. Cohen uses the term as the heading for this section in his table of contents and defines it, as noted earlier, as "similar to the new geometers, who for their hypereuclidean speculations have turned to the terms, coined by Leibniz, of metageometry. In this conception, which aims to complete and realistically ground the study of the state, 'society' [*Gesellschaft*] is the actual reality, which comparatively makes the state and its logical foundation, the law, into a mere abstraction."[57] Through a new definition of the "sciences of society," and new social sciences themselves, a metapolitics in the modern era would not take the state, law or any of their constituent parts as given or obvious objects for politics. Instead, it could study and cultivate new forms of organization. For instance, the social democratic concept of workers associations, and the legal concept of "association" more generally, could in Cohen's view thus hold a key role in future political structure, as could presumably new applications of the methodologies developing in fields such as sociology, anthropology or history.

In defining the state or law as comparative "abstractions," Cohen was by no means denigrating them or suggesting they are not real: the term is used literally to suggest that they are one legitimate abstraction or means of organizing a complex set of forces from an initial set of premises among many. Indeed, Cohen's ultimate argument is partly nationalistic, and aims to modify, not remove, the concept of the state. His emphasis is rather that in the future there may be many other varieties of meaning or forms of political abstraction for defining human collectives and human interactions. Ultimately, any science – whether within the human, social or natural sciences – holds metapolitical implications, as it defines or makes assumptions about one aspect of the totality of interactions within humanity.

August von Schlözer's original project of metapolitics held that nascent fields such as statistics or economics were to act as a counterweight to older forms of cameralism, definitions of state hierarchies and established interests.[58] Schlözer took the field of statistics, for instance, to be inherently democratic since it provided a true insight into the total activities of a society rather than being based on abstractions or archaic definitions within one aspect of it.[59] Similarly, Schlözer wanted to demonstrate the metapolitical value of phenomenon usually not recognized as political at all for providing the foundation for more directly "political" activities. Within history, for instance, he argued that by moving away from a traditional focus on nobility, territory and succession, a new emphasis on other motives of history would become evident: "One will be liberated from the taste for the stories of wars [*Mordgeschichten*] [...] and awake from the slumber in which our education has steeped us to realize that the present perfection of our loaf of bread, piece of printed paper, pocket watch, bill of exchange, planet globe and hundreds of other things has been the result of discovery after discovery over thousands and thousands of years."[60] In Schlözer's definition, metapolitics defined the material and structural preconditions for a current politics, and the basis for a future development of new political form.

For Cohen, the key modern field of metapolitics is economics, which defines the preconditions of any form of state or law.[61] "Economics," he writes, "is the operational field of circulation in which society implements itself."[62] Cohen modernizes Schlözer's term, however, by placing it in the context of the profound transformation in the concept of science that had occurred in the nineteenth century, a turn that also highlights the difference of Cohen's theory of a redemptive future from Schlözer's earlier and more linear notion of Enlightenment progress. Cohen emphasizes that the term metapolitics should be considered parallel to the use in contemporary geometry of the term "metageometry," as used to describe projective and non-Euclidean geometry, both terms finding a common heritage in Leibniz.[63] The implication is that just as the traditional definitions of geometry as a metrics of space was redefined, and placed in a broader context as a study of pure relations by metageometries such as projective geometry, so the same could occur in a metapolitics. A field such as projective geometry, for instance, started from observations within perspectival drawing and ultimately defined point and line as definitionally identical: a bizarre innovation that nonetheless had numerous real-world applications, such as in cartography (or more recently spatial recognition in robotics). The definition of the political or legal subject, and of ultimately of humanity, would by implication similarly be open to further redefinition and reinscription within the unfolding meanings of metapolitics. Unlike Schlözer, although Cohen imagined a world of progress, it is not a linear progress but one capable of surprising transformations in which, rather than having "perfected" the various forms of the day, society and science are open to unlimited future development.

The focus of Cohen's metapolitics is that both the individual and humanity as a whole are part of an incomplete project that is only partially bound and understood by the contemporary sciences that investigate them. Unlike traditional political systems or contemporary claims by biology, the definition of the individual person, the state and humanity cannot be assumed. "Humanity," Cohen writes in summary, "is not a natural reality."[64] Humanity is rather a present and future possibility and ideal goal. It is for this reason that a metapolitics is closely connected with a general critique of the limits of all of the sciences, since the misplaced faith in the natural sciences – such as Cohen thought was occurring in empirical psychology and presumably in other fields – could claim to define "human nature" and with it society.

In this context, Cohen's "Introduction" places two surprising but closely related limitations on a socialist metapolitics: the first is that it not collapse into materialism, and the second is that it needs to retain a horizon of theism or, as Cohen puts it, "the God-idea."[65] The former guarantees that there is no illusory "natural" ground to the assumption of what society or humanity is, the latter the possibility of a redemptive change in humanity beyond our present imagining – it promotes the "unending future of humanity."[66] "With atheism," Cohen writes, "socialism loses its peak, its roof; with materialism it loses its basis, its foundation."[67] In his attack on atheism, Cohen grounds his work directly on the older foundation of a negative theology, one that develops from the "God-idea" of Maimonides and later Salomon Maimon. The "God-idea" for Cohen functions as a guarantee that both humanity and its future are unknowable, that is have no single substantial essence, yet through hope in the future, and work on the present, both are capable of unlimited further development.[68] As Cassirer later put it in a summary of Cohen's ideas to the Jewish Student Society at Cambridge University in 1935, Cohen held the "very strange and bold" concept that "to believe in humanity, in its highest and ethical sense, and to believe in God means one and the same."[69]

The middle ground in which society can be transformed without making assumptions about the substance of humanity, by contrast, was through the relation of knowledge to the organization and structure of society, through the manner in which institutions could cultivate and be vivified by new patterns of behavior, belief and perception. The normative goal of this system was an operative ideal of equality and a notion – again Leibnizian in origin – of this equality as based not on the similarity but on the radical difference of individuals. On this basis Cohen argues that the legal basis of equality is also the foundation for individual difference. "Without laws there is no freedom," Cohen writes, "and without the society that forms through law there can be no free personality."[70]

Neither state, nor laws, nor even personality are given entities for Cohen, but rather all had to be understood as part of a "double reality" in which on one hand they are respected as the "normal reality" of a given historical period, while on the other they are understood as historical "products," as "transformational forms of ethical ideals."[71] Cohen envisioned the shaping of these transformational ideals as occurring through the institutions of Germany, foremost the university but also in practical institutions such as unions, professional societies and associations.

Cohen's Psychology and Aesthetics

The culmination of Cohen's critique of knowledge and the sciences in the "Introduction" is his description of "psychology," a field that was to be the capstone of his philosophy. Cohen's three-volume *System of Philosophy* was intended to culminate in a fourth volume, never completed, that defined psychology as the reciprocal determination of individual consciousness by the totality of the forces of his society. Only a "systematic psychology" linked to a wider philosophy of epistemology, ethics and aesthetics can address consciousness, Cohen held, and this compact occurs under the title of individual "psychology" as it reflects the totality of "culture" – a term that for Cohen encompasses the articulations of natural science and with it nature, as well as all cultural and social phenomena. "Psychology," Cohen writes against the empirical psychologists, "has to do with the consciousness of culture, and indeed with the unity of this consciousness. Culture, however, is a problem of history, and not as such with the norms and pathologies of the nervous system."[72] In particular, psychology would demonstrate "the illusion of a closed unity of consciousness" by demonstrating the necessary relation of consciousness with the infinite field of relations that structure it.[73]

Psychology would thus become a more complete and historicized form of Kant's "system of experience" in that it would explain how different forms of knowledge structure the continuous unity-in-difference of consciousness.[74] The basic assumption of Cohen's psychology is Leibnizian: the psyche is the differential in relation to the play of forces in society, and the relation of these two defines humanity as an emergent reality, to use a later terminology, without a set definition. It is in this sense that Cohen writes, "In psychology the unity of the substance of consciousness means the manifold [*Mannigfaltigkeit*] of its directions." Unity is not intended in this case as a totality either for the individual or for culture, but as a plenum connecting the two through the derivative, so to speak, of its "directions." It is a field of possibility that is nonetheless interrelated, so that Cohen can write, "This unity of consciousness [*Bewusstseins*] however is a great problem. It is not only a difficult problem for psychology, but an enormous and wide-ranging problem in the best sense of metaphysics, namely that of systematic philosophy."[75]

Cohen summarizes psychology in the "Introduction" as "the macrocosm of humanity in the microcosm of the cultural human."[76] Psychology describes the individual in an open dialectic with forms of society and nature. Thus Cohen writes that the "new branch of the system of the philosophy is formed by psychology, which depicts the meaning of humans in the unity of their cultural consciousness [*Kulturbewusstseins*], as well as the development of this unity, and the genetic relation of all of its branches and seeds [*Keime*]."[77] The term psychology as used by Cohen is obviously at nearly complete odds with its present meaning, and was meant in part as a polemical attack on the ascendant field of empirical psychology, but also as a return to first principles on the reciprocal relation of *psyche* and humanity.[78] By making psychology part of his system, Cohen writes, it would unify "culture, science and customs in their cultural manifestations and art," and would furthermore orient "logic, ethics and aesthetics" – which is to say Cohen's system itself.[79]

To explain the unity found in the "manifold of directions" of consciousness, Cohen uses Henri Bergson's then popular concept of duration (*dureé*), that is the sense of the inner consciousness of time and motion when it is not confused with spatial determination. Both the similarities and differences of the two projects illuminate Cohen's wider argument. Cohen's project is similar to Bergson's in that it is based on the immediacy of phenomenal experience grasped by the concept of duration, the "flow" of experience. Indeed, the primacy of this experience defines both Cohen's aesthetics and psychology. Whereas Bergson places duration in a dichotomy with problems of form and knowledge, however, Cohen sees them as a continuum defined by the infinitesimal idea and the concept of unity as a plenum. In this regard, he writes, Bergson's popular works need to be "sentence for sentence resituated and illuminated from our perspective."[80] Cohen held that the basic phenomenon of psychology was not any discrete moment of experience available through experiment or even reflection, but consciousness itself understood through the problem of "expressive motion" through which we grasp meaning in and as motion. "Expressive motion" will define a common plane of meaning equally encompassing both an "internal" sequence of "emotions" and an "external" change of states.[81] "Movement and a feeling for movement," Cohen writes, "are an elemental and fundamental factor in the structure of consciousness itself."[82] Expressive motion is then a term for the changing relation of a derivative, the moment of consciousness, to a larger differential ensemble or structure of claims.

In an earlier passage on psychology in the third volume of his *System of Philosophy*, the *Aesthetics of Pure Feeling*, Cohen summarizes the goals of his psychology, and intimates its ethical and political dimension, when he writes it would form a "hodegetic encyclopedia of the system of philosophy."[83] "Hodegetic" literally means an "indicator of the way" in Greek and was most frequently used in Cohen's time to connote that the method of a particular process is itself "the way" to further understanding its content.[84] In demonstrating the process of interactions of all forms of society, the process and meaning of consciousness would be revealed. The term thus suggested the revaluation of "psychology" at the heart of Cohen's work: psychology was to define the manner in which the psyche develops through forms of society, not claim access to a pre-existing substrate of the psyche itself. It is in psychology that Cohen finds the central premise of his philosophy, that the human spirit "produces itself in its productions."[85] Outlining these productions through a historical critique of knowledge revealed the directionality of consciousness of culture both individually and collectively.

Cohen apparently worked on this project until the end of his life: his very last lecture, canceled due to illness, was on the topic of the "Encyclopedia of Culture."[86] Much of Cassirer's later work turned to the terminology of "culture," but it often did so on the basis of Cohen's culminating definition of psychology as a unification of natural, social and humanistic fields, not in either of its standard definitions as a field opposed to nature or as a "higher" hierarchical feature of society. Interestingly, Cassirer's personal notebooks from Cohen's 1899 lectures on psychology appear to be one of the very few – perhaps the only – extant versions of Cohen's psychology, albeit at an early stage of development, and Cassirer clearly followed this aspect of Cohen's work very carefully.[87] Even Cassirer's late *Essay on Man* (which will be addressed in our conclusion) can be considered closely related to Cohen's planned psychology, and is only fully understood on the basis of the philosophy of the differential in which it was grounded.

The meaning of Cohen's psychology is illuminated by the last complete study of his *System of Philosophy*, aesthetics, which formed the topic of the third volume. Cohen's aesthetics developed a critique of narrative form, particularly through the opposition of comedy and irony, and was understood by Cohen as a paramount theme of his work and methodology. Whereas Kant's system culminated in his third critique in the aesthetic problem of the beautiful and sublime, in Cohen's system aesthetics culminated, somewhat improbably given his reputation for rationalism, in an open form of comedy that recognized both the infinite and its relation to human limitation.[88] The present era, Cohen held, was dominated by a Romantic concept of the sublime and an intimately related, but apparently opposed, concept of irony, and this affected both the form and substance of nearly all cultural production, and with it all individual perception. The danger of irony for Cohen is that it blocks the path to a wider skepticism by claiming tacit access to the "truth," and thus blocks both the specific modes of questioning of a systematic philosophy and the recognition of the infinite task of human striving in his psychology.[89]

For Cohen, comedy represents the awareness of the infinite task and development in the finitude of human experience – it thus counterbalances the sublime, which in Cohen's view often inadvertently assumes a sense of the infinite against which it is counterposed.[90] That is, when Kant suggests that the sublime "is to be found in a formless object," represented by a "boundlessness," say, in the mathematical sublime or the dynamical sublime of thunderclouds, this is easily mistaken as a Romantic sense of unity with the sublime.[91] Schopenhauer's work and the reception of Kant by the Romantics represents the clearest case of the misreading of the concept of the sublime for Cohen. In Cohen's reading, the modern comic style allowed the development of an aesthetics in which the ideas of both nature and culture as well subject and object could be correlated as an infinite difference open to permutation within a larger philosophical system. Cohen's principal example was Goethe's *Elective Affinities*, in which the amorous fate of the four protagonists also works analogously to the contemporary understanding of chemical processes.[92] The comic is the foundation of both a vision of humanity founded on future redemption – there is always more possibility for fortuitous forms than appears possible – and one grounded in a sense of the plenitude of present experience, particularly the hidden possibilities of the given human. "Man's love of nature thus receives its most profound foundation in this respect for man's dignity in all men" – and this curiously occurs foremost through the beautiful found in comedy.[93]

The ironic mode of contemporary philosophy indirectly assumed in Cohen's reading a "realistic" perspective and false "naturalism" in two different but closely related forms, either that of the Romantic claim to have access to the absolute and infinite, or the diametrically opposed naturalism found in the absolutism of materialism and "dogmatic" definitions of natural science.[94] These two antipodes represented the twin mistakes of the modern period for Cohen, those of neo-Romanticism and materialist positivism. His philosophy – first of natural science, then of ethics, aesthetics and finally psychology – was intended to provide an alternative path to understanding the problem of nature and reality as *always* the result of figuration.

Cohen's aesthetics was largely developed through the study of the specific art works of a period. The structure of surprise or incongruity in comedy highlights its complex structure in artistic form. Comedy leads us to expect reversals in structural form, and to pay attention to this horizon of meaning in any dialogic form of meaning. Cohen holds that the general form of comedy reveals a universal form of logic at work in all human activity; it is for this reason not only applicable to comic artists, but to the general form of art and meaning altogether. Even fields such as architecture can have the character of humor, so that Cohen describes Michelangelo's architecture as "unifying the two moments, the sublime and humor, in such a way that we take it as the unity of the beautiful."[95] What Cohen appears to mean here is that Michelangelo's work reveals a play of expectation and immediacy in a manner that both implies the infinitely great, and thus the sublime, while canceling obvious conclusions and frustrating expectation in the manner of the humorous or the infinitely varied.

Cohen's examples of *Elective Affinities* and Michelangelo's architecture suggest his ultimate aim in emphasizing the primacy of the comic within the beautiful: through it humans recognize themselves in a world in which complex webs of meaning on the one hand overwhelm any totalizing view of events, even as on the other they do not resolve into the equally simplistic "chaotic being" of Romanticism. This lack of closure includes human consciousness itself, so that the capacities of the human and human determination are also beyond expectation. In this manner Cohen's aesthetics provides a formal foundation for understanding his psychology. The goal in both is to avoid the "illusion of the closed unity of consciousness," which entailed the false correlates of closed definitions of subjectivity and objectivity, culture and nature.

Cohen's gifted student Cassirer arguably went beyond the formal analysis of historical works of art to use the principles of Cohen's aesthetics as the foundation of his own rhetorical and stylistic form. Throughout his life, Cassirer held that the very center of his philosophy was dominated by Goethe's claim that "style is rooted in the deepest foundations of knowledge, in the very being of things."[96] In his earliest work, Cassirer seems most influenced by Cohen's aesthetics, not in the content of his studies, but in the form of his approach to writing history. Cassirer's writing is comic in its emphasis on a history of problems that leads to fortuitous new creations, as Hermann Bahr had noted. The importance of this approach is two-fold. First, it demonstrates how in the form of his work Cassirer intended to advance Cohen's project just as much as he did in the content of it. Aesthetics is the one branch of Cohen's system that is not intended as a science; rather, it is a study of form as it protects the ineffability of experience. Cassirer's use of Cohen's model of comedy in his history of problems is in this regard consistent with Cohen's basic model. Secondly, the comic form of Cassirer's works will suggest that even as Cassirer could hold a realistic, and I will suggest even dark, view of his own time and human nature, he would always try to recast the elements of this time in a comic and optimistic manner, a project he considered similar to Kant's cosmopolitan history. Rather than reflecting his putatively irenic personality, or his lack of a critical relation to his own time period, Cassirer's formal style is most productively read as reflecting the highest principles of Cohen's philosophy.

Later readers have noted this aspect of Cassirer's work, even if they have mistaken its origins. The historian Hayden White in his study of nineteenth-century historical style, *Metahistory*, noted in passing Cassirer's use of the narrative mode of comedy, but mistakenly connects it solely to a Hegelian dialectic, rather than Cohen's dialogic theory of comedy. Discussing the difference of Hegel and Herder's form of comedy, White writes,

> A late Herderian and an exponent of the same Synecdochical intelligence that Herder represented in his own time (and, moreover, one who advanced his philosophy in a similar spirit – that is, as a way of transcending the Irony of his own age), Ernst Cassirer, said that Herder "broke the spell of analytical thinking and the principal of identity…" History, as Herder conceives it, Cassirer wrote, "… brings forth new creatures in uninterrupted succession, and on each she bestows as its birthright a unique shape and an independent mode of existence."[97]

Cassirer's own history and philosophy had a similar goal and valance "of bringing forth new creatures in uninterrupted succession," or as Cassirer puts it, transforming history into some thing that is "productive" in the sense of highlighting new possibilities for the future by highlighting novel structural relations in the past.[98] This style is rooted in Cohen's attacks on the ironic mode of the age and is developed from a Herderian "metaphysics of history," which, Cassirer argued, developed directly from "Leibniz's central doctrine" and ideas, a geneaology Cassirer himself first outlined in *Freedom and Form* and recapitulated in *The Philosophy of the Enlightenment*.[99] Cassirer's reading is not reducible to a Hegelian unfolding of the Idea, since it is based on a multivariant development of possibilities, one modeled on group theory, and on a correspondingly different "open-ended" form of logic, a logic that Cohen grounded in the problem of the so-called infinite or limitative judgment. Thus Cassirer's own work recasts Herder's philosophy through the same open-ended logic, and tacit comic form, that Cassirer had developed since *Substance and Function*, and links it directly to the historicist project that he thought was coequal with the role of science in functionalist philosophy. Herder's philosophy and its relation to history is thus described by Cassirer in terms broadly similar to those of his own *Substance and Function*: "It is from such complete heterogeneity that real unity emerges, which is conceivable only as *the unity of a process not as a sameness among existing things*. The first task of the historian is, then, to suit his standards to his subject and not, conversely, to make his subject fit into a uniform, stereotyped pattern."[100]

By developing structural relations and harmonic relations between themes, Cassirer's work demonstrates not only the content but form and methodological consequences of Cohen's philosophy. Far from being an escape from his time and its responsibilities, Cassirer's work was a translation of Cohen's immanent critique into a new and more approachable style. Although Cohen's philosophy of the natural sciences was the necessary starting point of his philosophy, and means of overcoming the naïve biases of the era, Cassirer saw that the core insights of Cohen's work lay in the later parts of his system – notably, in an ethics in which neither the individual nor "humanity" was fully knowable, an aesthetics based on a comic profusion of meaning and a psychology based on the relation of experience to the system and history of human knowledge. Each was based in its own way on the open-ended logic that Cohen, like his predecessor Leibniz, found epitomized in the infinitesimal problem, but none was even remotely reducible to a mathematical scheme. To understand the political implications of this project, we turn to the changing context of the Marburg school as it was reflected in Cohen and Cassirer's relationship.

Notes

1 Cohen, Hermann. *Das Prinzip der Infinitesimalmethode und seine Geschichte: Ein Kapitel zur Grundlegung der Erkenntniskritik* (Hildesheim: Olms Verlag, 1984).
2 Kuntze, Friederick. *Die kritische Lehre von der Objektivität: Versuch einer weiterführenden Darstellung des Zentralproblems der kantischen Erkenntniskritik* (Heidelberg: Carl Winter's Universitätshandlung, 1906), 249.
3 Frege, Gottlieb. "Translation of 'Rezension von: Cohens Das Infinitesimalmethode und seine Geschichte 1885,'" trans. Black et al., ed. B. McGuinness in *Collected Papers on Mathematics, Logic, and Philosophy* (Oxford: Basil Blackwell, 1984 [1885]), 108.

4 Cohen, Hermann. *Hermann Cohens Schriften zur Philosophie und Zeitgeschichte*, ed. Ernst Cassirer and Albert Görland (Berlin: Akademie-Verlag, 1928), vol. 1, 245ff. I will use Cassirer's edition of Cohen's works in order to retain Cassirer's selection of editions and corrections.
5 Cohen, *Schriften*, I: 265.
6 Cassirer defines all of Cohen's late work through the theme of the infinitesimal in the multiple editions of his "Neo-Kantianism" article in the *Encyclopædia Britannica*, 14th edition, XVI: 214–15.
7 Kant, Immanuel. *The Conflict of the Faculties (Der Streit Der Fakultaten)*, trans. Mary J. Gregor (Lincoln: University of Nebraska Press, 1992).
8 Meyer, Thomas. *Ernst Cassirer* (Hamburg: Ellert and Richter, 2007), 36.
9 Cassirer, Toni. *Mein Leben mit Ernst Cassirer* (Hamburg: Felix Meiner Verlag, 2003), 91.
10 Cassirer's critique of science was also a study, as Fabien Capeillères has noted, of the potential for philosophy itself to "become a science," which Capeillères argues occurs in Cassirer's later work through the model of energetics. See Capeillères, Fabien, "Philosophy as Science. Function and Energy in Cassirer's Complex System of Symbolic Forms" *Review of Metaphysics* 61, 2 (December 2007): 317–77, here 318. This development of philosophy itself as science is not, however, the theme of this book, as our emphasis is rather the necessary limits of the particular sciences and the political implications of these limits. Cassirer's break with Cohen in his early work can in this reading be defined through his departure from the concept of philosophy as science in favor of a project that emphasizes history, difference, and a Humboltian agonistic liberalism. Capeillères general reading of the model of the integral and differential of forms in this project, as derived from SF but applied to Cassirer's later work, is in accord with key themes developed here for the earlier period (ibid., 321). With the later development of a comprehensive philosophy of symbolic forms, particularly in relation to myth, it seems plausible that Cassirer may have returned to a strong definition of philosophy as science and unity. Ironically, then, precisely what is often taken as Cassirer's most evident "break" with Cohen in his turn to myth would be a return to the central theme of Cohen's project, a project he had earlier broken from in works such as *Substance and Function* and *Freedom and Form*.
11 KI 32; ECW9 33.
12 KI 32; ECW9 33n61.
13 KI 32n1; ECW9 33n61.
14 Cassirer, "Neo-Kantianism."
15 SF 300.
16 The basis of Cassirer's analogy is found in the "first fundamental theorem of calculus," covering the relation of the integral and differential of a function. Since this formulation will reappear throughout Cohen and Cassirer's works, it is worth parsing in some detail. Integral calculus is most readily characterized by its use to measure the area under an irregular graphed line using what we could initially imagine as infinitesimally small "strips." These strips are defined by the differential, and their "summation" provides the integral, the area under the graph. Thus, Cassirer writes of Leibniz's understanding of this phenomenon, "'Continuation' is here with other words the methodological expression of integration as the continuous 'summation' of infinitesimal moments" (L 170). Differential calculus at first appears to have a completely different application, describing how functions change as their inputs change, such as in the graph of a line. The derivative of such a function defines the behavior of the function near a certain point, such as the particular slope of a line at a given instance. Differentiation is the means of finding this derivative at a certain point, and second order differentiation is the rule of change for this development, the rate of its rise or fall, for instance along an axis. A further example taken from physics would have the differential define the instantaneous velocity as the derivative with respect to time of an object's distance from an original position, while acceleration is the second derivative of an object's velocity, that is, the rule for its rate of change in relation to time.

The heart of Cohen's wider claims for the calculus, and Cassirer's summary, appears to be the so-called "first fundamental theorem of calculus," which claims that as different as the application of integral and differential may at first appear, the two functions are precise and intimately related opposites of one another. This relation will typify what appears to be Leibniz's maximal description of the monad as defined by the relation of integral and differential. The first fundamental theorem of calculus holds that the overall functions of integral and differential calculus illuminate one another: the area under a line can be extrapolated forward using what is called an indefinite integral (also called the antiderivative or privative) so as to calculate out along another function (say, time), as it were, the varying growth and undulations of this area. In this way, however, the indefinite integral leads to something like the function for a line defining its space, and this can be differentiated. In terms of our example from physics, this would lead back to the point of instantaneous velocity of this function. Conversely, however, if we take the differentiation that is led to this derivative of instantaneous velocity and apply indefinite integration to it, we get the distance/area traveled. Technically, the fundamental theorem of calculus is thus expressed: "The definite integral of a function over an interval is equal to the difference between the values of an indefinite interval (or antiderivative) evaluated at the endpoints of the interval."

The import of the first formula of calculus is nicely summarized by David Berlinski at the end of his popular account of calculus, *A Tour of the Calculus*, presumably with no direct reference to Leibniz's philosophy, much less Cohen's. Berlinski summarizes a "coarse and vulgar" sense of the fundamental theorem in nontechnical terms that precisely evokes its philosophical implications for Leibniz, Cohen and Cassirer:

> The derivative of a function concentrates the mind at a point; the landscape it reveals is local. The integral of a function allows the mind to contemplate a region of space; the landscape over which it lingers is global. And this is what part one of the fundamental theorem reveals about the inner nature of continuity: from a continuous global description, a local system may be recovered, and from a continuous local description, a global system. In some ways, this remarkable exchange between global and local descriptions is the fundamental maneuver by which we come to understand reality, the very dance of life itself. […] The world impresses itself upon us as the single hot, charged center of consciousness. But the world is something that transcends the limits of my consciousness. […] The miracle of calculus is that in the realm of the real numbers, the passage from local to global and back again is both possible and necessary. (Berlinski, David. *A Tour of the Calculus* (New York: Vintage, 1995), 287)

17 EL 6.
18 I 19, 69.
19 HCS2 321–2.
20 Thus in regard to Leibniz, but in a manner that would apply equally to Cohen's analysis, Cassirer writes that the logic at the base of Leibniz's system was "not the result of Leibniz' studies of the problem of objectification of appearance in mathematics and dynamics" but rather "the condition of first being able to understand them" and the basis for Leibniz's understanding of fields as diverse as biology, ethics, aesthetics and law (L 478).
21 For a reading of this development within the general evolution of theory and the social sciences, see Hunter, Ian. "The History of Theory," *Critical Inquiry* 33 (Autumn 2006): 90. On Cassirer and Cohen: 89–92.
22 Kagan, M. I. "Hermann Cohen (4 July 1842–4 April 1918)," in *The Bakhtin Circle* (Manchester: Manchester University Press, 2004), 201.
23 K 426.
24 Cassirer, "Neo-Kantianism."
25 HC 253; ECW9 120.
26 Ibid.

27 HC 256; ECW9 122.
28 EK1 13.
29 Edward Skidelsky's reading of Cohen is the most recent of such interpretations of Cohen's work as scientistic. Skidelsky, 47–51.
30 HC 254; ECW9 121; EK3 125ff., 396ff.
31 HC 254; ECW9 121.
32 Elsewhere Cassirer would single out Gustav Fechner (1801–1887) and Du Bois-Reymond (1818–1896) – the former for overextending scientific explanation in the direction of "mathematizing of the psychic" and the latter for his critique of materialism – as both leading to arguments so muddled, Cassirer notes in his later history of the period, that "often both thesis and antithesis stood on the same ground." "Skepticism and dogmatism" crossed over into other, since even Du Bois-Reymond's attempts to reject mechanistic cosmology "could not help at every turn affirming just what it explicitly sought to oppose and deny" since any other model of science was "still far in the future." Cohen's philosophy of science was the first step towards such a new model of science that would avoid the related twin threats of simplistic realism and its purported opponent of mystical intuition or immediacy (EK4 87, 285).
33 HC 254; ECW9 121.
34 HC 260–61; ECW9 126ff.
35 Ibid.
36 HC 257; ECW9 123.
37 HC 259; ECW9 125.
38 Ibid.
39 Ibid.
40 HC 260; ECW9 126.
41 Ibid.
42 HC 257; ECW9 123.
43 EL 91.
44 EL 44.
45 EL 111ff.
46 EL 4, 116.
47 EL 124.
48 EL 27–8.
49 EL 104ff.
50 EL 111ff.
51 HC 253; ECW9 120.
52 EL 58–9.
53 *Dritten Internationalen Congress für Psychologie in München vom 4. bis 7. August 1896* (München: J. F. Lehmann, 1897), 68ff., 146–64, 218ff. Cassirer attended the congress and provides a summary of the *Psychologismusstreit* (STS 161–64).
54 EL 38.
55 Cohen on Cassirer in EL 26.
56 On "metapolitics," EL4 116; SP2 310. Cohen's use of it appears to derive from Schlözer, who saw the term "metapolitics" as defining the conditions for the possibility of integrating "prepolitical" aspects of humanity and the liberal state. See: Bödeker, Hans Erich. "On the Origins of the Statistical Gaze: Modes of Perception, Form of Knowledge and Ways of Writing in Early Society Sciences," in *Little Tools of Knowledge: Historical Essays on Academic and Bureaucratic Practices*, ed. Peter Becker and William Clark (Annarbor, MI: University of Michigan Press, 2001), 175ff. Becker summarizes the initial concept of Schlözer's metapolitics in political writings: "Treating of the pre–civil rights of the individual and the conditions precipitating a member of the 'society of citizens' to become a member of the state" (175). On Schlözer, see Cohen, Hermann. *System of Philosophy*, vol. 2 (SP2), 294; on metapolitics: Warlich, Bernd. *August*

Ludwig von Schlözer 1735–1809 zwischen Reform und Revolution: ein Beitrag zur Pathogenese frühliberalen Staatsdenkens im späten 18. Jahrhundert (dissertation, Erlangen-Nürnberg, 1972), 190ff.; Kern, Horst. "Schlözers Bedeutung für die Methodologie der Empirischen Sozialforschung," in Hans-Georg Herrlitz, *Anfänge Göttinger Sozialwissenschaft: Methoden, Inhalte und Soziale Prozesse im 18 und 19 Jahrhundert* (Göttingen: Vandenhoeck & Ruprecht, 1987), 55–71.

57 EL 116.
58 Lindenfeld, David F. *The Practical Imagination: The German Sciences of State in the Nineteenth Century* (Chicago: University of Chicago Press, 1997), 41–5; Reill, P. Hans. *The German Enlightenment and the Rise of Historicism* (Berkeley: University of California Press, 1975), 46ff.
59 Reill, *German Enlightenment*, 146.
60 von Schlözer, August Ludwig. *Vorstellung der Universal-Historie*, vol. 1 (Göttingen: J. C. Dietrich 1775), 33–4.
61 EL 117.
62 Ibid.
63 EL 116.
64 Ibid.
65 Ibid.
66 Ibid.
67 Ibid.
68 Cassirer, Ernst. "Cohen's Philosophy of Religion," *Internationale Zeitschrift für Philosophie* 1 (1996): 94. For Maimonides, incidentally, the designation of "negative theology" needs a qualification, as noted to me by Peter E. Gordon: certain attributes of action are revealed by God to humanity directly, and thus are not covered by the ban on predication, and these are exactly those in which God is revealed to humanity as capable of emulation.
69 Cassirer, "Cohen's Philosophy of Religion," 100.
70 EL 118.
71 Ibid.
72 SP3 II 429.
73 SP3 II 426.
74 Kant, Immanuel. *Critique of Judgment*, trans. Werner Pluhar (Indianapolis: Hackett Publishing, 1987), 402, section 19/A 215.
75 EL 43.
76 SP2 637.
77 EL 11, 43.
78 EL 53ff.
79 EL 11.
80 EL 51.
81 SP3 I 127. As Cassirer notes in an earlier work, the re-reading of the problem of motion earlier culminated in Cohen's replacing all spatial relations of "next-to-each-other" (*Beisammen*) with temporal relations of "after-one-another" (*Nacheinander*), particularly as epitomized by calculus. From "Die Begriffsform im mythischen Denken" (1922) in ECW16 48.
82 SP3 I 179, no. 2. The area to which Cassirer directs us makes it quite clear that "movement" in this sense is the key to Cohen's general use of the concept of origins, linking both his notion of "origin" in physics (through Galileo's law of falling bodies and its relation to Leibniz's notion of continuity) and to the movement of expressive wholes.
83 SP3 II 432.
84 For instance, as defined in Duden's *Fremdwörterbuch*, 10th edition: "Hod-ege-sis/Hod-ege-tik: Anleitung zum Studium eines Wissens- od. Arbeitsgebietes. nochwas: Hodegetria 'Wegführerrin': stehende muttergottes (auch als halbfigur) mit dem Kind auf dem linken arm (byzantinischer bildtypus); hodos=Weg steckt z.b. auch in method."
85 EL 44.
86 Kagan, "Hermann Cohen," 193.

87 See Cassirer, Ernst. "Cohen – Psychology," Ernst Cassirer Nachlass, Beinecke Library, Yale University, gen. ms. 98, box 56, envelope 1108. Andrea Poma claims that at least for Cohen's later classes on "psychology as encyclopedia of philosophy" from 1905–1908 there exists no documentation (*The Critical Philosophy of Hermann Cohen*, trans. John Denton (Albany, NY: State University of New York Press, 1997), 148); Cassirer describes his own project in the following terms: "It was one of my principal aims to convince the reader that all the subjects dealt with in this book are, after all, only one subject. They are different roads leading to a common center – and, to my mind, it is for a philosophy of culture to find out and determine this center" (E vii–viii). "The following investigations undertake to show that a coherent unity obtains. […] The nature of this coherence can only be designated and made known by following its growth" (PSF3 41).
88 SP3 I 9ff., 347ff.
89 SP3 II 48; SP2 507ff.
90 SP3 I 271ff.
91 Kant, *Critique of Judgment*, section 23, 244–5; section 28, 98–9, 119ff.
92 Through using humor as the basis of understanding the beautiful, and the beautiful as the means to understand nature and culture, is created a "fundamentalen Korrlation von Subjeckt und Objekt" (SP3 413).
93 SP3 I 233.
94 SP3 I 9–10.
95 SP3 I 237.
96 LC 83. Although my reading differs significantly from Barbara Naumann's, her work is perhaps the best guide to a reading of Cassirer's style as it developed from his reading of Goethe. See Naumann, Barbara. *Philosophie und Poetik des Symbols: Cassirer and Goethe* (Berlin: Wilhelm Fink, 1996), 21–71.
97 White, Hayden. *Metahistory: The Historical Imagination in Nineteenth-Century Europe* (Baltimore: Johns Hopkins University Press, 1973), 74–5, citing PE 231. White describes Herder's comic philosophy as holding that he "not only saw the plan of the whole historical drama as a Comic plan, but saw every act of that drama as a Comic play in miniature" (73).
98 FF 386.
99 In the continuation of the passage section referenced by White on Herder, Cassirer refers the reader back to the discussion of Leibniz and Herder in Freedom and Form (PE 231n40), itself a development of his Marburg interpretation of Leibniz, notably in L xii, 374.
100 Ibid. My emphasis.

Chapter Two

CASSIRER AND THE MARBURG SCHOOL IN THE ADMINISTRATIVE AND POLITICAL CONTEXT OF THE *KAISERREICH*

The Development of the Marburg Critique from F. A. Lange to Cassirer

That Cohen's work would principally be received as part of a broad reform project and not, as his detractors would have it, solely as a recondite theory of physics or, conversely, even a form of "mysticism" (Nelson) is evident if we place it in the earlier context of the Marburg school.[1] The same reception history also casts light on the meaning of Cassirer's works, which were written and presumably read during his life with this context in mind, but which in themselves rarely alert the present-day reader to this background. Against a reading of Cassirer or the Marburg school as aloof from practical concerns, their work proves to be consistently inflected by politics and social pressures. In confronting them, Cohen and Cassirer apparently acted in close consultation with one another.

The theoretical foundations of the Marburg school began with Lange's influential *History of Materialism* (1861), by far the most – indeed, in many ways the only – popular text of the Marburg school. The young Nietzsche, for instance, was characteristic of his intellectual generation in reading Lange's work numerous times. He went so far as to write a friend that Lange's was "the most important philosophical work of recent decades" and that all the young philologist needed for his education was "Kant, Schopenhauer, and this book of Lange's."[2]

Arguing against the dominant trend of scientific materialism of the 1850s, notably that of Ludwig Buchner and Karl Moleschott, Lange held that materialism was itself an idealist fantasy. There was no such entity as "matter" in itself, and it could only be understood as an ideal concept. The original Democritian basis of materialism in the natural sciences nonetheless provided a crucial foundation for critical thought in explaining the world. By depicting the history of materialism as a form of critique that ultimately turned against its own premises, Lange sought to suggest a nonmaterialist yet realist basis for the sciences and a critical Kantian approach to experience grounded on radical skepticism. Critical idealism in the manner first represented by Kant presented the principle modern alternative to materialism and the excesses of speculative idealism. Lange concluded the *History of Materialism* by noting that neither subjective nor objective reality can ever be known in itself: "The transcendent basis of our [psychic and physical] organization remains therefore just as unknown to us as the things that act upon it. We have always before us only the product of both."[3] Critical idealism would investigate this common world of phenomenon, and act as a check to assumptions of scientific materialism and religious or Romantic mysticism.

Closely connected with this rejection of false "essential" elements in early aspects of science was a parallel criticism by Lange of society, particularly an attempt to break down earlier social hierarchies into a new form of social democracy, as first put forth in *The Labor Question in its Present and Future Significance* (1875).[4] In a manner that would be repeated throughout the Marburg school's history, the two themes of scientific and social reform were linked by the demand for pedagogical reform. Lange developed this argument from the beginning of his career in works such as *Concerning the Relation of the Educational System to the Dominant Worldview of Different Ages* (1855, republished in 1894 and 1928) and *The Role of the School in Public Life* (1866).[5]

While retaining Lange's political focus, Cohen redefined the philosophical foundation of this project in a manner that radicalized its critique of "subjective" and "objective" reality, removed Lange's focus on physiology (which he had taken as the basis of the Kantian *a priori*) and defined the Kantian *a priori* categories as capable of historical transformation. As we have seen, Cohen similarly took the general notion of producer or worker collectives as the distinctive feature of a future socialism. He then applied to these collectives the concept of the "legal person" as the site of juridical responsibility and determination. The legal person would reside neither primarily in relation to the physical individual nor in relation to the business corporation. Rather it would be developed as a concept primarily in relation to the producers' collectives as the basis of ownership and society.[6] In this regard, Cohen's pervasive criticism of the empirical ego, that is, the commonsense person, as the basis of either epistemology or politics takes on a distinctly political resonance – it allows political, economic and social processes to be described primarily through collective endeavors and institutions. A corollary of this, however, is the defense of the absolute individual, in the Kantian sense of the unknowable transcendental ego, from "external" or arbitrary definition by means of an evolving system of rights, a defense central to Cohen's ethics and aesthetics.

Treating "humanity as an end" means first and foremost recognizing "humanity" as an incomplete project, which in turn for Cohen depends on recognizing the individual as ineffable yet defined through different forms of collective association. Curiously, the emphasis on this correlation of absolute individuality and collective humanity in Cohen's work became through the work of Emmanuel Lévinas one of the most influential aspects of his program, even as it was in the process largely removed from the social and institutional critique Cohen intended for it.[7]

By the time of his first 1896 introduction to Lange's *History of Materialism*, Cohen would closely link his re-reading of Kant with ethical socialism and the claim that "Kant is the true and actual creator of German socialism."[8] Similarly, in his *Ethics of Pure Will* (1904), Cohen argued that Kant's philosophy of ethics "contains the moral program for a new era and the entire future world history. [...] The idea of the priority of humanity as an end becomes the idea of socialism, which defines each human as an end in itself, as purpose."[9] The moderation of socialist politics under the Erfurt program, crafted in the wake of the new parliamentary legislation by Karl Kautsky and Eduard Bernstein and introduced in 1891, was developed in close connection with the philosophy of Lange and, to a lesser degree, Cohen. Eduard Bernstein's widely influential *The Preconditions of Socialism and the Mission of Social Democracy* (1899) in part replaced the Marxist foundation of the socialist party with a Kantian foundation, and this reading of Kant was often credited to Lange's work.[10] The direct application of Cohen's philosophy to the social democratic movement continued in the work of figures such as Karl Vorländer, Kurt Eisner and Ernst Reuter.[11] Particularly in Vorländer's reading, presented in *Kant and Marx: A Contribution to the Philosophy of Socialism*, it was Cohen's work that established a new philosophical basis for the Social Democratic Party.[12]

Throughout its history, the Marburg school's approach to the critique of science and its political implications had a descriptive and normative aspect. Descriptively each of the sciences would be revealed through historical analysis and conceptual criticism as a functional system open to maximal transformation, rather than as a system "copying" a pre-given reality. The beginning of this critique was found in Lange's attack on materialism, as well as in his own attempt to define a broader functional system in his late work, *Logical Studies: A Contribution to the New Foundation of Formal Logic and Cognitive Theory*.[13] In its more developed form in Cohen's *System of Philosophy* and Cassirer's *Substance and Function*, the Marburg critique held that all "metaphysical" claims of essence – such as matter, force or life – in the sciences had been progressively transformed into purely relational definitions, which were taken to be useful and "constructive" hypotheses. Similarly, within the human sciences, as most thoroughly argued in Cassirer's *Knowledge Problem*, ideas of subject and object would be considered open to radical permutation.

The normative aspect of the school was found in the claim that the functional description of the world provides in turn a basis for developing institutions in which the functional fluidity of society and nature, and with this the openness to future transformation, would be emphasized. For Lange, the rejection of materialism entailed the development of a new model of how science actually worked as a process, as well as a renewed attention to changing structures within society epitomized by a gradualist social democracy.[14] For Cohen, the model science was in many ways law, in which a maximal definition of human freedom was to be developed by providing the minimal and most flexible definition of rights, individuality and state. This project was initiated by Cohen's *Ethics of Pure Will* and carried forward in part by influential positive law movement of Hans Kelsen (1881–1973). For Cohen, the concept of personality of the state, and indeed the idea of the legal person at all, would be defined purely relationally and ideally, and rights would be understood as a limit case around which the ideal flexibility and possibility of the subject can be developed.

Both the normative and descriptive aspects of the sciences for the Marburg school fit within a larger aesthetic, ethical and political project that Cohen defined through the principle of the "redemptive" human future towards which activity had to be oriented. For Cohen, this future is grounded in a particular reading of messianism as a prophetic translation and projection of the future through present ideas and actions. Despite Cohen's religious terminology, the process through which the "future becomes the actuality of history" contains a necessarily secular aspect that Cassirer's reception of

Cohen will emphasize. Here too, the idealistic premises of the Marburg school have to be placed in their specific Wilhelmine context, for the school's apparently utopian promise of a transformative future was received within the tradition of the German "sciences of state," in which knowledge was understood to transform society through the intermediary of institutions and administration, as well as within the epochal transformation of the second Industrial Revolution. The Marburg school's vision of a redemptive future existed within a horizon suggested by real coordinates in the present, and indeed pitched against an equally dystopian set of prospects. In this context, Cohen saw "the conflict of the faculties" as resolved into a philosophical plan for pedagogical and institutional reform that would develop a contemporary critique of science and religion. Far from being limited to a "scientism" or attempting to level all of the German humanities and sciences to the form of natural science, the goal was to liberate Germany from previous false substantive definitions of science and to open a vision of radical human creativity in future society.

Self-Censorship and the Reception of the Marburg School

The historical challenge of reading the Marburg school is in part to recognize limitations that may have been evident to contemporaries on the writings and speech of its authors, in order that we can understand how such readers would interpret the implications of their philosophies. These limitations changed over time, even as certain fundamental aspects remained fixed. Four widely researched themes defined these aspects and inflected the school's reception. Having addressed these, we can then look in more depth at the reception of the school's "critique of knowledge" in the sciences and politics. In both cases, assumptions about the nature of German society that prevailed in the Weimar period and later in the liberal democracies to which Cassirer emigrated simply did not pertain in the Wilhelmine context.

The first limitation on communication in the Marburg school was the role of its members – along with all German academics – as state bureaucrats, with all of the attendant government oversight.[15] Despite the freedoms of the academic research university, the pressures for conformity were strong. As one state minister declared in 1898, it was an axiom of Prussian policy that professors were both to teach and "to fill youth with respect for the monarchy and the constitution and [...] our state institutions."[16] The Marburg school's progressive political orientation within a highly conservative national faculty meant in this context that it was taken as both the epitome of *Kathedersocialismus* ("lectern socialism") and that it potentially had to overcompensate for this by demonstrating the nationalist value of its philosophy. Cohen's often strident nationalism might be attributed in part to this motive, even as it was also no doubt an earnest commitment. Cassirer's more muted and cosmopolitan stance similarly anchors itself in German tradition as a means to broaden its accessibility.

The second influence on reception arose from the bureaucratic definition of the professoriate, and this was the role of the state, particularly the cultural ministry, in determining the direction of academic fields and universities. In the case of the Marburg school, it was of the greatest importance that the head of the cultural ministry from 1882 to 1908 was Friedrich Althoff, who had a profound interest in steering the German academic system away from its emphasis on the classics and humanities and towards the new realities of the German state in the age of the second Industrial Revolution.[17] The Marburg school's goal of reconciling natural science with elite culture fit surprisingly well within this framework, and Cohen's use of scientific terminology in particular can be read in this context.

This favorable proclivity towards the Marburg school was balanced by a countervailing third theme affecting its fate, namely the conflict of classical philosophy with the new field of experimental psychology that was arising in its midst. Cohen, as we have seen, took empirical psychology as perhaps the gravest threat to the Marburg school and a practical threat to German academic culture more generally. Particularly in Cohen's criticism of psychophysics and polemics over the role of the infinitesimal in philosophy, the context of the rise of empirical psychology within the Althoff era was of decisive importance. The conflict of philosophy and psychology indeed became one of the larger academic battles of the era, and was epitomized by Cassirer's inability to gain Cohen's chair in 1912 in favor of the empirical psychologist Erich Rudolf Jaensch.[18]

Finally, Cohen and Cassirer were in the difficult role of being Jewish German philosophers, both active in their faith, within a German context that favored Protestantism, positivist agnosticism or conversion. Cohen's utterly anomalous position in this regard is suggested in Thomas Nipperdey's history of the period: "For the unbaptized the situation [in the universities] was and remained particularly difficult, despite the tremendous [gewaltigen] success of Cohen in Marburg."[19] Characteristically, Cassirer and Cohen's approach to this theme was markedly different. Although Cohen had initially spurned Cassirer as a student, thinking he was both converted and assimilated, they developed a warm relation once Cohen recognized Cassirer's active commitment to Judaism.[20] Whereas Cohen made Jewish thought and philosophy increasingly central to his work, however, Cassirer largely kept Judaism a private matter, despite his active participation in the Jewish community and an increasing number of anti-Semitic insults and threats to his self and family. With the initial rise of the National Socialists in the early 1930s, however, Cassirer began more actively publishing on Jewish themes and in Jewish intellectual journals, and on emigrating he would seek out and speak to Jewish groups.

The Critique of Knowledge and Mandarin Culture

Although the internal and external limitations placed on the Marburg school by Wilhelmine society are evident in retrospect, less obvious are the means through which the school sought to transform German society within these bounds. In its attempt at a transformation of society from above, the Marburg project represents a high-water mark for "Mandarin" culture, in Fritz Ringer's famous term for the Empire's elite intellectuals as an ideal-typical category.[21] The "Mandarin intellectuals" were, in Ringer's wording, "the social and cultural elite which owes its status primarily to educational qualifications, rather than hereditary rights and wealth."[22] Characteristic in Ringer's view of a distinct phase between "the primarily agrarian level of economic organization and full industrialization," they attempted to reinvision state power through an academic elite, and thus had as their principal concern "the educational diet of the elite."[23] In itself, the Marburg school's ambition to transform this diet was not singular. What was remarkable about Cohen and Cassirer's work in the decade before the First World War, however, was the thoroughgoing philosophical redefinition of the *concepts* of knowledge and science themselves in an attempt to transform society. It is this peculiarity that makes their endeavor particularly telling for Mandarin culture and German academic politics more generally.

The Marburg school's vision of the reform of society through science is only comprehensible within the broad Wilhelmine meaning and application of the term "science" itself. Science, or *Wissenschaft*, had a far broader meaning in Wilhelmine Germany than it does in contemporary English, as it encompassed any form of organized knowledge in the natural, social or "human sciences" (the "humanities" in Anglo-American terminology). Even for the social and human sciences, collectively defined as *Geisteswissenschaften*, the term science took on an extremely broad meaning. A paradigmatic example is the Marburg-influenced journal *The Human Sciences* (*Die Geisteswissenschaften*), copublished by Otto Buek and Paul Heere, the former Cohen's student and Cassirer's coeditor on an edition of Kant's key texts before the First World War. The editors describe their goal as an internal critique of the human sciences, and as "the careful pursuance of the movements and directions of the disparate, and yet inwardly closely related, scientific areas to provide a creative and educational overview [of the whole]."[24] The project was thus precisely one of using a description of the present state of the sciences for a normative, or at least "educational," understanding of their further direction in the manner that Cohen and Cassirer's projects epitomize. The journal appears to have been intended as a centerpiece for the Marburg program of institutional change, and both Cohen and Cassirer contributed leading articles linking this critique of the sciences with the Marburg school's broader project – Cassirer's on the tendency of post-Kantian speculation and Cohen's on the nature of the human sciences themselves.[25]

Perhaps most telling is the definition of the human sciences as outlined in the subtitle of *Die Geisteswissenschaften*, which claims it as: "A Weekly for the Entire Area of Philosophy, Psychology, Mathematics, Religion, History, Languages and Literature, Art History, Law and Political Theory, Sociology, Economics and Statistics, Military Studies, Ethnology and Education."[26] Each term is defined by a compound of either "-wissenschaft" or the Greek equivalent of ordered knowledge, "-logy," and the inclusion of mathematics and military studies as well as law and statistics suggest the very wide range even *Geisteswissenschaft* alone could encompass. Or, perhaps, it suggests the necessary convergence of the human sciences with all other forms of science. In any case, within this range the reform of science initiated by the Marburg school could be imagined to affect nearly every aspect of German life.

Cassirer's reception of Cohen in the first decade of the century is consistent with the outlying liberal members of a generation that tended to be skeptical of the role of parliamentary and popular politics, even as they were optimistic about the potential for progressive and "scientific" reform from above.[27] The historian Kevin Repp has defined the peculiarities of a "generation of 1890," emblematized by Kaiser Wilhelm II's decrees in February 1890 of a vision of "social peace" through the technocratic state.[28] Followed by new legislation in 1891 that greatly modernized German working laws, and indeed made them among the most progressive in the world, the era seemed to some as one of general modernization and progress – albeit within constraints that in retrospect appear fatally narrow. Repp has usefully defined this style of reformist critique as an "antipolitics," that is, an attempt to develop a progressive political agenda apart from any actual parliamentary politics.

Rather than being apolitical, Cassirer's variety of "antipolitics" sought to develop the conditions for the possibility of political reform apart from actual political debate, to institute a sort of "second order" transformation of society through the matrix of science and history. Where Cohen advanced the idea of reform through a "metapolitics" of the sciences acting in concert with the social democratic movement, Cassirer envisioned "metapolitics" essentially acting on its own to form the conditions for the possibility of a new form of politics, one which would appear as each discipline developed new forms of critique and outlined new potentials for human liberation.

Cassirer as Disciple and Critic of Cohen

Cassirer's work in its early years, with its often understated political views and readable style, can be understood foremost as a reflection on and transformation of Cohen's demanding philosophy, highly public political stances, embattled situation and nearly incomprehensible style. Given the wide discrepancy between his close personal attachment to Cohen and his rare textual references to him, it has to be assumed that Cassirer often masked this relation in his public writings so he would not appear as an academic 'front' for Cohen or a mere mouthpiece for Cohen's ideas. As is clear from both their correspondence and Cassirer's wife's autobiography, Cassirer's early philosophy was nonetheless intended to act as a bridge to Cohen's theories and to the achievements of the Marburg school, and many of the important aspects of his work gain in depth when they are considered as a commentary on Cohen. He hoped specifically that his work would make clear the wider implications of Cohen's texts that were lost in Cohen's own occasionally overstated claims and in his limited reception. Far from appearing under the yoke of Cohen or his philosophy, as claims to Cassirer's "break" with Cohen's philosophy often stipulate, Cassirer clearly chose to work with Cohen and developed an exceptionally close relationship with him.

Cohen for his part undoubtedly saw Cassirer as his key successor, reflecting later that on first meeting Cassirer, "I felt at once that this man had nothing to learn from me," and continuing to defer to him, particularly in relation to his knowledge of the natural sciences.[29] Indeed, Cassirer and his colleague Paul Natorp appear to be the only individuals who Cohen endorsed as having fully understood his philosophy, and Cassirer was, along with Natan Gawronsky, the only student who Cohen apparently considered a close friend – both, for instance, were the only former students who acted as witnesses of his final will.[30] From the very beginning of Cassirer's relationship with Cohen, however, the two were in nearly constant, and occasionally humorous, debate and disagreement. As their letters attest, Cassirer in no manner simply reflected Cohen's views, and he very often contested them, but there was a real respect between the two.[31]

Cohen's relatively outspoken political views in the Wilhelmine environment made him the object of incessant academic and political harassment.[32] Mindful of Cohen's fate since the very beginning of his career, Cassirer recast key aspects of the Marburg program in a more tractable form, while clarifying and popularizing its philosophical meaning. Cassirer was closely connected with the presentation of Cohen's philosophy to the public: Cassirer's cousin, Bruno Cassirer, was the publisher of nearly all of Cohen's late works, and Ernst Cassirer wrote several key summaries of Cohen's work before the First World War. Cassirer also coedited the most relevant texts after Cohen's death in 1918 in the posthumous collection *Hermann Cohen's Writings on Philosophy and Contemporary History*.[33]

In her memoirs, Toni Cassirer described her husband's relation to Cohen, whom "he admired and loved extraordinarily," in conjunction with Cassirer's habilitation, or tenure publication, of the first volume of *The Knowledge Problem* (1906): "He explained to me the difficult situation that Hermann Cohen was in, since he was the only Jewish Ordinarius (Professor) in all of Germany, and gradually I came to understand that Ernst's habilitation was not really particularly important to himself, but that it should have the role of preparing the way for Cohen's own ideas, which everywhere encountered severe resistance."[34]

Cohen himself similarly viewed Cassirer as central to the Marburg school, seeing this early text and others as completing his own work to some extent. Cohen's estimation of Cassirer's *Substance and Function* as an extension of his own project is clearly in evidence in the correspondence of the two. "I congratulate you," he wrote to Cassirer after *Substance and Function*, "on your new and great achievement. If I shall not be able to write the second part of my Logic, no harm will be done to our common cause, since my project is to a large degree fulfilled in your book."[35]

Cassirer suggests his interest in presenting Cohen's philosophy from a different vantage point when he narrates his story of finding Cohen's philosophy through an earlier teacher, Georg Simmel. In an anecdote that he often repeated when speaking of Cohen, Cassirer writes that Simmel "emphasized how much he himself owed to the study of Cohen's books, but he immediately added that those books, in spite of their real profundity, suffered from a very grave defect. They were written, he said, in such an obscure style that as yet there was probably no one who had succeeded in deciphering them."[36] Cassirer notes that in his youthful enthusiasm he set to work voraciously reading all of Cohen's works before ultimately transferring universities to work with him.

The depth of relation – and mutual critique – between the two philosophers was remarkable. Cassirer's warm personal relation with the Cohens is described by Toni Cassirer as a "close, unclouded relation of friendship."[37] Speaking to the Jewish Student's Union at Cambridge University in 1935, Cassirer writes, "After my first visit to Marburg, I had not only become his pupil – he considered me and he treated me like an intimate friend. By this friendship, that, without any interruption, lasted up to the death of Cohen in the year 1918 I first got a true insight into the nature of Judaism, that was represented by Cohen."[38]

Despite his early success under Cohen, Cassirer at first had little interest in following his mentor into academics, largely because he was not interested in subjecting himself to the sort of political and anti-Semitic persecution that Cohen had experienced throughout his career.[39] Only with the success of his *Knowledge Problem* series after 1906 did Cassirer finally give in to Cohen's desire for him to stay in academia.[40] The conflict over Cassirer's succession to Cohen's Marburg chair in 1911 also appeared to be a turning point in

Cassirer's career, one in which even his well-honed diplomatic skills, so effusively praised by his colleague Natorp, failed to secure the Marburg chair.[41] Although Cassirer lost this battle, the same interpersonal skills undoubtedly played a key role later in his 1929 election to the rectorship of the Hamburg University by a drastically polarized faculty. At a time when the campus was already fraught with National Socialist agitation, Cassirer won the faculty vote to become the most prominent of the few Jewish German rectors in the history of the German university system. He began his position with what is now considered one of the few public defenses of the liberal and socialist Weimar state.[42]

The lives of both philosophers, then, can be characterized by highly visible public roles in German academic and political life managed in a nearly diametrically opposed manner. Cassirer's family was one of the leading industrialist families in Berlin, having gained a fortune in fields characteristic of the second Industrial Revolution: cellulose (used to make paper, textiles and cellophane) and the manufacture of cable.[43] As a scion of one of the most powerful families in Berlin, and someone brought up to be mindful of its economic operation, Cassirer was also singularly aware of the new economic and technological forces at work in Berlin and Germany. His educational and personal acquaintance with such figures as Walter Rathenau, the head of the German economic war effort and future foreign minister of Weimar Germany, and Aby Warburg, of the powerful Warburg family of bankers, further informed his reading of the complexity of this situation.

Even as Cassirer, like his friend Aby Warburg, adopted an intellectual life as a "higher calling" than his family's extensive business dealings, he was so thoroughly enmeshed in the Berlin of his family that he could not but be aware of its political, social and economic reality. Indeed, far from being naïve, when we look over Cassirer's life we see a success that evidences extreme canniness. In academic politics he was eminently successful up to 1933, and managed to help many of his students find work in exile. His public defense of the Weimar Republic was both courageous and, in retrospect, necessary. Yet he was also one of the very first prominent intellectuals to understand the lasting danger the National Socialists represented, and he was one the very first German Jews to leave the country immediately upon Hitler's enabling act of 1933, an act precipitated tellingly by his recognition that the change in German law transformed the entire nature of the country.[44] Similarly, in managing his own and his student's fate in exile, he was one of the more successful German émigré figures. Although Cassirer's caution and circumspection could be faulted, for the historian the assumption certainly cannot be that Cassirer was aloof from his surroundings.

Rather than being diametrically opposed or completely aligned with Cohen, Cassirer's work can be more productively defined by how he at once promoted Cohen's philosophy while also fundamentally critiquing and ultimately transforming it. This critique became a source of intense disagreement between the two philosophers, and was understood by Cohen as a break with some of his core assumptions, notably in relation to the problem of the infinitesimal, as early as 1912. Nonetheless, Cohen clearly understood Cassirer's work as presenting the most advanced development and complete prehistory of the Marburg school, and Cassirer as the best guide for its past and future.

Cassirer's Wartime Crisis

If the immediate prewar years mark an apogee of hopes for the Marburg school's reform of German society, the onset of the war quickly cast doubt on the premises of the entire project and for Cassirer led to a profound personal crisis. The culmination of Cassirer's early career was *Freedom and Form* (1916), which he developed during this crisis. Cassirer wrote *Freedom and Form* at night after his work, which he rapidly came to view as ethically dubious, as a chief censor in Berlin responsible for redacting French newspapers.[45] This work gave him a singular – and singularly painful – understanding of the war. As Toni Cassirer describes it, already by 1915 it quickly "robbed him of every illusion about the war and the 'German cause.'"[46] Due to his position, he could see the actual course of the war more clearly than most of his colleagues and he could discern the destructive role propaganda and xenophobia were having in all involved countries. As his wife's journal notes, he initially found this job both patriotic and interesting, but as the destruction of the war accelerated he found it dehumanizing and deeply problematic.[47] It is in this context that *Freedom and Form* explicitly transforms the earlier Marburg project into an argument for liberalism and against the centralized state, even as it extends the functionalism of his earlier argument in the social and state sciences.

Cassirer "felt himself transformed into a machine in his office to the highest degree," his wife writes, "and robbed of his personal responsibility just like every other soldier; he knew two years before the end of the war that Germany had lost."[48] As singular as Cassirer's resistance to wartime enthusiasm often appeared to later commentators, it may be less heroic or farsighted than it at first appears. Unlike most of his colleagues – notably his acquaintances Thomas Mann and Georg Simmel – he was aware of Germany's actual situation and thus partially immune to the "blind patriotism" of the era. Although it would be tempting to credit Cassirer with a particular ethical insight during this time, his position as a censor put him in a drastically different relation to the war than nearly all of his friends and colleagues.

Cassirer's job was not simply to censor, but also to excise words so that the meaning of French texts appeared conducive to Germany's position. He was, given his own skills in productively finding patterns of historical meeting, predictably good at this task, and rose through the ranks with what Toni Cassirer describes as "breathtaking" speed.[49] From this position, at once privileged and tormented, he perhaps could see how the war and the mode of self-understanding that went along with it were gradually eclipsing the plausibility of his own political project. *Freedom and Form* was written, Toni Cassirer says, as a means of attempting "to save the Germany […] that was no longer recognizable underneath the ashes that covered it."[50] The text was, however, much more than this, since Cassirer creatively combined key moments from German thought into an entirely new form, developing a literary pastiche that was presumably aimed in part at undoing the work he did during the day.

In the first pages of *Freedom and Form*, Cassirer described the response of elites in Germany to the horrors of the war through the discourse on culture as "one of the most bizarre aspects of the general image of our time."[51] He presumably had in mind the contrast of the ideal of culture in most German wartime writings – not least Cohen's – and the reality of industrial–bureaucratic warfare. For most German citizens, gaining an overview of the host of new catastrophes raised by the war was difficult due to the staggering scale of the conflict, its new technological horrors and the new combination of mass economy and "total war." A very few writers, notably Max Weber, noted at the time that the literature of "culture" obscured the larger technocratic and bureaucratic realities of the war. The "plain fact of universal bureaucratization," Weber wrote, "really stands behind the so-called 'ideas of 1914,' [as well as] behind that which the literati euphemistically call the 'socialism of the future.'"[52] Although Cassirer's characteristically measured response to the war may now seem inadequate to its realities, it is useful to remember that *Freedom and Form* was published during the war, and thus was one of the handful of published texts to argue for cosmopolitanism and against nationalistic neo-Romanticism.

Cassirer's perspective on the war placed him, as his wife accounts in her memoirs, in almost a completely isolated personal position. As she notes, "everywhere the blindest form of nationalism seemed to appear. Every trace of objectivity was gone."[53] Cassirer's own complicity in this blindness must have been particularly devastating. He had, moreover, direct contact with critics of the war, notably his cousin Paul. One of the few friends the Cassirers had during this time, Paul Cassirer was a leading figure in German publishing and gallery work. He also is notable for his role in German politics.[54] His journal *Bildermann* proved, in historian Roger Chickering's account of Germany during the First World War, "the most important source" of criticism – albeit still veiled – of the war in Germany.[55] Whether Cassirer discussed the war with Paul is unknown, but since they were by all accounts fairly close it seems likely that they would have at least exchanged views on the gravity of Germany's situation.[56] In any case, Cassirer clearly intended *Freedom and Form* to address the looming catastrophe in his own idiom, informed by his work and by the criticisms raised by his cousin.

The Cassirers' isolation was made even more stark by their increasing distance from their predominantly Jewish German friends, who, Toni Cassirer writes with some bitterness, were "more German than the Germans" and who "no one surpassed in patriotic shortsightedness."[57] This in turn was particularly disorienting since the Cassirers found themselves increasingly alienated as Jews from German society. Nowhere was this new alienation more clearly brought home than in Cassirer's attempted defense of Cohen, who was under a series of anti-Semitic attacks in the "Bauch affair" in the last two years before his death in 1918. Cassirer's response to anti-Semitic attacks on Cohen's philosophy during the First World War suggests his keen sense of the direct political relevance of his philosophical work to issues such as nationalism and racism, and are worth considering in some detail.[58]

Notably, we can see in Cassirer's drafts of responses to the affair – they were never published – the direct political implications of the form of social and human science he developed in works such as *Freedom and Form* on the basis of his and Cohen's prewar work. Precisely in their private form, they suggest one of the motivations for Cassirer's antiessentialism in a broad range of fields that was partially muted in his public writing. They also suggest, however, Cassirer's growing awareness that the Marburg model of changing society through its elite institutions could not confront the key challenges of the new era, and that nonrational forms of discourse would need to be addressed in a new philosophical idiom.

At the height of the war hysteria in 1916, the prominent philosopher Bruno Bauch wrote an article in *Kant-Studien* defining German nationality as a matter of birth, and claiming that only Germans could understand German philosophy.[59] In a letter to a separate nationalist/*Völkish* journal, *The Panther*, laid out bluntly what he meant: Jewish German philosophers, including Hermann Cohen, could never properly understand German philosophy since they were not part of the German nation.[60] Cassirer was understandably furious at this attack, and wrote several drafts of a long response to it.

Cassirer's answer to Bauch emphasized the use of visual typologies of nations as the most complete "naturalization" of the concept of culture. Given the frequent depiction of Cassirer as a straightforward humanist or proponent of Enlightenment, one imagines that he would criticize Bauch's lack of philosophical acumen and his crude empiricism in judging people by their heritage or how they look. Cassirer would presumably adopt an idealistic stance that we should judge all people as being inherently equal. The direction of his argument is, however, in nearly the opposite direction. For Cassirer, what is marked about Bauch's racism is its formalism and idealism. It first assumes an "essence" to people, around which it then constructs its visual and material object.[61] In this sense Bauch is making a turn towards "substantialism" of the form Cassirer decried in *Substance and Function* in relation to the sciences. Indeed, Bauch's argument is presented as parallel to the manner in which materialism abstracts the notion of a single "matter" from phenomenon or a set of causal material relations underlying experience.

Cassirer argued that Bauch creates ideal categories of "German" and "non-German," around the theme that "the Jew is a guest in the German house," and that this ideal categorization then determines from the outset how he would see all of the facts about the appearance of German Jews. "Since Jew and German, really or hypothetically, are taken in their 'essence' [*Wesen*] as being fully different," writes Cassirer, "for this reason in the end all of the inner, spiritual/cultural [*geistigen*] relations between the two are concealed and explained as mere appearance."[62] In short, it is Bauch who is allowing ideal forms to blind himself to more complex patterns of development.

Although specific here to the problem of anti-Semitism, the blindness Cassirer highlights in the inability to perceive more complex patterns and relations in "mere appearance" is in fact indicative for him of broad set of problems in contemporary popular history and culture – problems that were ever more pressing in the course of the war. *Freedom and Form* develops a similar critique of any description of "national" character or of the "intuitive" cultural truths, albeit in a more oblique reading cast through classical German authors. In both cases, Cassirer's earlier critique of the natural sciences is developed into a critique of social and cultural essentialism, which he sees as part of one manifestation of the form of knowledge of the era. Far from being an abstract escape from politics, he clearly understands this "relational" critique as the center of redefining the political possibilities of popular themes such as nationalism and anti-Semitism, as well as the social and human sciences that could redefine society.

The "subjects" of Jew and German to which Bauch refers, Cassirer ultimately concludes, never actually existed: "They are ghosts."[63] Due to the close relation of ideas and intuition, however, the two form a vicious circle, so that unless carefully examined, intuition and ideas tend to confirm one another. For this reason, the first demand Cassirer made of Bauch is that he return to a study of history, which would reveal the abstraction of his categories in all of their emptiness. History has to proceed not on the basis of vague philosophical or typological generalizations, Cassirer writes, but through a study of the "force and details of an individual historical situation and its particularities with which historians bring themselves and others to a living concrete intuition of a situation."[64] In avoiding "mere appearances" and then attempting to draw out patterns of relations within them, a more complex set of dialogs in history could be highlighted. Through a "living concrete intuition" more true forms of understanding, and with them patterns of relations, could become visible within the play of forces in society. Already in Cassirer's study of Leibniz in 1902, history is defined as creating ever new forms of understanding through which to recast and redefine the infinite "forces" at work in the world. Far from being Panglossian abstraction, this insight was for Cassirer part of a realistic critique not only of popular phenomena such as nationalism and anti-Semitism, but also of the entire mode of understanding that dominated the era.

Although of immediate concern in the Bauch affair, this approach to the politics of functional history was indeed at the center of Cassirer's work. Cassirer had developed a form of this argument since the very beginning of his *Knowledge Problem* in 1906, but here we see it newly applied to affective historical forms such as those of the "image" of peoples, rather than the sense of reality of objects of science or the "subjects" of different forms of politics. Cassirer's *Freedom and Form* itself defined German national character through a history of functional relations rather than claims to the "essence" of a people, but Cassirer had not directly noted the problem of the affective or "expressive" meaning in this regard. This new emphasis would be the key to Cassirer's symbolic form project, which he had already conceptualized by 1917. Understanding this project, however, rests entirely on the functional critique that preceded it and established Cassirer's broader understanding of reality and appearance.

The inspiration for Cassirer's symbolic forms project occurred and was developed, according to Toni Cassirer, during a tram ride home from his work as a censor in 1917, and this public setting is perhaps characteristic of Cassirer's turn of mind during this period.[65] Around the time he had his epiphany about symbolic forms, Cassirer was responding to the Bauch affair and the issues raised by *Freedom and Form*. Although Cassirer's intellectual shift had a number of other sources, his response to Bauch usefully highlights a common caesura in all of his work from this period that the symbolic forms project would answer. Namely, it is notable that Cassirer's demand for a return to functional history in his response – a demand already present in *The Problem of Knowledge* – only responded in part to the problem raised by Bauch's anti-Semitic attack on Cohen or the problem of nationalism. While a functional history might reveal the actual "relations" between peoples and events, it did not describe why a particular image of a nation or people would take the form of an intuitive or affective reality in the first place. It could not provide a phenomenology of the claims to the essence or reality of other people in sympathy (Scheler), in love or hate, that animated antisemitism, nationalism and racism – forces that surrounded Cassirer during the war years. Similarly, it could not explain why a narrative such as myth, whether in the forms defined by contemporary anthropology or developed as a modern technique of nationalism, presented a direct intuitive sense of a reality different from a mathematical or linguistic description.

Part of what was new in Cassirer's turn of thought to symbolic forms – which we will only intimate here through one of its aspects – was not the reciprocal relation between intuition and reality – for this had in fact long been a centerpiece of his philosophy, particularly in relation to science – but the addition of the element of affect, of love and hate, and the "tonality" or "physiognomy" of an experience as its "most essential" component. Far from superseding his earlier concerns, however, this new interest with "expressive" reality developed directly out of them. Cassirer's emphasis on function had always had the problem of reality as its key theme, both in terms of how science reveals "real" objects of study and how social forms such as the legal system develop "real" subjects. Only by first understanding how these apparently "obvious" features of an era were determined as real by its system of knowledge could one also understand how expressive phenomenon were not epiphenomenal but equally – indeed in Cassirer's ultimate reading – more foundationally "real." The problem of reality, however, and the problem of form had long been at the center of Cohen and Cassirer's philosophy, and had been developed most clearly through the deceptively esoteric problem of the infinitesimal problem – to which we will now turn.

Notes

1 The claim for Cohen promoting mysticism is found in Leonard Nelson's attacks, as noted by Cassirer in KI 32.
2 *Nietzsche Briefwechsel: Kritische Gesamtausgabe*, ed. G. Colli and M. Montinari (Berlin: DTV, 1975), vol. 1, part 2, Briefe an Hermann Mushacke (Nachschrift), November 1866, 184; for further on Lange see ibid., 160 (Briefe an Carl von Gerdsdorff, August 1866). On the literature of Lange's influence on Nietzsche, see George L. Stack's *Lange and Nietzsche* (Frankfurt: De Gruyter, 1983) and its critics including: Breazeale, D. "Lange, Nietzsche, and Stack: The Question of Influence," *International Studies in Philosophy* 21 (1989): 91–103; Stack, George. "From Lange to Nietzsche: A Response to a Troika of Critics", *International Studies in Philosophy* 21 (1989): 113–24; Stack, George. "Kant, Lange, and Nietzsche: Critique of Knowledge," in *Nietzsche and Modern German Thought*, ed. Keith Ansell Pearson (New York: Routledge, 1991); Wilcox, J. T. "The Birth of Nietzsche out of the Spirit of Lange," *International Studies in Philosophy* 21 (1989): 81–9; Pletsch, Carl. "Young Nietzsche: Becoming A Genius" (New York: Free Press, 1991), 82ff.; Salaquarda, J. "Nietzsche und Lange," *Nietzsche-Studien* 7 (1978): 236–53.
3 Lange, F. A. *Geschichte des Materialismus und Kritik seiner Bedeutung in der Gegenwart*, vol. 3 (Iserlohn: J. Baedeker, 1873), 218.
4 Lange, F. A. *Die Arbeiterfrage* (Berlin: Buchhandlung Vorwärts, 1910).
5 Lange, F. A. "Über den Zusammenhang der Erziehungssysteme mit den herrschenden Weltanschauungen verschiedener Zeitalter. Bonner Antrittsvorlessung 1855," in *Material zum Arbeitsunterricht an höheren Schulen* 71, ed. O. A. Ellissen (Bielefeld: Velhagen & Klasing, published originally in 1855 and 1894) and *Die Stellung der Schule zum öffentlichen Leben. Festrede, gehalten bei der Schulfeier des Geburtstages Sr. Majestät des Königs den 22. März 1862* (Duisburg: F. H. Nieten; reprint in 1862, 1975).
6 SP2 320, 615. Further on Cohen's socialism is contained in Van der Linden, Harry. *Kantian Ethics and Socialism* (Indianapolis, IN: Hackett, 1988), 221ff. and Keck, Timothy Raymond. *Kant and Socialism: The Marburg School in Wilhelmine Germany* (PhD dissertation, Madison: University of Wisconsin, 1975).

7 For a recent study summarizing the earlier literature, see Batnitzky, Leora. "An Irrationalist Rationalism: Lévinas's Transformation of Hermann Cohen" in *Leo Strauss and Emmanuel Lévinas Philosophy and the Politics of Revelation* (Cambridge: Cambridge University Press, 2006), 75–93.
8 EL 112.
9 SP2 320.
10 Bernstein, Eduard. *Die voraussetzungen des Sozialismus und die aufgaben der Sozialdemokratie* (Stuttgart: J. H. W. Dietz, 1899), 187ff.
11 Sieg, Ulrich. *Aufstieg und Niedergang des Marburger Neukantianismus: Die Geschichte einer philosophischen Schulgemeinschaft* (Würzburg: Königshausen & Neumann, 1994), 225ff.
12 Vorländer, *Kant und Marx: Ein Beitrag zur Philosophie der Sozialismus* (Tübingen: Mohr, 1926 [1911]).
13 Lange, F. A. *Logische Studien. Ein Beitrag zur Neubegründung der formalen Logik und der Erkenntnistheorie* (Iserlohe: J. Baedeker, 1877).
14 Thus Lange sees the rise of materialism from the 1830s to the 1850s as an ultimately unproductive form of resistance to the church and state, whereas he sees his own approach to realigning the forces of modernity as leading to progressive reform: "The breakdown in German idealism, which we date from the year 1830, turned gradually into a struggle against the existing state and church. [...] The whole character of the time began to incline towards materialism" (Lange, F. A. *Geschichte der Materialismus und Kritik seiner Bedeutung in der Gegenwart*, vol. 1 (Leipzig, J. Baedeker, 1902), 86).
15 Ringer, Fritz K. *The Decline of the German Mandarins: The German Academic Community, 1890–1933* (Hanover: Wesleyan University Press, 1990), 7, 34ff.
16 McClelland, Charles E. *State, Society, and University in Germany 1770–1914* (Cambridge: Cambridge University Press, 1980), 294 – citing *Deutsches Zentralarchiv* II, Merseburg, Papers of the Prussian Ministry of Churches, Education and Medical Affairs (Repetorium 76, Va, Sek. 1, Abt. IV, Nr. 42, Bd. II).
17 Sieg, Ulrich. "Althoff und die deutsche Universitätsphilosophie," in *Wissenschaftsgeschichte und Wissenschaftspolitik im Industriezeitalter: Das "System Althoff" in historischer Perspektive*, ed. Bernhard von Brocke (Hildesheim: Edition Bildung und Wissenschaft, 1999), 304; Sieg, *Aufstieg und Niedergang*; Ringer, *German Mandarins*, 51ff.
18 Sieg, *Aufstieg und Niedergang*, 357–72.
19 Nipperdey, Thomas. *Deutsche Geschichte 1866–1918, Band I: Arbeitswelt und Bürgergeist* (Munich: C. H. Bech Verlag, 1990), 402.
20 Gawronsky, Dimitry. "Ernst Cassirer: His Life and His Work," in *The Philosophy of Ernst Cassirer* (La Salle, IL: Open Court Publishing, 1949), 8.
21 Ringer, *German Mandarins*.
22 Ibid., 5.
23 Ibid., 6.
24 Otto Buek was a student of Hermann Cohen and later an influential conduit of the Marburg school to his native Russia and St Petersburg. He translated Cassirer's Kant selection into Russian and was an outspoken pacifist, being one of only of three signatories to Georg Friedrich Nicolai's "Aufruf an die Europäer." Paul Herre was professor of history at the University of Leipzig and author of both constitutional histories of Prussia and more popular works. He would become increasingly more reactionary following the First World War and ultimately wrote a peon to the fascists in 1941 as *Deutschland und die europäische Ordnung* (Berlin: Deutscher Verlag, 1941).
25 KP; and Cohen, Hermann. "Die Geisteswissenschaften und die Philosophie," *Die Geisteswissenschaften* 1 (1914): 3 reprinted in *Hermann Cohens Schriften zur Philosophie und Zeitgeschichte*, ed. Albert Görland and Ernst Cassirer (Berlin: Akademie-verlag, 1928), 525ff.
26 *Die Geisteswissenschaften* 1 (1914): 1.

27 Repp, Kevin. *Reformers, Critics, and the Paths of German Modernity: Anti-Politics and the Search for Alternatives, 1890–1914* (Cambridge, MA: Harvard University Press, 2000); Pieter M. Judson, *Exclusive Revolutionaries: Liberal Politics, Social Experience and National Identity in the Austrian Empire, 1848–1914* (Ann Arbor: University of Michigan Press, 1996). Cassirer summered in Vienna and rural Austria for most of his life so his connection with both German and Austro-Hungarian liberalism was equally vibrant. For an understanding of the centrality of the Austrian "Sommerfrische" intellectual scene to the Viennese intelligentsia, of which Cassirer appears to have been a part, see Deborah Coen, *Vienna in the Age of Uncertainty: Science, Liberalism, and Private Life* (Chicago: University Of Chicago Press, 2011).
28 Repp, *Reformers*, 19ff.
29 Gawronsky, Dimitry, "Ernst Cassirer: His Life and Work," in *The Philosophy of Ernst Cassirer*, ed. P. A. Schilpp (Evanston, IL: The Library of Living Philosophers, 1949), 7.
30 Cohen, Hermann. "Das Testament von Hermann und Martha Cohen: Stiftungen und Stipendien für jüdische Einrichtungen, herausgegeben von Ulrich Sieg," *Journal for the History of Modern Theology* 3 (1996): 263.
31 Toni Cassirer notes, for instance, the humor of this relation in regard to their dispute over the value of Bergon's philosophy. (Cassirer, Toni. *Mein Leben mit Ernst Cassirer* (Hamburg: Felix Meiner Verlag, 2003), 73.)
32 Sieg, 235–347.
33 Cohen, Hermann. *Hermann Cohens Schriften zur Philosophie und Zeitgeschichte*, ed. Albert Görland and Ernst Cassirer, 2 vols (Berlin, Akademie-verlag, 1928).
34 Cassirer, *Mein Leben*, 40.
35 Cohen to Cassirer, 24 August 1910, St Moritz. Ernst Cassirer Nachlass, Beinecke Library, Yale University, gen. ms. 355, box 2, folder 34, part 3; Gawronsky, "Ernst Cassirer," 21.
36 HC 222–3. Compare with the quote from the introduction that Cohen's works were "of unequalled value" but haunted by the "grave flaw" that they were written in a style that was nearly incomprehensible, "wrapt in such a mystery of philosophical style and technical language," as he quoted his earlier teacher Georg Simmel as saying, "that nobody as yet could understand them perfectly." From Cassirer, Ernst. "Cohen's Philosophy of Religion," *Internationale Zeitschrift für Philosophie* 1 (1996): 90. Reprint of discussion from June 1935 to the Oxford Jewish Society.
37 Cassirer, *Mein Leben*, 91.
38 Although Cohen was by several accounts an irascible character, there is no evidence for Skidelsky's claim of his overbearing demeanor with Cassirer, or for any break between the two of them. (Cassirer, "Cohen's Philosophy of Religion," 91.)
39 Gawronsky, "Ernst Cassirer," 15.
40 Ibid.
41 Sieg, *Aufstieg und Niedergang*, 344; Holzhey, Helmut. *Cohen and Natorp, Vol. 1: Ursprung und Einheit: Die Geschichte der "Marburger Schule" als Auseinandersetzung um die Logik des Denkens* (Basel: Schwabe, 1986), 16ff., 45.
42 Habermas, Jürgen. "Die befreiende Krafte der symbolische Formgebung: Ernst Cassirers humanistisches Erbe und die Bibliothek Warburg," in *Vom sinnlichen Eindruck zum symbolischen Ausdruck* (Frankfurt am Main: Suhrkamp, 1997), 9; Cassirer, Ernst. "Die Idee der Republikanischen Verfassung: Rede zur Verfassungs Feier am 11. August 1928," ECW 17 290ff.
43 Brühl, Georg. *Die Cassirers: Streiter für den Impressionismus* (Leipzig: Ed. Leipzig, 1991) 29–51; Paret, Peter. *The Berlin Secession: Modernism and its Enemies in Imperial Germany* (Cambridge, MA: The Belknap Press of Harvard University Press, 1980), 69; Krois, John Michael. "Ernst Cassirer, 1874–1945," in *Die Wissenschaftler: Ernst Cassirer, Bruno Snell, Siegfried Landshut*, ed. H. W. Eckardt (Hamburg: Verlag Verein für Hamburgische Geschichte, 1994).
44 Cassirer, *Mein Leben*, 195.
45 Ibid., 118ff.

46 Ibid., 119.
47 Ibid.
48 Ibid.
49 Ibid.
50 Ibid., 120.
51 FF 388. Cassirer removed the comments on the First World War from the third edition of the book.
52 Ringer, *German Mandarins*, 198; Weber, Max. *Gesammelte Politische Schriften* (Paderhorn: Salzwasser Verlags, 1921), 149.
53 Cassirer, *Mein Leben*, 119.
54 Tilla Durieux, wife of Paul Cassirer, discusses this in *Eine Tür steht offen: Erinnerungen* (Berlin-Grunewald: Herbig, 1954), 223–38; Mommsen, W. J. *Bürgerliche Kultur und künstlerische Avantgarde* (Frankfurt: Ullstein, 1994).
55 Chickering, Roger. *Imperial Germany and the Great War, 1914–1918* (Cambridge: Cambridge University Press, 1998), 139. As Chickering writes: "Given the censors' vigilance, open opposition was difficult, but veiled criticism of the war surfaced in the pages of several cultural journals, of which Paul Cassirer's Bildermann was the most important."
56 Yet it should be noted that neither they nor their spouses' writings discuss their meeting during wartime at all, although in the atmosphere of the postwar myth of the "stab in the back" this is an ambiguous fact, since neither would have an interest in highlighting any such exchange (Chickering, 70).
57 ML 99.
58 Ernst Cassirer Nachlass, Beinecke Library, Yale University, gen. ms. 98, box 52, folder 1057, 14. On Bauch's context and later career under the National Socialists, see Sluga, Hans. *Heidegger's Crisis: Philosophy and Politics in Nazi Germany* (Cambridge, MA: Harvard University Press, 1993), 82–5, 92–5, 164–7, 210–14. The episode was triggered by Bauch's anti-Semitic letter to a journal, followed by his claim that only Germans could understand German philosophy. Bauch, Bruno. "Brief," in *Der Panther: Deutsche Monatsschrift für Politik und Volkstum* 4 (June 1916): 742–6; and Bauch, Bruno. "Vom Begriff der Nation. Ein Kapitel zur Geschichtsphilosophie," *Kant-Studien* 21 (1916): 139–62.
59 Bauch, "Vom Begriff."
60 Bauch, "Brief."
61 Ernst Cassirer Nachlass, Beinecke Library, Yale University, gen. ms. 98, box 52, folder 1057.
62 Ibid., 22, see also 19.
63 Ibid., 19.
64 Ibid., 22.
65 Cassirer, *Mein Leben*, 120.

Chapter Three

"THE SUPREME PRINCIPLES OF KNOWLEDGE": CASSIRER'S TRANSFORMATION OF THE TENETS OF COHEN'S *INFINITESIMAL METHOD* (1882) AND *SYSTEM OF PHILOSOPHY* (1902–1912)

Few texts summarize and at the same time compound the challenges of their author's general philosophy as much as Cohen's *The Principle of the Infinitesimal Method and its History* of 1881. Cohen begins his text by writing that it is crucial to understand calculus "because it is the basic idea of the mathematical sciences," and there is no doubt that he wanted to convince his readers of its pivotal philosophical importance in this regard.[1] The work however almost immediately suggested a far broader philosophical agenda. Both in its initial form and its later adumbrations, the basic themes in Cohen's reception of the infinitesimal theory led to his split with key premises of Kantian philosophy and to the foundations of his own late philosophy. As Cassirer states in his "Neo-Kantianism" article for the *Encyclopædia Britannica* in 1912, Cohen's early readings of Kant were the "high point of Neo-Kantian thought" – with the implication that his late work moves in a new definition that might not fit under this rubric.[2] Cohen's late philosophy, he continues, develops directly from the *Infinitesimal Method* and the infinitesimal anchors Cohen's central claim that "reality is never 'given' in any sense, in neither sensation nor in mere intuition."[3] Reality is both "constructed" and "revealed" through forms of judgment following the unfolding of thought, and understanding the middle voice in which this occurs is the central challenge of Cohen's philosophy. From this starting point, Cassirer continues, Cohen traces through the entirety of human activity the "various ways and directions in which thought moves in [its] 'production of [its] objects,'" particularly in his later volumes on ethics and aesthetics.[4]

The riddle of Cohen's project for both contemporaries and to some extent for Cassirer was why the infinitesimal as a figure took on this importance for such a vast project, offering an explanation for nearly all phenomena and serving as a common theme that unified not only Cohen's philosophy but its claim to the unity of human experience itself. Cassirer's answer to this riddle on the one hand shattered Cohen's claim to a unitary explanation of all phenomena through the infinitesimal, but on the other confirmed many of Cohen's key tenets both within the philosophy of science and in his systems' development of a broader ethics, aesthetics and psychology. Cohen's emphasis on the infinitesimal, both men argued, was a return to basic problems of the limit and of logic developed in Leibniz's philosophy, which was now used as the foundation for a critique of Kant. Nearly all static claims in Kant's philosophy – for a faculty psychology, for settled *a priori* definitions of the categories and the intuition of time and space, even for a division between epistemology and ethics – would by this means be redefined as dynamic limit cases, open to permutation and skeptical reflection.

Ultimately, Cohen's use of the infinitesimal provided a means of critiquing and refining any definition of reality, particularly those revealed by science, even as it also provided a new means for an immanent turn towards experience as a process. Cohen's own terminology, particularly his emphasis on the quest for "pure knowledge" and "unity," led many commentators to mistakenly assume his was a purely and formally idealistic project: that thought constructs its object out of whole cloth, is fully separate from the possibility of sensation or even reality, and is concerned only with the development of a transcendental logic in the mathematical sciences. Cassirer's commentaries on Cohen's "critical idealism" completely reject this interpretation. The fulcrum of Cohen's work is the "problem of reality" and the consequence of it, in Cassirer's reading, is to emphasize a general theory of form that encompasses all experience.

The basic definition of form in Cohen and Cassirer's philosophy is Leibnizian. Leibniz, in Cassirer's historical reading, developed the foundation of the modern system of knowledge in two propositions. The first is the concept of a universe defined only by the play of relations, which entails the concept of force: "The foundation of all reality is a system of forces, of free-functioning [*freiwaltenden Tätigkeiten*] activities." Forces are always only understood through some kind of form, however, so that Cassirer continues, "But just in this, that these forces were grasped as forms, demonstrates that they carry in themselves a rule of inner connection and self-limitation."[5] A world of forces exists and these forces are given, but they can only be revealed through form. The root definition followed Leibniz's understanding of form as *any* constructive synthesis that demonstrates a rule whatsoever, starting with the various forms of topological space and encompassing the developmental forms of experience of the monad.[6] Practically applied, this meant that "objective spiritual forms," as Cassirer summarizes it in *Freedom and Form*, include those of "society and state, science and law," and encompass any local constellations of practices or knowledge through which a reality of forces is grasped, whether in the "forms" of sociation, political constitution, ideals of science or law. In this regard, Cohen's metapolitics and Cassirer's early work were in dialog with the contemporary use of "form" in the sociology of Cassirer's colleague Georg Simmel.[7]

Cohen held that both reality and sensation are fundamentally assumed, indeed assumed on the level of existential immanence, in experience, but that the particular presentation of reality or sensation are determined by forms or structures of knowledge. On its broadest level, form approaches the transcendental forms of understanding Kant developed in the *Critique of Pure Reason* and is indeed developed most clearly in relation to the natural sciences, as Cohen claimed in the first volume of his *System of Philosophy*. Cohen's definition of form is even more foundationally related to Kant's arguments in the third critique, however, and is involved in any relation of a particular to the whole process of nature (understood teleologically) and experience. The "whole" or "unity" involved is not the totality of experience as a closed unity of things but, as Leibniz had it, an infinite plenum of possibilities. The infinitesimal and limit figures are used by Cohen to weave together the Kantian schematism with the arguments of the *Critique of Judgment*, which then find new applications to ethics, aesthetics and psychology. Form reveals real continuities and apparently stable *a priori* features within experience, but it is also always open to permutation, and understanding these permutations and how they recast problems of reality, ideals or sense is the task of both science, in its broadest sense, and history.

Cohen's emphasis on the infinitesimal itself as a unitary problem within experience was characteristic of the faith of his era in science and in a single progressive unfolding development of human rationality. Both beliefs were fundamentally interwoven with concerns from Wilhelmine Germany that would partially vanish by the end of the First World War. Indeed, Cassirer's work will demonstrate one instance of this break from the earlier generation. To further frame the Wilhelmine context of this issue, and Cohen's enthusiasm for the infinitesimal, we can first look at the popular reception of calculus in the late nineteenth century as epitomized by the work of Gustav Fechner, before then returning to Cohen's technical reading of the problem and Cassirer's reception – and reframing – of the infinitesimal problem and its role in defining the sciences.

Popular Science, the Infinitesimal and the Human Sensorium in the Late Nineteenth Century: Fechner and Cohen

Cohen's *Infinitesimal Method* was formulated in terms of the specific history and philosophy of natural science at least in part to appeal to the widest possible audience and to address the most pertinent issues of its era, while at the same time fitting within the relatively narrow range of political acceptability of the German academic establishment. As Cassirer's 1912 retrospective on Cohen argued, Cohen's work focused on a key problem of late nineteenth-century Germany: the false naturalization and hypostatization of natural science. Even as he chose to address his philosophy through one of the most acceptable channels of concern in Wilhelmine Germany – the primacy of natural science – his goals were, however, quite different than those of mainstream scientific writers in Germany. Cohen sought to instil skepticism towards a deterministic or "substantial" definition of natural science, even as he hoped to explain the success and possibilities of all of the sciences on a new level – a level that he believed would open on to a new definition of the human future.

Cohen was of course not alone in the late nineteenth century in his fascination with calculus and its relation to experience. The popular reception of J. F. Herbart's (1776–1841) psychology and the psychophysics of Gustav Fechner (1801–1887) were part of a surge of interest in how the problem of calculus might explain the subtle perceptions of human beings.[8] The surprising popularity of such physiological explanations lay in the manner in which experience appeared to be convertible with measurement – qualitative perception to quantitative indices. The fascination in Wilhelmine Germany with the applications of calculus and the meaning of the infinitesimal epitomized the use and abuse of popular science, and it was not least for this reason that Cohen initially sought to "popularize" his work through a study of calculus.

Particularly in Germany, researchers and philosophers such as Fechner found in calculus a key means through which human perception could be redefined in connection to experience.[9] Cohen concluded his infinitesimal book with a study of the legitimate and illegitimate aspects of Fechner's work, and clearly saw his essay as playing off Fechner's immense popularity. Indeed, this may be a key reason why Cohen could look at the *Infinitesimal Method* as a "popularization" of his own work at all. The wide reception of Fechner's philosophy in not only the natural but social and human sciences was important enough, for instance, that by 1905 the prominent historian Karl Lamprecht could claim that Fechner's basic insights had been further developed by the founder of modern empirical psychology, Wilhelm Wundt, so that they now presented the possibility for a "psychic mechanics" that could be applied to history.[10] The same assumption had been suggested, and rejected, twenty years earlier by Wilhelm Dilthey when introducing his own argument for the particularity of history in the *Introduction to the Human Sciences* (1883). He praised psychophysics and its "first highly gifted elaborator," Fechner, but had to caution again any application of his work to history.[11] The promise of any psychological or historical study of psychophysics became in Dilthey's view impossibly ramified in the complex interactions of society and the nature of consciousness: "A waterfall is composed of homogenous falling water droplets; but a single declaration [...] shocks the entire living population of a continent through a play of motives in purely individual human beings: so different is the mutual effect appearing here, namely, motivation arising out of thought, and from every other form of cause!"[12] Nonetheless, Dilthey imagined that at least the possibility for imagining social complexity lay in calculus, even if its object was not tractable in the manner Lamprecht suggested. The same suggestion was made by no less a critic of historical explanation than Leo Tolstoy. The modern theory of calculus, Tolstoy writes in an aside to *War and Peace*, "unknown to the ancients," could yet be the basis for a future science of history, albeit one characteristically grounded in individual experience: "Only by taking infinitesimally small units for observation (the differential of history, that is, the individual tendencies of men) and attaining to the art of integrating them (that is, finding the sum of these infinitesimals) can we hope to arrive at the laws of history."[13] Like Dilthey, Tolstoy did not think this was possible, but the fact that calculus was the obvious model for such an endeavor is suggestive of the themes' vast import to the nineteenth century.

Fechner's work presented the most important reception of calculus, both as a quantitative method and as a theme for reflection in the sciences. Cassirer's own history of the period describes Fechner's *Elements of Psychophysics* as taking "the final step toward opening the way to the mathematizing of the psychic, since the relation of stimulus to sensation had been ascertained to be a quantitative one."[14] Fechner first proposed the term "psychophysics" to describe the uniting of the world of sensation and perception, and the movement proved to be a foundation for the nascent field of empirical psychology. Fechner's work led to a central law of modern empirical psychology, the "Weber-Fechner Law" that holds that the physical magnitude of the stimulus must increase geometrically in order for the sensed magnitude to increase arithmetically.[15]

Even as the problem of determination in calculus opened onto quantification and strict mathematical determination, in the reception of Romanticism it also opened on to a realm of speculation on holism and instinct within the field of aesthetic determination. If the relation of individual perception to the whole of experience could not yet be calculated, it could nonetheless be felt, and the best model for this problem would be the infinitesimal. Despite his advances in detailed research, Fechner himself ultimately represented perhaps the most intriguing use of Romantic themes in relation to infinitesimal philosophy.[16] As Cohen's colleague Lange commented, Fechner's combination of detailed research in empirical psychology with lofty speculations on its importance for the social and human sciences, and for spirituality, led him to "become a living example that even the most extreme religious enthusiasm [*schwärmerei*] does not always poison the spirit of true research." Lange and Cohen saw distilling the legitimate aspects of Fechner's work as a central problem.[17]

Fechner essentially held to a form of Spinozism in which material and bodily events were of one whole with consciousness and ultimately divinity, even as they were perceived as two different sides of meaning. Since the relation of excitation to sensation, and sensations to consciousness, could not be clearly delimited, Fechner reasoned that it was legitimate – or, one sometimes feels, at least not boring – to assume a universal link among them. Fechner distinguished the limited "night-view" (*Nachtansicht*) of materialism from the "day-view" (*Tagesansicht*) of consciousness of an infinite world. Both were epitomized by calculus. In the night-view, the relation of integral to differential is carefully calculated to explain the particular, as epitomized by the Weber-Fechner law itself, and in this way it promises to ultimately "mathematize" the world. The night-view not only encompasses materialism, however, but any attempt to understand the world as a "closed" system, whether in the epistemology of Kant or the system of ideas in Hegel.

Fechner's day-view, on the other hand, sees the same process of relating integral to differential as focusing on the holistic nature of this relation: intuitively our present experience *is* the experience of the whole. "The kernel and germ, the punctum saliens" of his theory, Fechner writes, is that "our relation to God is not an external one [...] but an internal one as a part of a whole, or as a step to stairs. For in this case God's being is not entirely ungraspable [*unfasslich*] to us; we ourselves are a breath, a small fraction, a small stage and experiment of it."[18] Rather than being understood as limit, the differential is taken as a given whole, and intuition is precisely the process of grasping this wider reality.

Thus, although the night-view remained closely linked to materialism, the day-view allowed for far broader speculative leaps in Fechner's reading, and these leaps were linked to a fully realistic sense that they could well provide a window into the "real" world. Although Fechner's more extreme thoughts occasionally appeared ridiculous, they often seem to be advanced in a playful manner consistent with both an earlier career writing satire and his belief that the world is undoubtedly far more complex than we yet grasp. Wilhelm Windelband suggests the popular appeal, and staggering ambitions, of Fechner's cosmology in his widely read *History of Philosophy* from 1898:

> As the sensations which correspond to the excitation of particular parts of the nervous system present themselves as surface waves in the total wave of our individual consciousness, so we may conceive that the consciousness of a single person is in turn but the surface wave of a more general consciousness, – say, that of the

planetary mind: and if we continue this line, we come ultimately to the assumption of a *universal total consciousness in God*, to which the universal causal connection of the atoms corresponds.[19]

As Michael Heidelberger has noted, Fechner's deism was in fact connected with an important innovation in realism: he intends his day-view to escape both subjective idealism and materialism to reveal the grander connections within the incarnate world as immediately given, not to be part of a secondary quality of experience. In this manner, Heidelberger contends, Fechner "sketches a new sort of epistemology, explaining the reality of the mental and the organic, bridging the cleft that separates nature and consciousness, reality and perceptual appearance, and combining science with direct human experience."[20]

Within this worldview, Fechner was open to hypothesis – and indeed hypothesis connected with immediate reverie – concerning speculative realities. Thus, Fechner reasoned within the day-view that individuals might readily live on after death, since even within materialism it was known that there was a complete change of all physical material in the body every decade or so, yet continuity of consciousness still existed from youth to adulthood.[21] Similarly, if in the night-view humans did not really understand how they detected consciousness in other people or creatures other than through the subtlety of physical signs, why in the day-view should consciousness be denied to plants? "Why should the beautiful structure and bejeweled form of the pure plant appear less worthy of a soul," Fechner writes, "than the inelegant [*unförmliche*] shape of a dirty worm?"[22] Such conclusions are what led Lange to consider this aspect of Fechner's work mere "enthusiasm," but it nonetheless presented a fascinating and popular understanding of natural science. At its base, however, was the model of calculus, which introduced the reality of the infinite into the present day-view even as it also determined, and translated, the precise figures of the night-view of psychophysics and empirical psychology.

Cohen saw his philosophy as presenting an alternative to Fechner's metaphysics and worldview, one that could however be similarly understood through the key problem of the infinitesimal. Fechner's night-view and accommodation with the exact natural sciences would be maintained, but Cohen – rather astoundingly, given his reputation as an arch-rationalist – argued that the "day-vision [*Tagesansicht*] of our own critical knowledge idealism" [*erkenntniskritischen Idealismus*] presented a new version of this aspect of Fechner's work that did not collapse into Romantic mysticism. Rather, it would explain how a form of infinite intuition is shaped and formed by ideas and is open to criticism. That Cohen chose the infinitesimal as the means to develop this argument made sense in the context in which it was made, even as it opened his argument to criticism from both sides: as arid "scientism" on the hand and, as Leonard Nelson had it, a "mysticism of the infinitesimal" on the other. Understanding how Cohen in fact understood the infinitesimal, and how it related to his wider philosophy, requires a closer look at its meanings in his system.

Cohen's *Infinitesimal Method*, Part One: The Paradox Calculus

With the broader context of the reception of calculus in mind, we can now work back to clarify the several, at times conflicting, meanings of calculus as they develop from Cohen's early *Infinitesimal Method*. The argument for the Marburg school's critical idealism, as Cohen and Cassirer often repeat, is that a misplaced essentialism of "being" or "substance" can corrupt our understanding of experience and ability to understand its transcendental foundations. Indeed, Cohen's starting point was to claim that experience itself is already misunderstood as simply given in the manner of a substance or something simply appearing. As Cohen describes this central problem in his *Kant's Theory of Experience*, "The content of experience, which Hume took for granted, is here what is put into question."[23] The specific mathematical theme of calculus comes to the fore as the key example demonstrating the "constructed" nature of immediate experience. The importance of mathematics is that it allows us to demonstrate the proper balance of thinking (*denken*) and sensible intuitability (*Anschauung*) in philosophy, a balance that will then affect nearly every other feature of Cohen's work.[24] "The definition of the infinitesimal method," as Cohen says, "is conditioned by the establishment of the boundaries of intuitability and thought."[25] The challenge of understanding the *Infinitesimal Method* both as a text and in the inherent difficulty of its subject matter will lie, as Friedrich Kuntze argued in 1906, in the manner in which this balance occurs outside the direct purview of either thought or intuition.[26]

The counterintuitiveness of the infinitesimal, which initially provoked so much resistance to its adoption, suggests the role of what Cassirer will call in his defense of Cohen its "relative being."[27] Bishop Berkeley summed up the central paradox of the infinitesimal in his *The Analyst, or a Discourse Addressed to an Infidel Mathematician*, when he wrote of Newton's fluxions (the derivatives of continuous functions): "And what are these fluxions? The velocities of evanescent increments? They are neither finite quantities, nor quantities infinitely small, nor yet nothing. May we not call them ghosts of departed quantities?"[28] As products of mathematics that define "reality" yet cannot be directly intuited as insular or discrete elements of being, the infinitesimal epitomizes the relation of thought, intuition and the transcendental imagination which in Cohen's view is to characterize all of modern science. The infinitely small is not defined simply by thought, since it relates to real change and phenomena in the world, as Galileo had noted.[29] Yet it is not given immediately to intuition other than through the relation of the transcendental imagination striving to grasp a functional relation of change. In this regard, the problem of the infinitesimal is in itself the core problem of first understanding calculus. Berkeley's mistake lay in imagining that the infinitesimal can be recognized in the discrete or empirical impressions of continuity, just as for others it lay in trying to find it under the substantial category of "thing" at all – approaches that lead to Berkeley's joking conclusion that fluxions must be supernatural "unthings," that is, "ghosts."[30]

The inability of Berkeley to grasp the reality of fluxions as "ghosts" is symptomatic of the broader philosophical confusions surrounding the concept. What was lost in these readings was in Cohen's view nothing less than the "creative positivity" of the limit concept.[31] One was left with a characteristic opposition of extension and nothingness, or in temporality of the past and the future, but not with the constructive quality of the infinitesimal or the fluidness of the present understood as becoming.

In Cohen's reading, calculus provides on this basis the key tool for avoiding the mistakes of modern empiricism and sensualism, as well as classical idealism. It demonstrates that we actively know the particular only through its generative context, and we know this context only through the basis of structural ideals and concepts that the transcendental imagination strives to relate to a particular reality. It is simply a tool through which we gain aspect to particular determinations of experience, not something that reveals or uncovers an absolute or settled given reality.

As such, however, calculus is not the exception, but the rule that reveals the basic manner in which any idea or method determines its objects within experience. If Zeno of Elea's paradoxes reveal limitations to "commonsense" modes through which we grasp problems such as motion or change in terms of static definitions of time or space, calculus provides a new model avoiding these pitfalls. It does so, however, by providing a means for the understanding the appearance of reality not reducible to either direct intuition or direct thought. Partly on this basis, Cohen argues that, far from being limited only to mathematical or scientific knowledge, the same process at the heart of the infinitesimal lies in all forms of perception and thought, forming the basis through which reality can be understood relationally and developmentally as a system open to ever new determinations.

In the attempt to discover the relation of motion to mathematics – and thus the changeable to what, in Plato, was the realm of the unchangeable – both Newton and Leibniz developed related forms of the calculus. As Cassirer notes both in his *Knowledge Problem* and in a later popularization of the problem for a North American audience, it is telling that the two thinkers reached a similar method of balancing thought and intuition in calculus from the opposed philosophical directions of scientific realism and logical idealism.[32] Similarly, both will in the process abandon the naïve definition of space and time as immediately given, with Newton removing this immediate reality to an "absolute" divine time and space, and Leibniz dispersing it to a purely relation definition of time and space.[33] Both developed what Cohen took as the purest model of a relational system in the calculus, which is both a system of "signs" allowing for a new perception of change, as Leibniz held, and the actual correlation of real motion, as Newton held. This functional model has no set "essential" definition of "being" beyond its "relative" being as an indication or revelation of change, yet it first allows a proper understanding of real change in the world. By presenting a historical synthesis of Leibniz and Newton's positions in the *Infinitesimal Method*, Cohen hoped in turn to refine the philosophical synthesis made of their work by Kant. In this manner a proper understanding of the infinitesimal problem and calculus would catalyze an entirely new philosophical interpretation of Kant's work.

Cohen's *Infinitesimal Method*, Part Two: Calculus and the General Problem of the Determination of Particulars

Cohen's philosophical reading of problem of the infinitesimal is principally situated in terms of his reading of Kant's reading of space and time, a problem that had been central to Cohen's work since his intervention in a debate on the topic between Kuno Fischer (1824–1907) and Friedrich Adolf Trendelenburg (1802–1872). Cohen considered a proper understanding of the Kantian problem of space and time as a specific topic that opens out onto the Kantian project generally as a "question of the principles of knowledge," raising the core question: "Is the nature of things grounded in the conditions of our mind? Or must and *can* our thought be confirmed by the law of nature?"[34] Cohen's hybrid answer to this broader question ultimately hinged for both himself and, he argued, Kant on a proper understanding of the infinitesimal problem.

Kant initially followed Newton in holding space and time to be absolute, but transformed this "objective" absolute into the forms of intuition of space and time. Cohen recast Kant's reading so that it is closer to Leibniz than Newton (a tendency he took as already present in Kant's initial ideal definition of space and claimed appeared with renewed clarity in Kant's later work). Essentially, Cohen claims that although space and time depend on transcendental rules that approach absolute definitions, their form nonetheless varies depending upon the mode of knowledge through which they are constructed.[35] They might always be described – or better, revealed – by a yet higher-order form as had notably occurred in the metageometries first outlined by Leibniz and later developed by Felix Klein and others.[36] Time and space are thus ultimately relational, "orders of coexistences and [...] successive existences" as Leibniz had it, even as within a particular form of knowledge they can appear to have absolute qualities.[37] In this way the legitimacy of non-Euclidean geometry and topology are guaranteed, and Cohen opens the path for an understanding of "reality" – and with it for him the core dilemma of the Kantian project – that is not substantial but rather develops from relations themselves.

In Cassirer's defense of Cohen's theory of the infinitesimal he notes that the real starting point of Kant's critical method, and by implication the best path to understanding Cohen's re-reading of Kant, can be found in Kant's transcendental dialectic, where he demonstrated a set of arguments for removing false definitions of being.[38] The transcendental dialectic explains away illusions of understanding that arise from false notions of "being" that plague our understanding of experience. In Cassirer's reading, the most relevant aspect of the transcendental dialectic for understanding Cohen is the antinomies, and particularly the so-called "cosmological antinomies."[39] Antinomies for Kant are a pair of propositions that follow from the same assumptions but prove apparently contradictory.[40] They are resolved by either showing a false basis in the initial assumption or the false nature of their opposition. Kant's analysis of each of the antinomies debunks a key element of false understandings of being. The first, the antinomy of rational psychology, unseats the idea of "substantial" ego; the second, of cosmology, that of a "substantial" definition of the world; the third, of rational theology, that of the idea of our ability to conceptualize a "substantial" God.

It is the cosmological antinomies that are in Cassirer's view the best starting point for understanding Cohen's re-reading of Kant and with this understanding his emphasis on the infinitesimal. Kant finds four antinomies of cosmology, relating roughly to the categories of quantity, quality, relation and modality. The second antinomy can stand in for the others as an example, since it relates most closely to the problems of calculus. Its first premise states, "In the world every composite substance is composed of parts; nothing exists anywhere except it is either simple or is composed of simple parts." Its opposite states, "In the world no composite things consists of simple parts and there exists nowhere in the world anything simple."[41] The solution to this antinomy, as to all of the cosmological antinomies, ultimately lies in recognizing the false substantiation of the concept of "world" as epitomized by the very article in "the" world. In fact, as Cassirer notes, "Experience as a whole is never given to us as such, as a rigid, closed entity; it is not a result lying behind us, but a goal lying before us."[42] Our particular experience in the present place and time never gives us access to "the" world as a whole, so we can neither have things "closed" off from it nor totally encapsulated "in" it. Instead, we have for Kant a particular set of transcendental logical rules, the categories, that lead us "from one conditioned thing to another" and we have a regulative idea of totality by which we place our particular experience in the horizon of a potential world of experience.[43]

Cassirer's reading of Kant's "process of determination" (*Bestimmung*) suggests why this process of experience, once freed of the substantial bias of the concept of "the" world, begins to closely engage the core ideas of the infinitesimal method. In analyzing our perception of experience, we find that experience is at its very base relational or functional and fundamentally temporal. Cassirer's own summary of Kant from 1919 highlights this relation:

> An individual member of a particular series always points to another that precedes it, without our ever succeeding to a last member, but also, when we grasp each series as a unity, the moment we wish to indicate how it coordinates with other series and depends on them, the result is a nexus of ever new functional connections, which, when we try to follow it out and express it, leads us straight into the indefinite distance. What we call "experience" consists in such a set of progressive relations, not in a whole of absolute data.[44]

The substantial or "completed existent" element is removed from immediate experience "The idea of totality is 'regulative,' not constitutive, because it contains only a prescription as to what we are to do in the regress, but does not determine and anticipate what is given in the object prior to any regress."[45] We constantly investigate or perceive our experience so as to redefine our definition of it, but never is it simply given without this process having taken place. Cohen's broadest use of the infinitesimal problem developed on this foundation: it assumed that "'experience' consists in such a set of progressive relations," but that for this reason any particular experience is only defined in relation to this infinite set. The relation of the particular experience to the entire infinite set was thus analogous to the relation of the integral and differential.

Several consequences follow from this approach. We never have experience of any *unconditioned* thing, and thus of any absolute being, for if we did it would stand in no relation to the rest of our experience, and would thus be incomprehensible. Nor, of course, do we ever experience a *fully* determined object, since this would require a consciousness that understands the infinite relations in time and space affecting any particular – a possibility, perhaps, for a different kind of consciousness and experience, but not for ours. Finally, the horizon against which the particular is related is always by definition one of a single plenum, since if any particular set of relations did not relate to other elements of the plenum, they too would not be intelligible in our world. Our immediate intuition of a particular does not guarantee the "reality" of an object in this plenum, rather thought and intuition together are guided by the ideal category of "reality" and the particular judgment of the "Real" – a problem to which we will return. Even as we have a general sense of reality, no particular reality is ever directly given. Rather we understand objects as hypothesis through the ideal category of the "Real," which allows imagination to unify thought and intuition so as to achieve the greatest probable level of "determination" of the particular against the "background," so to speak, of reality.[46]

In actual experience, simply stated, particular moments or events are determined in relation to the horizon of experience. Far from leading to an "abstract" definition of reality, for Cohen and Cassirer this was the basis for the immanence of perception, which connects thought and intuition under the guidance of the category of "reality" with the aid of the transcendental imagination. Reality is neither a "product" of the subject nor a "reflection" or copy of a pre-given world, but is an illumination of part to (infinite) whole as mediated by the transcendental imagination. To be sure, for both Cohen and Cassirer ideal categories and forms constantly inflect and enable this perception, but its key moment is the immanence of the present itself. In this regard, Cohen's philosophy follows Leibniz in holding that the same basic principle that is epitomized by calculus underlies our definition of present experience as a series, or as Leibniz puts it, in terms of temporal form, that the "present contains within itself the past, and is pregnant with the future."[47] The basic problem of determination of the particular object or event within an infinite context will serve for Cohen as the "'presupposition' of Leibniz's calculus even as the calculus will serve as its 'concrete demonstration.'"[48]

Cohen's *Infinitesimal Method*, Part Three: Calculus as a Universal Schema for Modern Natural Science and Experience

Even within the description given so far, it can be seen why Cohen would hold this basic approach to the problem of experience as critical not just for the particular case of calculus, but to understanding all of modern mathematics and physics. In a polemic against Charles Renouvier's (1815–1903) law of definite number, which claimed that all being had to be considered finite, and in favor of Cohen's theory of the infinitesimal, Cassirer raises contemporary theories of number, particularly the mathematician Richard Dedekind's (1831–1916) idea of the cut [*Schnitt*] of a particular irrational or real number as a confirming example. Whereas Renouvier had claimed that the idea of an "infinite" particular was illogical, and that the world of being could only be understood as a set of finite entities, Cassirer claims that such infinite particulars are the norm:

> Whether or not something is in the sense of pure logic a legitimate "object," does not in fact depend on whether we can imagine its parts individually realized in the imagination, but rather it depends on whether we can make fully determined judgments about it that are separable from every other "object." By this basis the problem of infinite contents [in irrational numbers, etc.] are not a problem: they are the true logical-mathematical "individuals," which through the particular rule of the contents are separated from each other with absolute clarity. The logical being that we give to these contents entails certainly no absolute, but rather a relative

Being [*Sein*]: it suggests in its fundamental concept simply a "being-different than" [*Unterschiedensein*].[49]

The particular is determined in its horizon of experience in a manner that is again neither purely "thought" nor "intuition," but is rather a limit ideal or "tool" based on its interrelation in the world, guided by imagination (*Einbildungskraft*) and the regulative ideal of reality.[50] Although defined purely through relations and delimited negatively, the numbers or objects revealed in this process nonetheless have a positive quality of being "real," as is suggested by both calculus and Dedekind's theory of the "cut." In *Substance and Form*, Cassirer clarifies the mathematical critique of the "traditional doctrine of the plurality of things" through the work of Dedekind in terms of a "series principle" parallel to Cassirer's earlier logical critique.[51] Dedekind argued in "Continuity and Irrational Numbers" (1872) that numbers should be understood as, in Cassirer's words, "the expression of a complex whole of relations…" in the "determinateness of division itself."[52] Dedekind's "cut" argument provides a demonstration of the primacy of this definition by resolving the problem of how all numbers could be defined so as to resolve the opposition of the continuum of the number line and the discreet nature of numbers themselves. Put simply, we can define the number line into two parts around a number, say the real number e, such that there is a set B which is greater than the number, but does not include it, and a set A which is less than a number but does not include it. The "cut" in which the number appears is in neither set but is defined purely by its relational position, thus allowing for all irrational and real numbers on the number line.[53] In this manner, even numbers can in a sense be recontextualized or reinscribed within a new definition of relations, say, by finding a further "cut" in the number system and thus discovering or locating a "new" number, such as the real number e, itself as developed by Jacob Bernoulli, Leibniz and Leonhard Euler. The conditions of the possibility of mathematical experience are the same for Cassirer as the conditions for the possibility of all experience. Cassirer thus writes in response to Renouvier, "Removed from the necessary relational forms of experience, there would be no phenomenal 'content' more: for the 'appearances' are as empirical objects only given in experience and do not exist at all outside of this experience."[54] It is in this regard that we can recall Cassirer's summary statement to Nelson, quoted above: "The infinitesimal is not a thing [*Ding*] but a condition [*Bedingung*], not any sort of reality at hand, but an instrument of thought for the discovery and construction of true being [*wahrhaften Seins*]."[55] The same will be true for Cohen and Cassirer for any scientific hypothesis, method or event properly considered.

In Cohen's reading, the theory of the infinitesimal is exemplified by its history. The initial historical attempts at understanding the infinitesimal involved comparison of two extensive entities, one much smaller than the other, such as in the method of exhaustion.[56] Only once the bias in favor of the necessary intuition of extension in this process was dropped could the infinitesimal properly be understood as a conceptual diminution to the infinitely small that was real yet graspable only as a function in combination with imagination.[57] Even mathematically the infinitesimal is only a "relative-being," it nonetheless correlates or provides access to "true experience" in a manner demonstrated by its practical application.[58]

As Cohen acknowledged, the first thinker to place the insights of Newton and Leibniz in the Kantian format was not Cohen but Kant's contemporary Salomon Maimon (1753–1800). Cassirer's reading of Maimon in the third volume of his *Erkenntnisproblem* is singular in almost the entire series for its direct interpolation of Maimon's philosophy as a predecessor to his own work and that of Cohen, and as such it can be read as an aspect of Cassirer's exegesis of Cohen's work.[59] The critical component of Maimon's treatment of calculus is its use in describing the Kantian "thing-in-itself," and with it any element of "fact" or "object" whether of science or of perception.[60] Maimon claimed that the thing-in-itself is a formally contradictory concept that could never be known or usefully theorized in itself, even as the term functions as a signifier for this very contradiction. This contradictory nature is found in the very definition of the thing-in-itself as "something" from which the characteristics of thought as we know it are removed; the term has a function similar to an imaginary number (such as the square root of a negative number) in its suggesting a turn of thought which is in itself not open to intuition.[61] Nonetheless, noumenal reality does find a place in Maimon's work, and it occurs through his understanding of calculus. Here he writes that "differentials of objects are the so-called *Noumena*; the objects that develop from this set of relations are the so-called *Phenomena*."[62] Although somewhat unusual in his use of the Kantian terminology, what Maimon appears to mean by this is that the differentials of objects are noumenal in the sense of connecting up with an unknowably infinite horizon of connections. In this definition, which was to prove influential on Kant's own changing readings of the problem, both the thing-in-itself (*Noumena*) and immediate experience (*Phenomena*) have ceased to be defined substantially, and are now defined only as idealized poles of a single functional relation.

The appearance of objects here functions, Cassirer writes, "so that understanding cannot simply have 'objects' appear without taking them as 'appearing,' that is, without taking them as flowing."[63] Our entire definition of nature is no longer understood as an "objective being" but as "the entirety of relational determinations."[64] These relational determinations themselves, however, stretch between the noumenal fully unknown and unknowable, on the one hand, and the presentation of the phenomenal object on the other. Maimon's thought, at least in Cassirer's reading, thus demonstrates how the dualisms of noumenal and phenomenal, and similarly dualisms such as inside and outside, subject and object, can be recast into unified limit problems. The appearance of these forms ceases to be a given fact, but a consequence or construction of different forms of knowledge that can be explained and analyzed.

Maimon's turn of thought here, Cassirer suggests, is essentially the basis of the Marburg definition of functionalism: any particular "object," broadly defined, is shaped by the entirety of the relations, or functions, in which it stands, particularly as understood as part of a temporal or historical process. Even the basic dichotomies assumed in Kant's original system cannot be used to stabilize this process; instead, a quest for the "logic of pure knowledge" involved finding the generative principles and contradictions that leads to particular formations of an object. Maimon, however, in Cassirer's view ultimately remained skeptical about the possibility for his system to provide a description of physical reality, and to explain how thought and perception coincide.[65] The intervening century, however, had provided numerous new examples in which Maimon's basic template could indeed be modified to form the best possible schema for understanding the relation of knowledge and reality.[66]

Cohen's *Infinitesimal Method*, Part Four: The Principle of Anticipation and the Continuum

Cohen argued that Fechner's colorful combination of psychophysics and metaphysics, particularly in its application to empirical psychology, assumed a dualism of body and soul, inside and outside, of a "being" (*Wesen*) of consciousness and material, "that ultimately had no foundation in experience."[67] From the Kantian interpretation of the infinitesimal first proposed by Maimon, Cohen claimed to avoid such dualisms, as well as the dualism of quantitative exact science (Fechner's night-view) and qualitative intuition (Fechner's day-view). Rather than reducing qualitative understanding to quantitative, philosophy would demonstrate that all quantitative experience first derives from a single field of qualitative meaning. Cohen's technical arguments for this claim can at first appear daunting, but resolve into a fairly straightforward explanatory theme that will suggest how meaning unfolds out of a unified field of experience. Only in understanding this process can the fundamental unity of thought and intuition, form and matter, or concept and sense be understood. As we have seen, the question held immediate stakes for the Marburg school; the development of empirical psychology within philosophy – and with it the claims to the priority of quantitative information found in much of psychophysics – was one of the significant threats to the schools viability under the Prussian cultural ministry.[68]

Cohen's counterintuitive use of calculus to defend the value of philosophy and qualitative meaning hinges on his transfer of the center of gravity within Kant's *Critique of Pure Reason* from the *a priori* categories of reason and intuitions of time and space to the principles and schematism.[69] Synthetic principles describe, as Cassirer puts it in his functionalist summary of Kant, how "the function that characterizes a specific category relates to the form of pure intuition and permeates it in a synthetic unity."[70] Kant's ultimate explanation of why natural science was able to use mathematics to understand the world, and thus why *synthetic a priori* judgments were possible, was arguably that the rules for understanding the world were the same as the rules that governed our consciousness, since experience is, after all, one unified whole. Kant's explanation of this problem within the principles is found first in the problem of extension, an area covered by the so-called "general axioms of intuition." These axioms have as their foundation the idea that "all perceptions are extensive magnitudes."[71] Perceptions in any form of experience, whether in mathematics or everyday life, work through extensive magnitudes whose rules will be identical no matter where or how they are encountered.

In the "anticipations of perception," Kant made a parallel claim for quality and intension on the basis of calculus as the "axioms of intuition" made for quantity and extension on the basis of geometry. Kant's principle of the anticipation of perception is directly formulated as: "In all appearances the real which is an object of sensation has intensive magnitude, that is degree."[72] Our temporally changing definition of both the "more" or "less" of sensation, and, more importantly, of the relation of the ideal to the particular – say in Newton's ideal laws of motion in relation to actual events of motion – are defined by the anticipation of perception, which in turn in Cohen's reading forms the basic template of calculus. In Cohen's reading of Kant, calculus finds so many corollaries in the "real" world of natural science for the same reason as the axioms: its innermost rules are intrinsic to any experience whatsoever.

Cohen thought that the rules of the "anticipation of perception" prove of much greater importance than those in the "axioms of intuition," however, since they will describe the limitative relation of how any idea conditions a particular fact, how the horizon of experience inflects the particular moment of time and space. Thus, in Cohen's reading, the anticipation of perception plays a role in every aspect of Kant's philosophy, since everywhere that the categories and the intuition of space and time are involved there will be limitation in relation to the particular. The anticipation of perception, and with it the fundamental logic at the base of calculus, can thus can be applied to the entirety of Kant's philosophy as a sort of universal solvent, which will allow Cohen to recast the meaning of the whole.

It would thus be difficult to overstate the importance of this principle for Cohen. "The principle of anticipation," he writes, "contains in itself the problem of the critique of knowledge."[73] In his argument with the psychologists, Cohen will contend that the rules of intension can explain the process of extension, rules of quality can be used to derive quantities, but the reverse is not true. The basic rationale for Cohen's argument can be suggested by recalling Cassirer's argument with Renouvier over the nature of numbers and entities. The only definition of a "legitimate" mathematical or scientific object, Cassirer had claimed, was whether "we can make fully determined judgments about it that are separable from every other 'object.'"[74] The only means of doing this in a manner that explains all mathematical and scientific objects, however, was not by claiming legitimacy only for objects that can be intuited as finite entities or quantities, but through the method of determination – epitomized by Dedekind's "cut" – in which "the particular rule of the contents are separated from each other with absolute clarity." The key to this relation of the particular to the whole, however, ends up being founded on something like the qualitative relational model of the differential to the integral. Quality rather than quantity is for Cohen ultimately the universal foundation of scientific explanation and calculation.

The same basic qualitative role in the anticipations of perception will similarly provide the only comprehensive definition of the appearance of physical objects in space and time. Cassirer later pithily summarizes this relation in regard to space: "The consciousness of a single point contains reference to space as the sum and totality of all possible designations of position. [...] The 'integral' of consciousness is constructed not from the sum of its sensuous elements (a, b, c, d...) but from the totality, as it were, of its differentials of relation and form (dr_1, dr_2, dr_3, dr_4...)."[75] The basic concept of a "relational" matrix for space, and thus a qualitative foundation even for a quantitative designation – such as the Cartesian grid – is here identical to Cohen's definition. Even as we often retrospectively assume a quantitative definition of, say, a number or a point, or any quantitative object to be self-sufficient, it is always dependent on this broader qualitative embeddedness in "relations and form." While such a synthesis is needed, as Cassirer notes, to "produce concrete geometrical figures, it is absolutely indispensible in matters of specifying physical objects," and indeed in any experience whatsoever. It is the embeddedness in "relation and form," in its infinite imbrication, that Cohen and Cassirer claim presents itself precisely as the immediacy of phenomenological experience, of not merely the "form of perception," as Cassirer puts it, "but its content as well," which has "its subjective, psychological expression in sensation."[76] In this way, as Cassirer writes, the appearance of the "mathematics of intensive magnitudes," notably first calculus, entailed nothing less than that "the 'real' in appearance achieves its first scientific designation and objectification."[77]

Cohen further argued that the logic of the infinitesimal and calculus is omnipresent in what he took to be the root form of human judgment and logic, the so-called limitative or infinite judgment. An understanding of this form of judgment is pivotal for understanding Cohen's argument concerning quality, and indeed his philosophy as a whole. Using this form of logic Cohen claimed to show the priority of the anticipation of perception not only within the exact sciences, but for all human experience. In Kant's original reading, limitation itself is one of the so-called categories of quality (a term defined as a category in a related but slightly different manner than in its use in the anticipations of perception), which includes reality, simple negation and limitation.[78] The logical form of judgment relating to these categories are affirmative ("all men are mortal"), negative (it is not the case that "the soul is mortal") and limiting, or infinite, judgments (it is the case that "the soul is non-mortal [*Nichtsterblich*]").[79] Limitative or infinite judgments are for Kant of the form "x is non-y," so that they affirm a quality by negating it against something else. The importance of these judgments is suggested in passing by Kant when he notes, "Now by the proposition, 'The soul is non-mortal,' I have, so far as the logical form is concerned, really made an affirmation. I locate the soul in the unlimited sphere of non-mortal beings."[80]

Although Kant will largely drop this observation and conclude that this aspect of correlation between the table of categories and the older table of judgments was not particularly revealing, Cohen found in it a general form of thought that was at the same time the methodological core of calculus. Limitative judgment, in Cohen's view, has precisely the same form as the mysterious relation found in thinking of the infinitely small, or of the integral and differential in calculus.[81] In imagining the infinitely small, we have to strive to envision a series beyond the "smallest part of the smallest part." In calculus, a similar relation ensues in thinking of the particular as only defined relationally through a function against the infinite plenum of the whole of possibility, a relation that reveals a negative definition of both elements, part and whole, analogous to the limitative judgment.

In the history of philosophy and theology, Cohen found important precedents for his argument that explain why he thought it constituted a crucial means for redefining the problem of knowledge. Cohen was aware that a variant of limitative judgments, developed earlier in Aristotle and medieval Arabic, Christian and Jewish thought, provided an open-ended means of definition that was developed in negative theology in statements on the attributes of God – notably by Maimonides and Nicholas of Cusa, whose work is extensively thematized by Cohen and Cassirer.[82] In this variant, the emphasis is on the negation of what the medieval period would consider "improper" predicates of a subject – for instance, "The wall is non-seeing" – which form the core of "infinite" judgments.[83] Even more than in the Kantian definition, here it is clear that neither subject nor predicate are meant to operate as given substances, but rather act as a functional relation, a mode of directing the mind towards a continuum of concern. When applied to divinity, the effect is to suggest attributes without presuming them, such as in the statement that "God is not blind." The predicate of human sight would not relate to the divine, but by invoking its opposite a continuum of possibility is created about God's abilities without presuming divine attributes. In effectively "modernizing" the importance of this earlier version of infinite judgments, Cohen places the emphasis on the open-ended project of humanity, rather than the nature of divinity alone, since he will define all of experience and knowledge, including that of the ego, through this relation of logical striving. The logic of infinite judgments opens a path towards perception of the infinite itself and provides a means for avoiding essentialist simplifications in defining humanity.

For Cohen, limitative judgments precisely describe the relation of thought and intuition in the infinitesimal as leading to a hybrid form of "objective intuition" that is not purely sensible, but rather a *means* of knowledge.[84] As Cohen puts it in his original formulation in the *Infinitesimal Method*, this is nothing less than "the secret of the differential idea, which is revealed as the logical secret of knowledge."[85] Although the concept of infinite judgments had previously been ridiculed by Hegel and Lotze, Cohen holds limitative judgments to be the key to understanding the infinitesimal, and with it modern science. Although he does not fully have the logical tools to develop his use of limitative judgments, Cohen appears to have in mind something like the forms of logic later developed by George Spencer Brown and popularized through systems theorists such as Niklas Luhmann as a "calculus of indications."[86] In such a scheme, the initial assumption is an undifferentiated plenum of meaning in which one "draws a distinction" to begin to reveal certain forms of reality. By using the call of "draw a distinction" a division or operation within the plenum of experience is made, which then leads to ever-greater differentiation.

Lacking such a calculus, Cohen presents an absurd but useful example to illuminate the process of differentiation as applied in logic. We might understand the definition of human, he writes, by suggesting that "nonhuman" can be defined by the three terms "triangle, melancholy and sulfur."[87] In Cohen's view, the fact that the mind immediately develops a *sense* of the object of such limitative judgments demonstrates a central aspect of thought: for it demonstrates a union of thought and intuition using imagination to make sense of a reality without reference to substantive terms, and thus in a manner similar to the logic of the infinitesimal. Cohen's preference for the medieval definition of the limitative judgment over the Kantian is precisely because of this lack of substantive terms, since for the negative theologies the predicates under consideration don't exactly "match" their substance in the first place (God/Blind).

Precisely in regard to leaving open the definition of "objects" such as "human" or "consciousness," infinite judgments for Cohen have great value. Through them "we are able to form some concept by considering it in opposition to its opposites."[88] Presumably by redoubling this process through multiple terms, ever more refined judgments could be made about an object without either presuming to know the essence of the object or assuming that the chain of distinctions leading to a particular definition is the only possible one.[89] In defining consciousness (*Geist*) for instance, Cohen notes that we could neither call it material nor a "substance" at all, nor purely ideal, but that "it is part of an infinite group that is a third type (*tertium comparationis*)."⁹⁰ Just as in Cassirer's example of Dedekind's definition of "number," where numbers are defined most inclusively through a process of qualitative determination based on what they not, on their "being different than" as described by their "cut" in the line of numbers, so for Cohen in logic and experience any particular "object" will ultimately be most inclusively defined by the form of negation epitomized by infinite judgment. The medieval mode of these judgments represents an ideal case of "pure" negation to which modern logic can aspire. In this manner, Cohen's philosophy thus defined itself as at the opposite extreme of all attempts – notably those of the empirical psychologists – to "quantify" human spirit and find a substantial definition for humanity, even as he shared with the psychologists the language of the infinitesimal.

The final key to the universal applicability of limitative judgment and calculus for Cohen is found in its relation to the problem of continuity. Continuity is traditionally defined through the idea that *natura non facit saltum* ("nature does not make leaps"). Essentially, limitation will reveal that continuity is a constitutive feature – indeed, for Cohen it is *the* constitutive feature – of all thought and with it of all experience. "Continuity," Cohen succinctly writes, "is the general basis of consciousness."⁹¹ The mind always places each particular in its wider context, and this context by definition – once it is seen to rest on infinite judgments – is an infinite continuum or plenum. Defining "nonhuman" by a particular set of predicates, such as "triangle, melancholy or sulfur," creates an infinite range of possibilities for how "human" can be defined and perceived, as would a more proximate set for delimiting "humanity," such as the claim that "nonhuman" is defined by the predicates eternal, mechanical or omniscient. Cohen argues that the continuity established by limitative judgments is characteristic of all consciousness, perception and affect. Its importance is that it presumes neither the primacy of subject nor object, psychology nor logic, but rather works from a series of determinations within a plenum of meaning.

Although the skepticism of his contemporaries, including Cassirer, is justified in many regards concerning Cohen's claims for the necessary legitimacy of his argument linking the infinitesimal to infinite judgments, its power as a mode of interpretation is profound. Despite the centrality of limitative judgments to his argument, however, Cohen never directly demonstrates the link of this particular form of negation with that found in calculus. Nor does he demonstrate how from its esoteric medieval use it could be applied to any judgment whatsoever, as he will claim to be the case, nor how these judgments can combine into more complex modes of perception. Nonetheless, Cohen's model of infinite judgments is invaluable for understanding what he takes the infinitesimal problem to be, and in Cassirer's work the consequences of this reading are broadly influential – even as it is also tacitly critiqued.

Cohen's *Infinitesimal Method*, Part Five: The Problem of Reality

The culmination of Cohen's argument concerning the logic of the infinitesimal and infinite judgment is found in his theory of reality, which he argued was first developed by Leibniz but partially obscured by Kant's reception of the polymath.⁹² Cassirer's *Leibniz's System* will, as we will see, extensively address this lacuna and in turn further illuminate this central aspect of Cohen's philosophy. That the definition of reality is the centerpiece of Cohen's late philosophy is made clear by Cassirer in his summary in the *Encyclopædia Britannica* (updated from 1911 through 1927, noted earlier, in which he describes Cohen's late work as being grounded in the initial establishment of the differential calculus as "the indispensable and basic intellectual means for any scientific cognition of 'reality,'" because, he continues, "reality is never 'given' in any sense, in either sensation nor in mere intuition."[93] It is this basic insight, Cassirer continues, that is then the foundation of Cohen's three-volume *System of Philosophy*, which is to say his entire late philosophy.

For Cohen, "reality" is not a terminus, but itself a function – and as such it is nothing less than the "great problem" of philosophy.[94] Here the infinitesimal model takes on its broadest definition: the diffuse sense of "reality" is the differential of all potential possibilities in the universe, but we connect it with a particular "object" – and set of assumptions about this broader reality – through the specific function of the Real (*das Reale*).[95] Cohen writes in the *Infinitesimal Method*, "If the differential makes of reality a valid constituting condition of thought [*konstituierende Denkbedingung*], so the integral demonstrates the Real as object."[96] Ultimately, this means that the ideal function of reality is how objects are defined out of the flux of all possible real relations: "In the [judgment] of reality the object has its foundation."[97] Although for Cohen any judgment concerns a specific relation of a particular to the whole of its context, there is also a judgment of reality at work in all of experience that finds its summation in our understanding of what is "Real" in a given instance.

Cohen considered two of his students to have directly developed this problem: Dimitry Gawronsky, who wrote a dissertation on the *Judgment of Reality* in 1908, and Cassirer, whose *Substance and Function* will develop this theme in a different manner.[98] The basic theme is fundamental to the Marburg reception of Kant. Cassirer thus at one point summarizes Kant's work by writing that "the most general consequence of 'transcendental idealism' can be described as follows: that the concept of reality [*Wirklichkeit*] is not a one-sided, pre-given, set terminus, but rather that it contains various content and various meaning according to consciousness-function [*Bewusstseinsfunktion*], to which it serves as the correlate."[99] Both Cohen and Cassirer potentially go beyond Kant, however, by claiming that this is not solely a function of "consciousness" but also of experience itself that cannot be reduced to subjective experience. The "correlation" – a central Marburg term that will reappear in Cassirer's study of Leibniz – of the particular Real to the overarching plenum of possibilities in reality is the basic function through which Cohen's critical idealism is defined.[100]

Although Cohen's terminology is a bit vexing, its application appears to be fairly clear. If, to take a later example from gestalt psychology, we are walking through the woods on a cloudy day and see a dabble of sun on the ground, we may at first be surprised, but then on closer inspection find that the light is actually caused by spilled lime. The general context of our perception is our given sense of reality, so in this case our delimited sense of the infinite plenum of possible relations connected with walking in the woods on an overcast day. When confronted with any particular phenomenon or object, it "finds its place" in this system through the specific "promise," or in Cohen's terminology "hypothesis," of the real object. Thus we connect this specific object – the more obvious feature of sunlight on the forest floor, rather than the less likely but on closer inspection "actual" phenomenon of spilled lime – through the *function* of the Real to the wider possibility of reality. Far from leading to an "abstract" definition of reality, the definition of reality developed by Cohen is meant to explain an immanent *process* evident in scientific thought and experiment, but also in aesthetic and ethical experience. As such, Cohen, following Leibniz and followed in turn by Cassirer, aims to intensify *present* perception and provide a maximally skeptical argument against any dogmatic claims to ultimate reality.

In experience, there is no initial separation of sense and intellect at all, nor between "intuition and thought" (*Anschauung und Denken*), rather both are always already combined and ordered by the critical category of reality.[101] This process is what is most directly experienced as physical perception, but occurs in any aspect of experience. For Cohen, it always involves the transcendental imagination linking sense and intellect, intuition and thought, through the ideal category of the Real. A good example of this process is suggested by the historian Amos Funkenstein, following Karl Pearson, in the history of theories of motion: the concept of different theories of motion entail different "facts," such as the supposition of rest as a natural state in Aristotle's physics to that of inertia in Newton, which in turn condition the sense of "reality" in a given period.[102] In perceiving the arc of a thrown object, intuition and thought are unified by the concept of the Real in a manner that means motion in the twelfth century both appears and in some sense *is* different than in the sixteenth or twentieth centuries. It is in this regard that we are to understand Cohen's frequent claims to reject the primacy of sense or stimulus. His point is not that concept totally overrides sense or stimulus, or that the mind or science creates reality of whole cloth, but that we always already have a general horizon of the possibility of "sense," but that any particular moment of sense is always already shaped by concept and by the structure of knowledge in which it is perceived.

This is clearly seen in the "culmination" of Cohen's system in his psychology, where he develops an idiosyncratic, but revealing, argument in relation to the "new physiology" of Johannes Müller.[103] The starting point of his psychology, Cohen argues in his "Introduction" as well as in his *Aesthetics*, is in Müller's argument concerning specific nerve energy, or in the phrasing Cohen prefers, "necessary preconditions" (*Notwendigen Vorbegriffe*). Indeed, this is technically the beginning of his lectures on psychology 1899, as written down by Cassirer.[104] Müller's notion of specific nerve energy held that perception is defined by the form or type of nerve over which stimulus is carried, not the stimulus itself. The classic example is physical pressure on the eye, which causes a form of light to be seen even as the stimulus is physical. The concept fascinated the nineteenth century since it suggested that different nerves were "specialized" to react to distinct phenomena, and indeed it was the basis of physiological interpretations of the Kantian *a priori* – such as Lange had initially put forth but that Cohen rejected – that claimed our physical organization determined our experience of the world.

In Cohen's interpretation, however, the theory of specific nerve energies resolves neither to a purely "subjective" nor mechanistic "objective" definition. Rather, for Cohen it suggests that "consciousness is not to be determined as a reaction to a stimulus; rather the primal resource [*Urbestand*] of consciousness must be taken to be its disposition to such a reaction. The stimulus can not be allowed to be the first; for hidden in the stimulus is the [concept of the] object, which cannot be assumed from the outset to be given."[105]

Cohen argues that the real basis of the concept of specific nerve energy is a philosophical one, namely the *disposition* understood as an anticipation of perception. The "disposition to such a reaction" is a form of continuum preceding the appearance of any particular object or subject, indeed of any form of experience at all.[106] Although the biological basis of this argument seems slight, the basic form of the argument is the same as Cohen's earlier claims for both the infinitesimal and limitative judgments, in that the "necessary preconditions" (*Notwendigen Vorbegriffe*) of nervous energy suggests a definition of a particular reality against a plenum or continuum of meaning, one which cannot be reduced to external stimuli or purely defined as "subjective reality." Cohen's argument developed from an initial form in Lange's writings in a manner that will be more familiar to some readers from its reception by Nietzsche: we do not first have an external stimulus, say the prick of a pin, and then a response, but rather we first have the response, and then retroactively work back to define the source of the stimulus.[107] In this way, as Cohen said, the "object" can be understood as constructed and not "assumed from the outset to be given."

In both Cohen's psychology and his general definition of reality, he took motion to be a predominant theme. For psychology, if stimulus is not first, and anticipation is the basis of nerve energy, the problem is one of a developmental series of anticipations or determinations. The basic form of "motion" is shaped by relations of continuity, of qualitative anticipation, of both the part to the potential whole to which it relates and, through this relation, of one part in a form of change to the next. This relation is of course epitomized by calculus, which combines both of these determinations in one form as a "direction" of movement.[108] Motion, Cohen argues, could not exist merely as a sequence of insular times or places in experience any more than in Zeno's paradoxes.[109] A telling detail of this development is found in his rejection of Herder's definition of touch as the most primary aesthetic and psychological sense in favor of temperature. For in the sense of temperature, and a number of other phenomena such as attention itself, we have a process of almost pure movement in which "our sense of temperature," Cohen writes, "establishes the stimulus to which it is itself the answer."[110] That is, we sense a "growing colder" or "growing warmer" that is fundamentally reflexive and transitional. As such it epitomizes the qualitative "motion" that Cohen considers the essence of consciousness, and which he thought was first clearly illuminated by the calculus, which combines both of these determinations in one form as a "direction" of movement.

Cohen's argument appears to be that from its most basic and primal forms through to the most complex apprehensions of natural science or aesthetics, the problems of "sense" and of "reality" develop through a growing immanent differentiation of elements within this form of motion. This should not be understood to occur "inside" or "outside," in a subject or object, but rather, these distinctions are themselves constructs of this one problematic. For Cohen, reality is thus fundamentally temporal: it is in the developing continuity of experience that different aspects of reality can become defined as objects, as real, so that Cohen defines this process as a "continual and regular generation of reality in time."[111] This temporal quality is central, since it allows the role of negation and limitation to occur consistently through process, and thus provides the horizon of possibility for the mediation of phenomenal and noumenal, first suggested by Maimon.

"For reality is continual fruition [*Erfüllung*] in time," as Cohen puts it, "this is however the limitative, which establishes reality through the restriction [*Einschränkung*] of the negative."[112] Objects as intensive forms are established within limit horizons of probability, or what Cohen defines as the *hypothesis* of the Real.

In Kantian terms, Cohen holds that in striving to make sense of the particular within the context of the whole through the "judgment of reality," the mind necessarily uses the productive imagination to combine intuition and thought, but this is neither a product of "fantasy" nor a process of construction from nothing – it is rather always determination within a pre-existing continuum of experience.[113] The principle of continuity established between the particular and the horizon of "possible" reality guarantee that this productive imagination is never merely fantasy. Rather it is conditioned at once by its "ideal" determinates and its possible real corollary in experience in a manner that cannot be considered simply inside or outside, subjective or objective. The world is understood as a constant play of movement, "becoming and being, movement and solidification," based on the role of negation in experience, specifically the negation of the limitative judgment as it defines particular objects of experience.[114] Again, neither reality nor sensation in this scheme has to be "added in" to experience. It is rather only because we are first given the fact of reality at all, or the undifferentiated premise of the sensorium, that we can have this or that particular reality and perception defined by the category of the Real. Far from being purely a subjective or ideal category, only this dialectic of an existential and ineluctable reality and the "Real" allows us to first grasp new aspects of experience.[115]

The importance of Cohen's argument about reality becomes evident in his reading of history. Each moment in the development of scientific concepts conditions the particular interplay of thought and intuition for a given historical period as a limit problem defined by its particular form of knowledge. The interrelation of different forms of limitative judgment define different problems, events or objects of science, and the dissonances of these relations already contain within themselves the material for the next "unfolding" of the concept. Cohen sees science, or any human endeavor, "unfolding" newly relevant questions and perceptions of reality out of the problems posed by old answers and theories using the open-ended relation of different forms of limitative judgment, which occurs as part of a system or dialectic of changes.[116] Concepts, although capable of possessing timeless truth in themselves, can always theoretically be reinscribed into a new structure – as he already thought had happened in the relation of non-Euclidean to Euclidean geometries, the development of the concept of number or contemporary physics. One of Cassirer's major innovations was to use group theory as the model for this form of transformation. Already in Cohen's work, however, the process leads to a result that has been compared to Kuhn's theory of paradigm shifts, yet is more dialogical: from what is "projected" by the possibilities and failures of the old system arises a new system, and this system is open to contingent or chance inputs.[117]

In Cohen's thought, reality as a whole, the Real in particular and the development of history through them always retain an open-ended, infinite and even mysterious character. It is in part this aspect of experience that allows Cohen to link his epistemology to his aesthetics, ethics and psychology – and ultimately theology. It is not surprising in this regard that Cohen would suggest already in the *Infinitesimal Method* that his "idealistic knowledge criticism" was presenting a "day-view" of reality similar to Gustav Fechner's, but superior in its critical relation to scientific development.[118] "Reality" as a whole, as we have seen, is itself a limit concept shaped by all possible forms of relation, and is thus, as Maimon had it, intrinsically linked with the concept of the noumenal.

The same limit quality, and relation to noumenal reality, is also true of the "Real" itself. In a dense and crucial passage, Cohen defines the "Real" as that which "demonstrates and will demonstrate what can only be described as becoming and what can only be thought of as becoming, in and for itself, without attending to the organization and ordering of its relation to others."[119] By Cohen's own definition, this means that the "Real" cannot be comprehended in itself, since it has no "relation to others" and thus no means of being perceived. The "Real" is thus itself an ideal limit function of the promise of change within experience. Despite both this paradox and Cohen's jargon, this definition is in part commonsensical: a present moment of experience is defined as "present reality" primarily because it is assumed to be open to change and defined by the principle of sufficient reason in a matter that is open to unlimited inspection. The "Real" in itself has at its core the promise, at once commonsensical and ineffable, of pure becoming. In everyday experience, this is borne out by our sense that if something is "Real" it can not only be further investigated, but indeed also infinitely contextualized in a manner open to surprising new meaning, an infinite contextualization that is arguably not the case in fiction. Even if we do not know them, the details of Shakespeare's life are infinitely open to discovery in a manner that those of Hamlet's fictional life are not.

As startling and speculative as Cohen's work may appear, it had a strong philosophical precedent in the work of Leibniz. Towards the end of the *Infinitesimal Method* Cohen writes that "every differential is like a monad," and that each moment of experience is therefore "*un monde entier plein d'une infinité*, a whole world full of an infinity."[120] Already at the end of *Kant's Theory of Experience*, Cohen argued that he used Leibniz as the key corrective to Kant, particularly in rejecting Kant's assumption of psychological faculties and emphasis on the subject, while conversely using Kant to remove the false assumption of the monad as a substantial entity connecting "unity and reality."[121] This is indeed the basic challenge of Cohen's philosophy: by avoiding the fiction of the monads, Cohen drops the one "substantive" element left in Leibniz's philosophy to promote a purely open play of relations within a single continuum of reality, one which redefines both unity and reality itself in a new manner. The Leibnizian foundation of Cohen's late philosophy cannot be emphasized enough, for without it readers consistently and falsely impute dualisms to this work that do not exist. Cohen himself went so far as to claim in 1912 that Leibniz did the "preponderance of philosophical work" that was absorbed by Kant, and that Kant should be considered, "despite his originality, principally first and foremost the student and successor of his German teacher [Leibniz]."[122]

Understood within the context of the limitative judgment and the problem of continuity, it can now be understood why Cohen would claim that he is presenting a particular version of the day-view of reality in Fechner's sense, that is, a maximally open definition of reality. By thinking of any particular through limitative judgments, that is, by considering any particular "in opposition to its opposites," Cohen allows for an infinite field of possibilities that can further be specifically delimited through different concepts.

In this regard, Cohen's definition of judgment can retain a relation to the unknown, affective and divine, but of a form different than Fechner's, since by considering any aspect of reality through limitative judgment, we progressively delimit its definition, and this process is open to criticism at every stage. The goal of "reality" approaches the noumenal both in the meaning of what the plenum of reality "is," which can never of course be known, and in how it inflects the particular moment of the Real as change, which can similarly never be fully known.[123] It is thus parallel to negative theology in assuming the existence of its objective, whether defining humanity or experience, but not presuming to explain this reality in absolute or substantive terms. Cohen thought more detailed forms of limitative judgment could, in a manner similar to Hegel's definition of force, precisely define any positive concept. The means through which Cohen imagines limitative judgments to define all forms of judgment is never fully revealed, however, and appears to be beyond the capabilities of the philosophical logic of his period.[124] The premises of Cohen's thought, however, allowed him to suggest how humanity and experience are defined negatively by structures of knowledge, yet at the same time are immanently given as an enigmatic presence. The numerous speculative and methodological leaps in his argument were left to his student Cassirer, by far the more authoritative writer on natural science and the better stylist, who attempted to demonstrate both the validity of Cohen's core argument and develop its central premises on a stronger footing. Even as Cassirer also developed a critique of key aspects of Cohen's work, the foundational concepts of Cohen's philosophy established the basic direction of Cassirer's work to the end of his career. This can be briefly suggested by surveying Cassirer's reception of the theme of the infinitesimal itself, which will then be situated in the broader framework of Cassirer's early work in the coming chapters.

Cassirer's Transformation of Cohen's Late Philosophy

In *Substance and Function* (1910), Cassirer develops Cohen's theme of the relation of the integral and differential in a key moment of his definition of representation and judgment as the foundation of epistemology and the sciences.[125] For his 1906 habilitation under Alois Riehl and Wilhelm Dilthey, Cassirer had "The Concept of the Limit and its Knowledge-Theoretical Meaning" listed as his first recommended theme for the sample lecture; his second was "Substance Concept and Function Concept," the full title of his later work.[126] Both Cohen and Cassirer recognized the continuity of the lectures, but the first in particular suggested the infinitesimal theme of Cohen's work, and the two related lecture topics were integral to Cassirer's written work during this period.[127] Thus Cohen wrote from vacation in St Moritz on first reading *Substance and Function* that even if he were not able to complete the second volume of his *System of Philosophy* this would not matter, since "the project is in its main outlines developed in this book."[128] This was particularly true, Cohen wrote, for the "general method of the leading thought and *the prevalence of the meaning of the infinitesimal.*"[129] Thus, even as Cassirer places far more emphasis on group theory and later aspects of mathematics than the infinitesimal proper, Cohen clearly saw the general meaning of the work as a direct development of his infinitesimal project. Cassirer's exposition, however, provided to Cohen "the most transparent clarity" and "liveliness" of exposition possible – both features Cohen was well aware his work lacked.[130] Cassirer had also succeeded, Cohen continued, in another area where Cohen found his own work fell short: integrating the Marburg project with the most recent developments in all of the sciences. Yet, within this praise Cohen also notes that their "unity is endangered" by Cassirer's significant shift of emphasis *within* the infinitesimal problem, particularly within its application to the judgment of reality, a problem to which we will turn in our extended reading of *Substance and Function* in Chapter Five.

A culminating statement of *Substance and Function* establishes what Cassirer calls nothing less than a "logical universality of the supreme principles of knowledge [*Erkenntnisprinzipien*]," as it has developed in a preceding survey of the contemporary sciences of geometry, physics, chemistry and psychology.[131] It is this "logical universality" that effectively recasts Cohen's reading of the infinitesimal problem. Summarizing it, Cassirer writes,

> The fact must be granted unconditionally, that the particular "presentation" reaches beyond itself, and that all that is given *means* something that is not directly found in itself; but it has already been shown that there is no element in this "representation," which leads beyond experience as a total system. Each member of experience possesses a symbolic character, in so far as the law of the whole, which includes the totality of members, is posited and intended in it. The particular appears as the differential, that is not fully determined and intelligible without reference to its integral.[132]

As with Cohen, the figure of calculus is here used as a model for a general epistemological assumption of a negative determination of the particular that at the same time suggests the reality of the particular. It demonstrates that "all that is given *means* something that is not directly found in itself," a theme that is applied not only to the exact sciences, but to the entire range of experience. Cassirer's early definition of "symbol" as it derived from Cohen and Leibniz is at the center of this definition: a symbol negatively defines the particular in relation to the whole, and allows a particular "reality" to appear within the set of all possible realities.

There is, however, a subtle but vast difference between Cassirer's statement of the role of the calculus in this example and Cohen's use of it. Cassirer begins by writing that: "The fact must be granted unconditionally [*ist unbedingt zuzugestehen*], that the particular 'presentation' reaches beyond itself."[133] Where Cohen claimed to *prove* the logic behind the infinitesimal and calculus, and then use it to connect all fields of knowledge, Cassirer with this statement suggests that this theme is rather a necessary first *assumption* and principle for a particular mode of philosophizing. By "unconditional" Cassirer means literally not conditioned by specific assumptions about intuition or thought, sense or concept, subjective or objective meaning, since this will be an initial premise *preceding* such definitions in a functional definition. With this one turn, Cassirer bypasses many of the systemic, logical and methodological problems raised by Cohen's *Infinitesimal Method*, while still retaining the basic argument that grounds the later *System of Philosophy*. Cohen presumably read "must be granted" as merely a claim of wholehearted agreement with his philosophy, but Cassirer clearly means it as a fundamentally different claim for a "logical universality."

Any number of philosophies, of course, from empiricism to many forms of idealism would not "unconditionally" grant this definition of representation, much less treat it as a "fact," but Cassirer will argue that his preceding chapters have demonstrated both its value and indeed, under his own definition of the term, its facticity in being the best explanatory model for the array of forms of modern logic, science and ultimately experience that he canvassed in his argument. Cassirer repeatedly emphasizes, for instance, that once philosophy draws itself within the solipsism of the Cartesian ego it is completely consistent for it to never be able to reconnect with the world.[134] As with a calculus of indications in later systems theory, the first assumption of Cassirer's "ultimate principle" is of an undifferentiated field of meaning, and the second assumption is that within this field distinctions can be drawn that order this field negatively on every level.[135] For this reason, Cassirer does not need to claim – as Cohen had – that his philosophy could be literally mapped out through all of the implications of calculus, as for instance Cohen had in the *Infinitesimal Method* by claiming that our understanding of how all forms of objects could be defined through higher orders of calculus.[136] Calculus is instead for Cassirer principally a superb model for understanding a general principle of consciousness that is open to nearly universal implication and creative reinterpretation.

For Cassirer, the claim of this use of the infinitesimal model accomplishes nothing less than to overcome the opposition of both the "psychological immanence of impressions" and "the metaphysical transcendence of things" – precisely as we would expect following Cohen's precedent.[137] Using the model of judgment epitomized by the relation of integral and differential is the best means of developing a logic of indications that can investigate how worlds are functionally constructed or revealed without first presuming their substantive natures. Cassirer demonstrates this process, to use an example we will further develop later, in the evolution of the concept of the atom in relation to chemical experimentation, through the models of Democritus, Lavoisier and Dalton through to Planck and Einstein. The "symbol" of the atom provides a specific schema for revealing its reality, and further investigating it, even as this symbol proves to change radically over time – neither the particular definition of "atom" as "smallest part of matter" or later "electrical field," for instance, nor the horizon of reality in which it exists, are simply given but rather they are only revealed within a web of relations of ideas and experimentation.

The importance of this "ultimate principle of knowledge" and model of representation for Cassirer's philosophy cannot be overstated, since this definition forms the starting point for his theory of judgment as well as of perception, and his epistemology as well as the horizon of this epistemology with his later critique of ontological philosophy and later theories of myth and language. Only through means of this precise definition and its use of the figure of the infinitesimal is Cassirer able, as he writes in *Substance and Function*, to overcome the "the kernel of all misconceptions among the various epistemological tendencies" that is found in the opposition of "subject and object" and its corollary in the "opposition of thought and experience."[138] Cassirer further notes that with this claim any transcendent reality, particularly the claim to a transcendental noumenal reality, is dropped, for transcendence would suggest a lack of relation to the immanent whole: "An absolutely lawless and unordered 'something' of perceptions is a thought, that cannot be realized [even] as a methodological fiction; for the mere possibility of consciousness includes at least the conceptual anticipation of a possible order, even though the details may not be made out."[139]

Cassirer provides a brief but powerful statement of how not only natural science but ultimately everyday perception will be affected by his functional and relation approach. Cassirer writes that both the mathematical and physical concept "cannot be comprehended, as long as we seek any sort of presentational correlate for it in the given; the meaning only appears when we recognize the concept as the expression of *pure relation*, upon which rests the unity and continuous connection of members of the manifold."[140] In a more poetic style, Cassirer notes that this does not preclude "real experience" being at the center of this system:

> All that the "thing" of the popular view of the world loses in properties, it gains in relationships. [It is] connected inseparably by logical threads to the totality of experience. Each particular concept is, as it were, one of these threads, on which we string real experiences and connect them with future possible experiences. The objects of physics: matter and force, atom and ether can no longer be misunderstood as so many new realities for investigation, and realities whose inner essence is to be penetrated, when once they are recognized as instruments produced by thought for the purpose of comprehending the confusion of phenomenon as an ordered and measurable whole.[141]

The world, in short, is a manifold of functional relations that are capable of unfolding with unlimited complexity and increasingly comprehensive viability.[142]

Cassirer later provides a particularly pithy and epistemologically focused summary of this problem in the first volume of his *Symbolic Forms* project, entirely consistent with the core idea of *Substance and Function*, even as it more explicitly broadens it from the conceptualization in science to all experience. Characteristically, Cassirer's description is considerably clearer than Cohen's and thus worth citing here at length. In it we can see how Cassirer, like Cohen, essentially retained a Leibnizian reading of the infinitesimal problem while rejecting Leibniz's assumption of the monad as the basis of unity. Although limited only to the epistemological problem of consciousness, Cassirer will elsewhere make it clear that this description applies to experience itself before any division of subject and object. Cassirer writes,

> The element of consciousness is related to the whole of consciousness not as an extensive part to a sum of parts, but as a differential to an integral. Just as the differential equation of a moving body expresses the trajectory and general laws of its motion, we must think of the general structural laws of consciousness as given in each of its elements, in any of its cross sections – not however in the sense of independent contents, but of tendencies and directions which are already projected in the sensory particular. This precisely is the nature of a

content of consciousness; it exists only in so far as it immediately goes beyond itself in various directions of synthesis. The consciousness of the moment contains reference to temporal succession; the consciousness of a single point contains reference to space as the sum and totality of all possible designations of position; and there are countless analogous relations through which the form of the whole is expressed in the consciousness in particular. The "integral" of consciousness is constructed not from the sum of its sensuous elements (a, b, c, d...) but from the totality, as it were, of its differentials of relation and form (dr_1, dr_2, dr_3, dr_4...).[143]

In this brief summary, Cassirer essentially presents a condensed version of the epistemological foundation of his work as well as its debt to both Leibniz and Cohen.[144]

Cassirer's early work uses the Leibnizian concept of symbol to define any moment of representation, and correlatively defines form as the particular relation of a unified experience to its component aspects. The two terms will later fuse into a new concept of "symbolic forms," but in Cassirer's early work each has a distinct and crucial meaning. Although form is broad enough to encompass what Kant would define as forms of understanding in the categories, Cassirer will also use it to define any particular constellation of knowledge whatsoever, such as in the "form" of the atom for Democritus or the form of legal representation in Roman Law. The later concept of "symbolic form" defines at once a narrower definition and a broader problematic. It is a specific study of how forms of knowledge articulate different modes of reality in language, myth and science. Cassirer explicitly links this project to *Substance and Function* and describes it as a "broadening" of the concept of theory to encompass not only "the scientific world view" but "the natural world view implicit in perception and intuition" and the "mythical world" that "disclosed relationships, which [...] are by no means without their laws and reveal a structural form of specific and independent character."[145]

It is not surprising that perhaps the most important moment of continuity between Cassirer's early and late projects will be the reading of judgment and reality through the figure of the infinitesimal. The continuity of Cassirer's later and earlier arguments in relation to Cohen's idea of the infinitesimal is perhaps clearest in Cassirer's analysis of the meaning of "*symbolische Prägnanz*" in the third volume of his *Philosophy of Symbolic Forms* on the "phenomenology of knowledge." Symbolic forms grasp very different modes of experience of the world, largely through the expressive, representative and significative function of the three principle symbolic forms, loosely correlating to the mythical, perceptive and scientific modes of knowledge. Linking these various modes of symbolic formation is the theme of symbolic prägnanz. Symbolic prägnanz describes how the symbol combines sensuous immediacy with meaning to form a given reality: it is then a direct development of Cohen's exploration of the process through which "reality" becomes immanent, but it has expanded to include mythological and expressive dimensions, as well as representational and significative ones. As has been noted by commentators, this is perhaps the most important philosophical theme in Cassirer's symbolic form project, but it is vexing precisely in its concision. The German term *Prägnanz* has no exact equivalent in English (although it is now an adopted term from gestalt psychology) and is only loosely translated by "pregnance." It combines, as John Krois notes, the German *prägen* – to impress, mint, give contour or characterize – with the Latin *praegnens*, which means "full of" or ready to give birth.[146]

Cassirer uses the theme of reality and the infinitesimal in the central part of his description of symbolic prägnanz to describe how any aspect of perception is both immediately "real" even as it is also shaped by its context. The core paradox of symbolic prägnanz is how it defines what was earlier termed the Real, that is, how "perception as a sensory experience contains at the same time a certain nonintuitive meaning which it immediately and concretely represents."[147] Reality is now clearly in the domain of "life" overall, not the scientific reality that was the focus of *Substance and Function*. Symbolic prägnanz exhibits how "perception itself [...] by virtue of its own immanent organization, takes on a kind of spiritual [*geistige*] articulation – which, being ordered itself, also belongs to a determinate order of meaning. In its full actuality, its living totality, it is at the same time life 'in' meaning."[148] Symbolic prägnanz will explain how matter and form, intuition and thought, are united in the immediacy of perception of the real to create not just different objects, but different forms of meaningful and particular experience as expressed in the "physiognomic" feeling of immediacy of a complex whole ("the physiognomy of the city") and "feel" of a moment.[149] As in Cohen's definition of the continuum of the "Real" with "reality," this can only occur because we always already have an inchoate sense of "reality" as an immanent form of life: "Reality could never be deduced from the mere experience of things if it were not in some way already contained and manifested in a very particular way, in expressive perception."[150] It is a "certainty of a living efficacy [...] primarily apprehended here [as] life as such far more than any individual spheres or centers of life."[151]

Cassirer provides several deliberately simplified examples of symbolic prägnanz that can help clarify the concept, and the manner in which it is effectively a capstone of his development of Cohen's concept of the infinitesimal. Consider the same "optical experience" of an undulating line, which could be considered through symbolic prägnanz in three vastly different manners. "Such an experience," Cassirer writes, "is never composed of mere sensory data, [but] as sensory experience it is always the vehicle of a meaning and stands as it were in the service of that meaning. But precisely therein it is able to perform very different functions and through them to represent very different worlds of meaning."[152] Taken purely as a "physiognomic" occurrence, and thus in the function of expression, the line would be present as a "particular mood [...] expressed in the purely spatial determination."[153] Such a reading is not merely subjective for Cassirer: "The form gives itself to us as an animated totality, an independent manifestation of life." The same line could also be considered as a "geometrical schema, a means of representing a universal geometrical law. Whatever does not serve to represent this law, whatever appears as an individual factor in the line, now becomes utterly insignificant; it has departed, one might say, from our field of vision."[154] Finally, if we take the line "as a mythical symbol," it might embrace "the fundamental mythical opposition between the sacred and the profane. It is set up in order to make a separation between the two provinces, to warn and frighten, to bar the uninitiated from approaching and touching the sacred."[155]

Cassirer uses the theme of the infinitesimal in the central part of his description of symbolic prägnanz to describe how any aspect of perception is both immediately "real" even as it is also completely shaped by its context in a manner that is epitomized by the infinitesimal relation. The core paradox of symbolic prägnanz is again the manner in which it defines how "perception as a sensory experience contains at the same time a certain nonintuitive meaning which it immediately and concretely represents."[156] Cohen had defined the paradox of this "certain nonintuitive meaning" as a purely negative ideal pole of the "Real," as that which was pure change in itself. In a central passage of his later philosophy, Cassirer redefines this concept within a phenomenology of symbolic prägnanz:

> The analysis of consciousness can never lead back to absolute elements: it is precisely pure relation which governs the building of consciousness and which stands out in it as a genuine *a priori*, an essential first factor. It is only in the reciprocal movement between the "representing" and the "represented" that knowledge of the ego and of objects, ideal as well as real, can arise. Here we feel the true pulse of consciousness, whose secret is precisely that every beat strikes a thousand connections. No conscious perception is merely given, a mere datum, which need only be mirrored; rather, every perception embraces a definite "character of direction" by which it points beyond its here and now. As a mere perceptive differential, it nevertheless contains within itself the integral of experience. This integration, this apprehension of the totality of experience starting from a single factor, is only made possible by definite laws that govern the transition from one form to another.[157]

In symbolic prägnanz, the earlier appearance of the particular "object" in the "ultimate principle of theoretical knowledge" is broadened to a phenomenology of experience of the becoming of any particular, but the basic scheme is analogous. Perception is a lived and immediate reality, here described as the differential, intimately interwoven with the whole of experience, here represented as the integral.

Cassirer defines symbolic prägnanz as a process through which the mythical, artistic or mathematical, in which an experience occurs, takes on "a certain nonintuitive meaning" as the *haecceitas*, the "thisness" in which it is "immediately and concretely" presented. Similarly, Cassirer notes, the process of imagining past or future time is always presented as itself a symbolic prägnanz in which the future, for instance, is not "jointed to the sum of present perceptions as given to us in the now" but rather "presents itself as a wholly distinct form of vision."[158]

Cassirer is essentially here restating, albeit in a far more approachable manner, Cohen's original objection to the separation of intuition and idea, reality and sense. We do not "first" have sense or perception, and then a reaction or meaning to it, but rather sense and perception always first "stands as it were in the service of that meaning." It is for this reason that Cassirer finds "matter" and "form" completely fungible: the aspect of the "matter" of line that appears simply given can completely change in relation to the form under which it is perceived, even as the moment of particularity of some aspect of "matter" will always be present.[159] A phenomenology of these experiences demonstrates that sensation is codetermined within the structure of meaning, even as meaning presents itself foremost as the immediacy of sense, and both exist as part of an already given world.

Against contemporary claims that continued to define his work as a straightforward extension of the Kantian project, Cassirer claims that the phenomenology of symbolic prägnanz leads his philosophy beyond the dichotomies presented by Kant (such as sensation/understanding) and Husserl (such as the hyletic/noetic distinction).[160] An actual understanding of this bold claim within his brief argument, however, appears to assume knowledge of Cassirer's earlier philosophy and its intrinsic relation to Cohen's late philosophy of origin, where precisely this transition beyond Kant (and to some extent Husserl) had already begun. Cassirer's footnotes to this section do indeed return the reader to these earlier sources in his work, linking the use of the integral in symbolic prägnanz to the earlier citation on epistemology, which in turn is introduced by a claim that the reader must return to *Substance and Function*. Only in this manner can the reader understand the means through which "the world of 'reality' is constructed for us [through] *functions* by means of which a particular form is given to reality."[161] Only here too can it be seen that this is never a construction from whole cloth or by an insular subject, but one that always reveals a reality that precedes any definition of subject and object. In this regard, symbolic prägnanz is not an entirely new aspect of Cassirer's philosophy, but a transformation in this new forum of his first principle and basic insight.

Notes

1   I 1.
2   "Neo-Kantianism," in *Encyclopædia Britannica*, 14th edition, XVI: 214–15.
3   "Neo-Kantianism," 214.
4   Ibid.
5   FF 52.
6   On comparison of Leibniz's "form" with that of Descartes, see L 155 (int); form as developmental, 186.
7   On the close relation of Simmel and Cassirer's theory of form, see the collected essays in the *Simmel Newsletter* 61, including my contribution, "The Problem of Physiognomy and the Development of Cultural Theory in Georg Simmel and Ernst Cassirer," *Simmel Newsletter* 61 (1996): 44–56; Skidelsky, Edward. "Simmel and Cassirer From Epistemology to Cultural Criticism: Georg Simmel and Ernst Cassirer" in *History of European Ideas* 29:3 (September 2003): 365–81; Blumenberg, H. "Geld oder Leben: Eine metaphorologische Studie zur Konsistenz der Philosophie Georg Simmels," in *Ästhetic und Soziologie um die Jahrhundertwende: Georg Simmel*, ed. H. Böhringer and K. Gründer (Frankfurt: Vittorio Klostermann, 1976), 121–34.
8   Schnädelbach, Herbert. *Philosophy in Germany, 1831–1933*, trans. Eric Matthews (Cambridge: Cambridge University Press, 1984), 78; EK3 364ff.
9   Schnädelbach, *Philosophy in Germany*, 78; I 10–12, 84–6, 156ff.
10 EK4 285; Lamprecht, Karl. *Moderne Geschichtswissenschaft* (Freiburg: Heyfelder, 1905), 16; Lamprecht, Karl. *Die Kulturhistorische Methode* (Berlin: R. Gaertner, 1900), 12ff.
11 Dilthey, Wilhelm. *Introduction to the Human Sciences: An Attempt to Lay a Foundation for the Study of Society and History*, trans. Ramon J. Betanzos (Detroit: Wayne State University Press, 1988 [1923]), 96.
12 Ibid., 96, 98.
13 Tolstoy, Leo. *War and Peace*, trans. Nathan Haskell Dole (Cambridge: T. Y. Crowell, 1889), book 3, part 3, ch. 1, 284. Similarly, he later writes somewhat more cryptically that history like the natural sciences has to "put aside the question of causation [...] and seek for laws. History

14 EK 4 285.
15 Fechner's Log Law is $S = c \log (I)$, where sensation (S) is proportional to the log of stimulus magnitude (I) and c is a constant of proportionality related to Weber's constant. Weber's constant – developed starting in 1834 to explain the difference threshold of a number of different senses (carried weights, sounds, etc.) that proved not to be constants – describes for a specific sense the relation of the difference threshold as a function of magnitude for a given sense.
16 On Fechner, see particularly Heidelberger, Michael. *Nature from Within: Gustav Theodor Fechner and His Psychophysical Worldview* (Pittsburg: University of Pittsburgh Press, 2004).
17 Lange, Friederich Albert. *Geschichte der Materialismus und Kritik seiner Bedeutung in der Gegenwart* (Leipzig: J. Baedeker, 1902), 193.
18 Fechner, Gustav Theodor. *Die Tagesansicht gegenüber der Nachtansicht* (Leipzig: Breitkopf und Härtel, 1904), 16.
19 Windelband, Wilhelm. *A History of Philosophy with Especially Reference to the Formation and Development of its Problems and Concepts*, trans. James H. Tufts (New York: Macmillan, 1926 [1893]), 644–5.
20 Heidelberger, Michael. *Nature from Within: Gustav Theodor Fechner and His Psychophysical Worldview*, trans. Cynthia Klohr (Pittsburg: University of Pittsburgh Press, 2004), 3.
21 Fechner, *Die Tagesansicht*, 92ff.
22 Fechner, Gustav Theodor. *Nanna oder Über das Seelenleben der Pflanze*, ed. Kurd Lasswitz, 3rd edition (Hamburg: Verlag von Leopold Voss, 1903), 10.
23 KE 374.
24 Cohen had noted this quality of mathematics as "hypothesis" as early as his *Platos Ideenlehre*: "Hypotheses serve as 'construction images' [*Konstruktionsbilder*] for the methodical handling of things. [...] Aristotle understood correctly: mathematics stands in the middle between sense and ideas." HCS I:366.
25 EL 6.
26 Kuntze, Friederick. *Die kritische Lehre von der Objektivität: Versuch einer weiterführenden Darstellung des Zentralproblems der kantischen Erkenntniskritik* (Heidelberg: Carl Winter's Universtitätshandlung, 1906), 249ff.; EL 2.
27 U 114.
28 Berkeley, George. *The Analyst*, ed. David Watkins (Dublin: Trinity College Dublin, 2002 [1734]), 18. Online: http://www.maths.tcd.ie/pub/HistMath/People/Berkeley/Analyst/Analyst.pdf (accessed 18 March 2013); see also Cassirer's critique of Berkeley's reading of calculus (EK2, 302ff.). Here Cassirer notes Berkeley stressing not the failure of intuition but rather thought: "No reasoning about things whereof we have no ideas, therefore no reasoning about infinitesimals."
29 Cohen claims the first modern recognition of the "real" correlate of infinitesimal motion is in Galileo, presumably either in the "First Day" of *Galileo's Dialogues Concerning Two New Sciences* (1638) where Salviati, Galileo's spokesman, observes that a curve is made of infinite indivisibles, or in his description of acceleration in falling motion, which assumes the actual process of falling as the correlate of the equation (EL 32, 44).
30 For an exhaustive study of the failure of Berkeley's arguments with Leibniz, and the latter's continuity with twentieth-century logic, see Mikhail G. Katz and David Sherry, "Leibniz' Infinitesimals: Their Fictionality, their Modern Implications and their Foes from Berkeley and Russell and Beyond," in *Erkenntnis* (forthcoming) 1–55. My thanks to Mikhail G. Katz for sharing this article with me.
31 I 30.
32 Cassirer, Ernst. "Newton and Leibniz," *Philosophical Review* 52 (1943): 366–91; EK3 149ff., 401ff.; ECW 4 125ff., 394ff.

33 Ibid, 385.
34 Cohen, Hermann. *Schriften zur Philosophie und Zeitgeschichte*, vol. 1 (Berlin: Akademie Verlag, 1928), 229.
35 EK4 35.
36 EL 129; EK4 37–54.
37 EK4 35; letter from Leibniz to Conti, *Opera Omnia*, Dutens, III, 446.
38 U 115–16; EL 55.
39 Kant, Immanuel. *Immanuel Kant's Critique of Pure Reason*, trans. Norman Kemp Smith (New York: St Martin's Press, 1965), 402ff. (A434).
40 Kant, *Critique of Pure Reason*, 328 (A340).
41 Kant *Critique of Pure Reason*, 402ff. (A434).
42 K 202.
43 K 204.
44 K 202.
45 K 206.
46 EL 27.
47 FF 35.
48 FF 34.
49 U 114–15.
50 EL 18.
51 SF 36; U 90 EK4 66.
52 SF 59.
53 "A partition of the rational numbers into two non-empty parts A and B, such that all elements of A are less than all elements of B and A contains no greatest element. The cut itself is, conceptually, the "gap" defined between A and B. In other words, A is every number between the cut and any number lower than the cut, and B is every number between the cut and a number greater than the cut. The cut itself is in neither set." Dedekind, Richard, "Continuity and Irrational Numbers," *Essays on the Theory of Numbers* (Dover: New York, 1924 [1901]), 13ff., 33; SF 124.
54 U 115.
55 KI 32n1.
56 EL 30.
57 EL 2.
58 EL 32, 44.
59 Kuntze, in *Die kritische Lehre*, appears to have first noticed this relation between Cohen and Maimon. Cohen recognizes the affinity in S 389, no. 12. Jakob Gordin's historical introduction to the infinite judgments similarly gives Maimon, not Kant, the central role as the predecessor to Cohen (*Untersuchungen zur Theorie des unendlichen Urteils* (Berlin: Academie-Verlag, 1929), 16ff.). Gordin emphasizes the variety of solutions Maimon provides to the problem before hitting on the central role of the judgment of reality, which he claims Cohen ultimately clarified (33–7).
60 EK3 80.
61 EK3 86.
62 EK3 95; Gordin, *Untersuchungen*, 14ff.
63 EK3 94.
64 EK3 81.
65 EK3 118ff.
66 EK3 80.
67 EL 157–8.
68 Sieg, Ulrich. "Althoff und die deutsche Universitätsphilosophie," in *Wissenschaftsgeschichte und Wissenschaftspolitik im Industriezeitalter: Das "System Altoff" in historischer Perspektive*, ed. Bernhard von Brocke (Hildesheim: Verlag A. Lax, 1991), 304ff.; Sieg, Ulrich. *Aufstieg und Niedergang des Marburger Neukantianismus: Die Geschichte einer philosophischen Schulgemeinschaft* (Würzburg:

Königshausen & Neumann, 1994); Ringer, Fritz. *The Decline of the German Mandarins: The German Academic Community, 1890–1933* (Hanover, NH: Wesleyan University Press, 1991), 51ff.
69 As Cassirer repeatedly notes, the "fundamental merit" of Cohen's Kant interpretation is that he "fully and clearly described this relationship" of the "system of synthetic principles" for the first time" (K 175).
70 K 175.
71 Kant, *Critique of Pure Reason*, 148 (B202).
72 Kant, *Critique of Pure Reason*, 201 (B207).
73 EL 28.
74 U 114–15.
75 PSF1 100, 104–5.
76 K 178.
77 K 180.
78 The difference in definition of "quality" is telling for the wider argument. In the anticipation of perception, "quality" is delimited to mean simply that "in all appearances the real which is an object of sensation has intensive magnitude, that is degree," as epitomized by the speed of a body or its temperature. As an overarching term for the triad of categories, quality is more broadly defined pertaining to the logical category being: as existence, nonexistence or – and this category will prove both the most important and the crossover term – limitation. For Cohen, however, the two definitions tie nicely together, since ultimately the application of *any* category to experience entails the principles, and thus the anticipation of perception. In this regard it, and its logical corollary in the category of limitation and infinite judgment, are the catalyst for redefining all of Kant's system.
79 Kant, *Critique of Pure Reason*, 88 (B97).
80 Kant, *Critique of Pure Reason*, 108 (B 97).
81 I 40ff.; SP1 210ff.; Funkenstein, Amos. *Theology and the Scientific Imagination from the Middle Ages to the Seventeenth Century* (Princeton: Princeton University Press, 1986); Funkenstein, Amos. *Perceptions of Jewish History* (Berkeley: University of California Press, 1993), 271–84.
82 EK2 20ff.; SP1 31ff.; see Maimonides, Moses. *The Guide for the Perplexed*, trans. M. Friedländer (New York: Dover Publications, 1956), 83.
83 Wolfson, Harry A. "Infinite and Privative Judgements in Aristotle, Averroes, and Kant," *Philosophy and Phenomenological Research* 8, 2 (1947): 186.
84 EL 18.
85 EL 29.
86 Brown, G. S. *Laws of Form* (New York: Julian Press, 1972); Baecker, Dirk. "Introduction" in *Problems of Form*, ed. Dirk Baecker (Stanford: Stanford University Press, 1999), 2ff.
87 EL 36.
88 I 37.
89 This argument is developed in Funkenstein, Amos. "The Persecution of Absolutes: On the Kantian and Neo-Kantian Theories of Science," *The Kaleidoscope of Science: The Israel Colloquium for the History and Philosophy of Science* 1 (1986): 329–48.
90 I 37.
91 I 34.
92 I 13. The key importance of Leibniz for the infinitesimal problem is this relation to the problem of reality, which Kant, Cohen claims, "developed so little, if not to say even left hidden away." Jakob Gordin's reconstruction of Kant's reading of the infinitesimal judgment similarly downplays Kant's role (Gordin, 4–14, 32ff.).
93 "Neo-Kantianism," in *Encyclopædia Britannica*, 14th edition, XVI: 214–15.
94 EL 43.
95 I 144.
96 Ibid.

97 SP1 138. On Cohen's development of the judgment of reality, see Holzhey, Helmut. *Cohen and Natorp: Band 1 – Ursprung und Einheit, Die Geschichte der "Marburger Schule" als Auseinandersetzung um die Logik des Denkens* (Basel: Schwabe, 1986), 246–67.
98 Cohen to Cassirer, St Moritz, 24 August 1910. See Dimitry Gawronsky's dissertation *Das Urteil der Realität und seine mathematischen Voraussetzungen* (Weimar: Hof-Buchdruckerei, 1910) for an alternative interpretation of *Substance and Function* on this topic.
99 FF 160.
100 Ironically, one of the only ways in which the neo-Kantian theme of the "Real" is recognized in contemporary thought is through its use in the psychoanalytic theory of Jacques Lacan, where it was developed on the model of psychotic patients into a general, and radically antirational, model of psychic structure. For Cohen and Cassirer, on the other hand, the model is the foundation for a probabilistic and rational theory of hypothetical means of engaging in the world, with the model being the function of hypothesis within a larger reality. On Lacan and Cassirer, see Lofts, Steve G. "L'Ordre Symbolique de Jacques Lacan à la Lumiere du Symbolique D'Ernst Cassirer," in *La pensée de Jacques Lacan: questions historiques, problèmes théoriques*, ed. Antoine Vergote, Steve G. Lofts (Louvain: Peeters, 1994), 83–106. On the role of probabilism in liberal thought of the turn of the century, see Coen, Deborah R. *Vienna in the Age of Uncertainty: Science, Liberalism and Private Life* (Chicago: University of Chicago Press, 2007).
101 ETK 423; I 3. For more on the relation of intuition and thought, see: Holzhey, *Cohen and Natorp*, 128ff.
102 Funkenstein, *Theology*, 52–3.
103 Cohen's reading of Müller stands in contrast to the "unquestioned dominance" in the late nineteenth century of the psychological reception theory of Müller's work and his "specific nerve energies," particularly as spread through the work of Heinrich Helmholtz and the early editions of Friedrich Albert Lange's stunningly popular *History of Materialism* (EK4 4); Schnädelbach, Herbert. *Philosophy in Germany*, trans. Eric Matthews (New York: Cambridge University Press, 1984), 97–8. For Cassirer's in-depth criticism of this phase of psychologism in relation to spatial perception, see SF 287ff.
104 SP2 156–7. "Diese erste Anlage zum Bewusstsein wird die Grundlage unserer Psychologies bilden. Die empfindung wird an die zweite Stelle treten, erst die zweite Stufe bilden müssen. Sie hat zur Voraussetzung jene erst Stufe der allgemeinen Anlage, des notwendigen Vorbegriffs, des Ursprungs" (ibid., 157). Cassirer, Ernst. "Student Notebooks: Cohen – Psychologie" n.d. (env. #138), gen mss. 98, series IV, box 56, folder 1108, 1.
105 SP3 133–4.
106 Ibid. Cohen himself notes that the term "disposition" is far preferable to the German term *Anfang*, and that it is intended as a translation of the Greek *arché* (SP1 79).
107 Nietzsche, Friedrich. *Twilight of the Idols: or How to Philosophize with a Hammer*, trans. Duncan Large (Oxford: Oxford University Press, 2009), 17ff. Nietzsche places this under the general rubric of confusing cause and effect, which leads to the conclusion that "the apparent world is the only one; the real world has just been lied on" (17).
108 Cassirer's reading of this problem is in large part a rereading of Cohen's use of calculus to, as Cassirer describes it, replace all spatial relations of "next-to-each-other" (*Beisammen*) with temporal relations of "after-one-another" (*Nacheinander*). From: "Die Begriffsform im mythischen Denken" (1922) in ECW16 48. On the central role of calculus in Cohen's work see SF 99 and I.
109 I 74. For a wider ranging study of this role of motion and its appearance in Cohen's work, see also Meyerson, E. *Identity and Reality*, trans. Kate Loewenberg (New York: Dover Publications, 1962), 229.
110 "Die Temperaturempfindung setzt den Reiz voraus, auf den sie die Antwort ist" (SP3 135).
111 ETK 426.

112 ETK 426.
113 ETK 423; I 28.
114 SP1 122.
115 The classical locus of this problem was, in Cohen's reading, Galileo's grasping the simple reality of the law of falling bodies (I 50). For a further description of this theme, see Giovanelli, Marco. *Reality and Negation: Kant's Principle of Anticipations of Perception: An Investigation of Its Impact On the Post-Kantian Debate* (New York: Springer Verlag, 2011), 189ff.
116 Funkenstein, Amos. "The Persecution of Absolutes"; Gordon, Peter. "Science, Finitude, and Infinity: Neo-Kantianism and the Birth of Existentialism," *Jewish Social Studies* 6, 1 (1991): 30–53; Kinkel, W. "Das Urteil des Ursprungs," *Kant-Studien* 17 (1912): 274–82; Gordin, *Untersuchungen*, 142ff.
117 Funkenstein, *Theology*, 40; PSF3 306.
118 "Und diese Materie ist uns, nach der Tagesansicht unseres erkenntniskritischen Idealismus, lediglich eine Inhaltsgruppe des Bewusstseins. Diesen Inhalt beschreibt die Materie, und zu dieser Beschreibung bedarf sie des Differentials" (I 228).
119 EL 89.
120 I 146.
121 K 426.
122 E 24.
123 KE 423.
124 To the degree that Cohen tries to ground his understanding of limitative judgment in Leibniz's infinitesimals it could be argued that Leibniz in fact provided a more substantial argument for the foundation of limitative judgments and continuity in all judgments than Cohen himself. On the modernity of Leibniz's theory of infinitesimals, which the authors see as consistent with the modern logic of Robinson's "hyperreals" and "the mathematical implementation of Leibniz's heuristic law of continuity," see Mikhail G. Katz and David Sherry, "Leibniz' Infinitesimals: Their Fictionality, their Modern Implications and their Foes from Berkeley and Russell and Beyond" in *Erkenntnis*, 55.
125 Tellingly, Cassirer introduces his notes to the planned conclusion of the "fourth" volume of the symbolic project with a restatement of this problem in relation to the whole: "We start with the concept of the whole: the whole is the true (Hegel). But the truth of the whole can only be grasped in a particular 'aspect.' This is 'knowledge' in the *broadest* sense – 'seeing' the whole 'in' an aspect, through the medium of this aspect. With this, the problem of representation becomes the central problem of knowledge" (PSF4 193). Cassirer goes on to describe this, the starting point of his philosophy, as opposed to Mach's work, the relation of the part to the whole in his description is "a development (negative) against the *falce* concept of an element: Mach, etc." (PSF4 193, no. 3).
126 Cassirer, Ernst. "Meldung zur Habilitation, 10 May 1906," in *Habilitanden-Buch* 12, journal no. 250, Bestand Phil. Faculty 1228, Humboldt University Archive, Berlin.
127 Cohen refers to extended conversations on this theme in Cohen to Cassirer, 24 August 1910, St Moritz. Ernst Cassirer Nachlass, Beinecke Library, Yale University, gen ms. 355, box no. 2, folder 45; Gawronsky, Dimitry. "Ernst Cassirer: His Life and Work," in *The Philosophy of Ernst Cassirer*, ed. P. A. Schilpp (Evanston, IL: The Library of Living Philosophers, 1949), 21.
128 Ibid.
129 Ibid. My emphasis.
130 Ibid.
131 SF 300.
132 Ibid.
133 SF 300.
134 SF 295ff.
135 Jakob Gordin appears to suggest already the potential of Cohen's work to form something like a calculus of indications that later formed the basis of systems theory: "The infinite judgment

should and can be the methodological fulcrum of philosophy, a guide through the structure of systems of philosophy" (Gordin, 133).
136 I 147.
137 SF 300.
138 SF 296.
139 Ibid.
140 SF 166.
141 Ibid.
142 Not coincidentally, Cassirer finds this insight most analogous to the work of Gustav Fechner, the founder of psychophysics and the champion of the model of the infinitesimal. Fechner, although plagued by "an inner obscurity in the definition of the objects of physics," nonetheless recognized the fundamental principle that "the constant reaching out beyond any given, particular content is itself a fundamental feature of knowledge" (SF 301–2). The further implication, that "the world of phenomenon is such that one phenomenon can exist only in and through another" has led many, as Cassirer quotes Fechner, to "deny real existence to phenomena in general" or to presume something "behind" phenomena that provides their "basis and kernal" (SF 301). Instead, reality is defined only by the relation of part to totality, or as Fechner phrases it with a characteristic mysticism: "The whole is the basis and kernal of whole and of all that is in it" (SF 301, citing Fechner, Ludwig. *Über die physikalische und philosophische Atomenlehre*, 2nd edition (Leipzig, 1864), 111ff.).
143 PSF1 104–5.
144 Thus discussing the problem of the infinitesimal method in relation to mechanical dynamics and then acceleration, Cassirer gradually broadens its description to nearly all of the sciences Leibniz considers under the model that "the relation of the elements to the construction [*Gebilde*] that arise from this problem of continuity are, as one can see, represented in their scientific generality by the relation of the differential and the integral" (L 171ff.; ECW 1 154ff.).
145 PSF3 xiii.
146 Krois, John Michael. *Cassirer: Symbolic Forms and History* (New Haven: Yale University Press, 1987), 53.
147 PSF3 202.
148 Ibid.
149 PSF3 73.
150 Ibid.
151 Ibid.
152 PSF3 200–201.
153 PSF3 200.
154 PSF3 201.
155 Ibid.
156 PSF3 202.
157 PSF3 202–3.
158 PSF3 202.
159 PSF3 9.
160 PSF1 193. Cassirer claims symbolic prägnanz forms the key moment within Kant's synthetic unity of apperception and Husserl's phenomenology. The transcendental unity of apperception guarantees the connection to the events of the world – including any mode of self-perception – into a whole and thus enables the fundamental possibility of perception. "Perception 'is' only insofar," as Cassirer notes, "as it takes on determinate forms" (PSF3 194). The dichotomies of both Kant and Husserl, Cassirer claims, can be overcome in his own theory of symbolic prägnanz, which will suggest the original phenomenology beneath distinctions of sensibility and thought, noetic and eidetic

stratums of relations (PSF3 198). The limitation of Kant's philosophy had been that ultimately, although "perception is only insofar as it takes on determinate forms," it cannot describe how "meaning exists in forms and not before them" (PSF3 194). This limitation had been concealed by a faulty use of eighteenth-century faculty psychology: "Thus here again 'receptivity' and 'spontaneity,' 'sensation' and 'understanding' might seem to be conceived as 'psychic faculties,' each of which exists as an independent reality but which, then – in their empirical cooperation, their causal concatenation – bring forth experience as their product" (PSF3 196). In doing so, however, they violate Kant's transcendental premise of understanding "not so much objects as our mode of knowing objects" as far as possible *a priori*. By assuming faculties as things he merely displaces the problem of meaning to the question of the "manufacture of forms." Conversely, from a position that at first appeared "diametrically opposed," Husserl ran into similar problems. He had argued that in fact all of consciousness is always already intentionality: "It embraces a distinctive direction, a determination *Bestimmung* of meaning," and meaning *per se* is the nature of consciousness itself (PSF3 196). This intention cannot be grounded on any other function and particularly not on a "thing," "image" or faculty that precedes it without being caught in a vicious circle. "We always arrive at antinomies [...] when we forget that the fundamental relationship of representation and intention is the condition of the possibility of all objective knowledge [...]. Consequently, no objective fact or occurrence belonging to the world of things which is first made possible by this relationship may be drawn into the description of it" (PSF3 197). Yet to retain the purity of his analysis, Husserl separated the strata of pure intentional consciousness, the *noetic* mode of phenomena, from that of all others, the material or *hyletic* (PSF3 19).

161 PSF3 203n13; PSF1 91n5.

Part II

CRITICAL SCIENCE AND MODERNITY

Chapter Four

LEIBNIZ AND THE FOUNDATION OF CRITICAL SCIENCE: *LEIBNIZ'S SYSTEM IN ITS SCIENTIFIC FOUNDATIONS* (1902)

"Leibniz's Metaphysic," Cassirer writes in *Leibniz's System*, "is the first complete expression of the consciousness of the modern era," but we could add that for Cassirer this consciousness had only haltingly been recognized.[1] Although presented as a historical exegesis, Cassirer's *Leibniz's System in its Scientific Foundations* (1902) presents an interpretation and critique of Leibniz's work as the foundation of the system of knowledge, one that refutes the dominant Cartesian philosophy of Leibniz's era and its contemporary successors in materialism and positivism. It is a strong interpretation – in both senses of the term – that was largely consonant with both Cassirer's own functionalism and the work of the Marburg school. As such, it provides a lucid and broad historical introduction to both Cohen and Cassirer's work, as well as to the Marburg school's broader vision of the role science in the development of German society. Cassirer effectively suggests how the logic that stands at the base of the infinitesimal method might be of relevance for all of the sciences and how the sciences in turn relate to ethics, aesthetics and psychology. In particular, the work lays out the foundation for a developmental and antiessentialist definition of humanity, and demonstrates how the progress of humanity could be instituted as part of a broad program of academic and political reform consonant with Cohen's program and "metapolitics."

As the historian Ulrich Sieg has noted, *Leibniz's System* can be read as both a "popular" description of Cohen's often arcane philosophy and an argument for its continuity with the German tradition of science and critique as it began with Leibniz.[2] Along with the first volume of *The Knowledge Problem* four years later, it can be described, as Thomas Meyer claims for the later work, as a "philosophical historical position paper" anchoring and summarizing the key positions of the Marburg school.[3] In stark contrast to Cohen's presentation, Cassirer developed a lapidary explanation of the core features of this argument, which as an expository history has the added advantage of being able to withhold any truth claims for Leibniz's philosophy. Tellingly, Cohen praised the work, along with the first two volumes of *The Knowledge Problem*, as establishing the historical foundations of German classical philosophy – and by implication the Marburg interpretation of that philosophy – and noted further that Leibniz's work was the only context in which Kant can be fully understood.[4] An initial understanding of the metaphysics and politics of Leibniz's system was a necessary precondition for understanding the late Marburg school, as Cassirer's early work evidences from *Leibniz's System* through *Freedom and Form*.

The reception of *Leibniz's System* suggests both the continuing challenges faced by the Marburg school and the perception of such a text as intrinsically linked to the Marburg project, despite its lack of any reference to it. Tellingly, the essay on Leibniz was awarded a second place by the Prussian Academy with high praise – even though no first place was awarded.[5] Although perhaps in part motivated by the rising anti-Semitic climate, particularly in relation to Cohen, and the reactionary dislike of the school's politics, there is some truth to the stated grounds that the text presents a brilliant but somewhat one-sided view of Leibniz – a view that essentially establishes his work as the foundation of the Marburg school itself.

The Centrality of Leibniz in Cassirer's Early Career and Thought

In Cassirer's reading, Leibniz's philosophy is the signal development that defines the modern trajectory of central European thought. Cassirer's early career is defined by his reception and attempts at popularization of Leibniz's work, and indeed it is telling that his fame initially rested as much on his Leibniz scholarship as on his neo-Kantianism.[6] In addition to *Leibniz's System*, Cassirer attempted to facilitate the reception of Leibniz by translating and publishing his central works in two popular and inexpensive translations entitled *Leibniz: Key Works for the Foundation of Philosophy*.[7] Even late in his emigration to the United States, Cassirer continued to stress the importance of Leibniz for his own work and for an understanding of German modernity. In an essay from 1943 entitled simply "Newton and Leibniz," Cassirer attempts to bridge the gap between his Anglo-American reception and his original arguments by suggesting the importance of Leibniz for his new North American audience.[8] Whereas the Anglo-American tradition is dominated by the philosophical trajectory from Newton through Locke, Cassirer writes, Germany's relation to the Enlightenment and Romanticism is only comprehensible through the work of Leibniz.[9] The two represent, Cassirer writes, a place where "modern thought had reached a parting of ways," yet – and Cassirer leaves this premise unstated – for most North Americans Newton's work effectively constitutes "common sense" while Leibniz's work remains a bizarre curiosity.[10] Yet from the perspective of *fin de siècle* science and functionalism, it was Leibniz's philosophy that was more philosophically durable, and for Cassirer Leibniz's contemporary quality extended to fields such as law and logic, social science and the humanities.

Cassirer's late attempt to define the divergence in the Western philosophical tradition signified by Newton and Leibniz may also have been an attempt to open the way for a reception of his own work – and that of Cohen – in North America, for without the awareness of Leibniz's "alternate" definition of modernity these philosophies are only haltingly comprehensible. Leibniz's philosophy changed markedly throughout his life, and many of its key elements are open to multiple interpretations in a manner that defies summary.[11] Cassirer's own reading of Leibniz, however, remained similar throughout much of his career – indeed, it can be read as forming a baseline for his philosophical endeavor. In what follows, we will summarize Cassirer's reading of Leibniz as much as Leibniz's own philosophy, with the understanding that this reading was only one approach to Leibniz and was specifically tailored to address Cassirer's own era. In the interests of simplicity, we will use not only *Leibniz's System*, but also Cassirer's own later congruent statements on Leibniz, such as his descriptions in the *Problem of Knowledge*, *Freedom and Form* and the *Philosophy of the Enlightenment*, since these are often written in a more abbreviated form.[12]

In Cassirer's histories, Leibniz's work is portrayed as synthesizing the negative theology of Nicholas of Cusa with Renaissance natural philosophers such as Giordano Bruno, Galileo and da Vinci, while also responding to the challenges of Hobbes, Spinoza and Newton.[13] The result is an unusual reading of Leibniz that emphasizes the continuous creative transformations of his thought, his focus on immanent processes as well as practical matters and – perhaps most surprisingly – the relation of his thought to skepticism. Cassirer, for instance, sees Leibniz's thought as influenced by Montaigne and Pierre Bayle, and as critical of his rivals Hobbes, Spinoza and Newton largely on the basis of the assumed metaphysics in their systems.[14] For Cassirer, it is characteristic of Leibniz that in his long polemics against Newton, his key point was not that he was opposed to the working of Newton's physics, but to the process "by which he, like the scholastics earlier did, places qualities at the base of things, when there is no reasonable calculation that leads to their foundation."[15]

As Cassirer depicts it, Leibniz's work pervaded the environment of eighteenth-century German culture, but did so at first through the popularized, systematized and deeply distorted version presented in the voluminous works of Christian Wolff.[16] This reception of Leibniz was followed by the reception of his aesthetics in the Romantic movement, notably in the Swiss aestheticians and in Kant's later work, that in Cassirer's view formed the countervailing tendency in German intellectual history to Wolff's rationalism.[17] After its initial problematic popularization by Wolff, Leibniz's actual philosophy and writings remained in a sort of chrysalis through the modern period.[18] Cassirer saw Cohen's work as first opening this chrysalis, and his own early scholarship as making Cohen's interpretation open to a wider public. As always in the Marburg school, in Cohen's words, "the systemic and the historical in philosophy are interrelated": the present fabric out of which we think is itself historical, and rethinking historical philosophy in turn redefines contemporary systemic issues.[19] From Cassirer's perspective this was the legitimate ground on which to develop the most philosophically powerful reading of Leibniz, even as certain aspects of Leibniz's original philosophy were downplayed or not mentioned at all.

Leibniz's contemporary importance, Cassirer argued, was redoubled by his new relevance for the exact sciences, which was in turn an intimation of his larger importance for redefining the nature of science itself. Throughout his career, Cassirer held that the basic insights of Leibniz's system were confirmed by contemporary physics in the successive variants proposed by Planck, Einstein and then by Paul Dirac and others in quantum electrodynamics. Thus, in his 1923 *Einstein's Theory of Relativity*, Cassirer anchors Einstein's theory in Leibniz's turn from defining "the truth of knowledge [...] from a mere pictorial to a pure functional expression," and argues that Leibniz's basic insights are born out by Einstein.[20] Cassirer notes that this change "appears in the setting of a metaphysical system, in the language of a monadological scheme of the world" that needed to be purged of its "unproven metaphysical assumptions," but he also insists that the modern definition of truth, function and reality effectively developed from this system. The recent work of Christiane Schmitz-Rigal on Cassirer's "open constitution" of knowledge in physics and Karl-Norbert Ihmig on Cassirer's invariant theory of truth in many ways confirm Cassirer's strong Leibnizian focus in his philosophy of science, a focus he presumably developed in part to confirm Cohen's late work.[21]

Leibniz and Modernity

Cassirer's broadest summary of Leibniz's modern worldview, noted earlier but worth repeating, states, "The foundation of all reality is a system of forces, of free-functioning [*freiwaltenden Tätigkeiten*] activities. But just in this, that these forces were grasped as forms, demonstrates that they carry in themselves a rule of inner connection and self-limitation."[22] The "rule of inner connection and self-limitation" will broadly define what Leibniz refers to as "functions," that is, possible rules of relation between forces. Forces are never known in themselves but only through forms. Forms could be initially defined as coherent sets of functions grasped as a whole, which precisely as sets are capable of redefinition and transformation. The clearest model of mapping "forces" through form is mathematics and mechanics, and calculus is the first modern historical example of how this is done. That is, within the multiple interlocking functional assumptions that shape assumptions about a moving body (i.e., is unified, is affected by gravity, etc.), the calculus describes a particularly clear relation between function and force.

As in Cohen's use of calculus, so Cassirer writes that for Leibniz the problem at the heart of calculus is however of far greater applicability than that of mathematical physics and constituted "a general critique of the 'given' [*Gegebenheit*] that went beyond its particular field to address a theme of general logical importance."[23] Cassirer notes at the beginning of the text that his historical interest in Leibniz arose from seeking the "logical foundations" of mathematics and mechanics, a project whose contemporary meaning he will address in *Substance and Function*.[24] On pursuing the problem, he continues, he realized that the entirety of Leibniz's philosophy was crucially linked with the modern sciences, particularly the problems of the infinite and dynamics.[25]

This relation also included the formation of the modern human sciences (*Geisteswissenschaften*), where Leibniz's thought defines the break of "ethics, history and aesthetics" from their "particular and earlier forms," and provides the foundation for nearly all of German classicism.[26] Cassirer depicts Leibniz's logic of science as equally important in the social sciences, where it effectively forms the catalyst for the break from early modern forms of politics and sociation. In Cassirer's interpretation, Leibniz's influence can moreover be detected nearly everywhere in German classicism. Herder, for instance, came to define the aesthetic moment of Leibniz's thought as the most important aspect of his own system.[27] Partially through Herder, Leibniz was in Cassirer's reading the key influence on modern German historicism.[28]

While Cassirer's reading of the natural sciences and logic appears to situate the book as a successor to Cohen's *Infinitesimal Method*, his reading of the human and social sciences links this project clearly with the later volumes of Cohen's *System of Philosophy* on ethics, aesthetics and psychology. Although Cassirer's analysis of Leibniz often highlights analogous moments in Cohen's philosophy, however, he is careful to avoid Cohen's bold attempt to definitively claim that the logic at the base of the infinitesimal – and each of the sciences – is that of the limitative judgment. He instead demonstrates the parallel non-Aristotelian logic of each science Leibniz develops, and notes a parallel of negative constitution of the individual or force by its context based on the problem of continuity. In his own work, as we will see with *Substance and Function*, this transformation is carried even further, as here the logic at the heart of the calculus and limitative judgments are clearly described as initial "suppositions" from which a powerful unified philosophical and methodological perspective can develop, not necessarily given truths in themselves.

The basis of Leibniz's insight into the negative relation of part to whole, and the infamous peculiarity of his system, is the concept of the monad. Monads are, in his familiar phrase, "the living mirrors of the universe."[29] This mirroring and fusion by the monad quite simply, but subtly, *is* its identity. From a human perspective – to take the most intuitive initial approach to the monadology – the basic experience of the monad is consciousness itself, which is a form of unity created entirely of difference, and which cannot be defined as a spatial "thing" or, in Leibniz's view, accurately reflect upon itself as a substantial ego. This "mirroring" does not reflect an "absolute object" as we might imagine in the analogy of an object reflected in a common flat mirror.[30] Cassirer notes that Leibniz's model is rather, in a characteristic figure for the Baroque period, a concave or anamorphic mirror, and the monad is the meeting point of this mirror's reflections – it is the limit point of all given interrelations of the whole from a particular perspective.[31] Moreover, as the theodicy emphasizes, this "mirroring" is not simply a reflection of a given state of affairs, but is a "living" part of a process in which judgment and imagination illuminates new connections, new functions and forms, within reality. Experientially, perception does not work out, so to speak, from a given something, but rather has a horizon of possibility shaped by what might be termed a constant imaginative guesswork making connections in the world. Leibniz defines his philosophy as shaped by the two problems of "infinity" and "unity" in this sense: a diverse infinity of relations is creatively unified as an open plenum, rather than a closed totality, from a certain perspective.

Neither Cohen nor Cassirer retain Leibniz's literal definition of the monad, but both see its intrinsic importance as a summary of Leibniz's logic and phenomenology of experience. As noted earlier, Cohen himself had argued at the end of *Kant's Theory of Experience* that his work used Leibniz as the key corrective to Kant.[32] Cassirer attempts to demonstrate this connection, and this priority for Leibniz, for the Marburg school tradition.

Cassirer's reading in *Leibniz's System* culminates in a long chapter entitled "The Concept of the Individual in the System of the Human Sciences [*Geisteswissenschaften*]," which suggests the vast implications for the human and social sciences of Leibniz's logic.[33] Cassirer's interpretation of Leibniz holds that the individual – which is to say the individual monad, which in the human sciences concerns the entelechy of human consciousness – is determined (*bestimmt*) by and through the forces of the human society, and that these can be grasped and potentially transformed by the various functional forms developed in both common sense and, on the next order, the human and social sciences. In each case, the individual is defined essentially as analogous to a "force" that is negatively determined by its place within a set of processes or functions as they coalesce through different forms As we will see in the next chapters, this Leibnizian understanding of modernity and form, and critique of its limits, dominates Cassirer's own work and translation of Cohen's thought from *Substance and Function* through *Freedom and Form* – texts, indeed, whose very titles are not intelligible without this Leibnizian background, as it is first defined in *Leibniz's System*.

Leibniz's Philosophy as Functionalist Critique of Cartesian Substantialism

Cassirer began his academic career by outlining the background of Leibniz's transition from a world of substances to a world of forces with the simultaneous publication of his essay *Descartes' Critique of Mathematical and Natural Scientific Knowledge* (originally written in 1899) and its continuation in *Leibniz's System* (1902).[34] A similar juxtaposition is used to conclude the first volume of *The Knowledge Problem* (1906). As a brief sketch of Cassirer's understanding of Leibniz, we can outline his reading of Leibniz's opposition to Descartes' philosophy through Leibniz's critique of Descartes' three fundamental substances of ego, extension and God. The consequences of Leibniz's rejection of Descartes' substantialism are profound, since it allows him to abandon the tacit but "inflexible set of fundamental ideas" at the base of Descartes' work.[35] Instead, Leibniz's close logical critique of these ideas allowed him to focus on experience as a set of "procedural-directions" (*Verfahrungsweisen*), so that the "claim for improvable givens is replaced by the reference to fundamental syntheses."[36] With this move, the problem of judgment and the focus on the conditions of experience (that is, reality) become the centerpiece of Leibniz's thought, and the search for ever more fundamental forms of judgment and synthesis the central priority of his philosophy. By using the contrast of Descartes and Leibniz as our guide, we can also contrast a traditional North American history of philosophy, which often places Descartes' substantialism at the starting point of Western thought and then follows it with the successive philosophies of Newton and Locke, with the German emphasis on a Leibnizian tradition. Such an approach can allow us to avoid mistaken tacit assumptions, frequently made in his North American reception, about the starting point of Cassirer's own philosophical endeavor on the dichotomy of subject and object, the "interiority" of self versus the "externality" of space, and the mundane world in opposition to the transcendence of God. All are foreign to the historical foundations of Cohen and Cassirer's philosophy in Leibniz, as even the fundamental divisions first assumed by the figures of the self, spatiality and divinity are woven together in his philosophy.

A. Descartes' cogito *and Leibniz's monads*

Descartes held that the *cogito*, or "I think," intuitively reveals the substance of the ego in the immediacy of thought given to reflection. The "given-ness" of the *cogito* is further valid because it appears to exist even if all else is assumed to be destroyed – a modification of the nominalist argument *toto mundo destructo* used for God's existence.[37] For Leibniz, the "ego" or "soul" of Descartes is transposed into the "monad," defined as a temporally unfolding entelechy. On closer inspection, Leibniz argued, it is not the unity of the abstraction of the ego that is given in reflection, as Descartes had it, but the multiplicity of its contents. The unity in difference of these contents is considered by Leibniz as a fundamental and ungraspable mystery of experience, never knowable as a substance or ego in any form.[38] "The concept of the I," Cassirer simply and directly states, as we noted earlier, "is foreign to [Leibniz's System]."[39] Leibniz's concept of the monad is for Cassirer "distinguished by the fact that it conceives of the relationship between unity and multiplicity, between duration and change, as a pure correlation. Genuine knowledge, accordingly, cannot be knowledge of the enduring or of the mutable; it must exhibit the correlation and mutual interdependence of these two elements."[40]

The correlation of the infinite plenum of relations with the evanescent movement of the particular moment in the monad is analogous "in its scientific generality [...] to the relation of a differential to its integral in calculus."[41] Just as in calculus the relation of integral to differential reveals a "real" phenomenon that cannot be described as substance, but only as a process relating a limit ideal of unity to difference, so too with monadic consciousness. The central role of the problem of continuity, the theme Cohen thought he had isolated through limitative judgment, for Leibniz was at the heart of both phenomenon, and indeed formed the catalyst for understanding all of experience anew. Calculus is however not meant to be hypostasized as some sort of "reality" that defines the content of the multiplicity of all possible correlations, and certainly is not supposed to suggest that all of reality is defined by a mathematical metrics.[42] Rather, it is a model that vividly enables a description of "change as a basic element of determination" through a new set of conceptual tools. Both the monads and calculus are for Leibniz "a well-founded fiction" (*fiction bien fondée*) that reveals with particular vividness a "new fusion of ideality and reality in which a new concept of reality is created."[43]

When transposed to human society, Leibniz's theory of monads reveals what might be considered a modern transposition of earlier negative theology and anthropology: the conscious entelechies of which Leibniz's universe is made cannot be described positively, or in the manner of "things," but are rather defined negatively by the forms of knowledge and activity in which they are placed. Although this at first appears counterintuitive, its consequences are immensely practical: the universe is understood as actions and relations of actions rather than things, and knowledge of such actions or forces is always provisional. "To the most central tenets of [Leibniz's] system," Cassirer summarizes, "belongs the idea that being, and particularly spiritual [*geistige*] being, is only revealed and disclosed in activity."[44] In rejecting Descartes' idea of the ego, Leibniz reveals the cogito as activity, and indeed essentially as action in the world. For Cassirer, this approach will transform the human sciences by transforming the focus of concern from stable patterns onto practices, actions and institutions.

Leibniz held that monads "have no windows" into the reality of other monads or consciousnesses, but in Cassirer's reading monads are also not capable of reflexive circumspection into their own infinity in a manner that would lead to anything resembling a substance in Descartes' sense. The idea of the "ego" in itself is only grasped in the same way as any other object, and as such has no priority over any other object epistemologically. "Thought can only be understood on the basis of the content in which it is represented," Cassirer writes of Leibniz, which entails that the "the unity of consciousness only allows itself to be constituted in the unity of the object."[45] This unity of the object as "ego" has no particular meaning for Leibniz beyond any unification in the minds grasping towards any object, or better objective, in the world.

Cassirer saw Kant's reception of Leibniz, and the necessary relation of their two philosophies, as helpful for clarifying this aspect of Leibniz's philosophy.[46] Leibniz's understanding of the monad was refined through the work of Kant to form the basis of the distinction of the transcendental and empirical ego. Kant's transcendental ego in Cassirer's reading defines simply the mysterious factum of connection in the world at all and is thus equivalent to Leibniz's monad as a unity of apperception, or "correlation" of unity and difference. The empirical ego is always shaped by contingent forces and accessible only within the mediation of these forces and is thus again equivalent to our knowledge of any other object.[47] Summarizing Kant's work, Cassirer writes, "Even the self is not given to us originally as a simple substance: the idea of it only arises in us on the basis of the same synthesis, the identical functions of unification of the manifold by which sense-content becomes the content of experience, impressions become object."[48] The "individual" of both the transcendental ego and of the "thing in itself" are equally unknowable, even as both drive the formation of knowledge and understanding. The conclusion of this aspect of Kant's thought in Cassirer's Leibnizian reading is that "the ego has no preeminence and no prerogative over and above the other facts attested to by perception and empirical thought. For even the self is not given to us originally as a simple substance."[49]

In place of any definition of substance, the prime focus of Leibniz's philosophy in Cassirer's reading becomes how different forms of knowledge define experience as "correlations" of actions themselves. Indeed, for this reason Cassirer's reading focuses on the value of Leibniz's work as a form of phenomenology, a focus only on the presentation of experience in correlation. The apparent exotic animism of Leibniz's monadology – the entire universe, from rocks to plants, from humans to stars, is assumed to be made of monads – is often taken as the weakest and most speculative aspect of his system. For Cassirer, however, it proves on closer inspection to be the basis of a profound skepticism and return to first principles: there can be no initial assumption of consciousness as a different form of unity from any other form of correlation.

The concept of the monad – although itself ultimately metaphysically problematic for Cassirer – performs the invaluable historical service of allowing all assumptions about the difference of animate and inanimate life, "internal" and "external" experience and numerous other assumptions of Western philosophy to be bracketed. Leibniz's philosophy avoids psychological and empirical assumptions, and avoids indeed any preconceived notion of reality itself. Since ancient Pyrrhonism, and with renewed interest since Montaigne's "Apology for Raymond Sebond," the difference of human and animal consciousness, and indeed animate and inanimate reality, has been a touchstone of skepticism: is there ultimately any way to know that a rock is not in some sense conscious? Far from obfuscating such problems, Leibniz's focus on the monad as correlation allows him to start anew investigating and categorizing aspects of experience. Whereas Descartes had a "double truth" of the world and human will, in which the basis for neither is discoverable in itself, Leibniz uses the basic logical form at the heart of calculus to establish a unified perspective on experience.[50]

The most basic feature of monadic experience, indeed the key assumption Leibniz makes about monadic correlation, is its principally anticipatory or "prospective" character as defined by the problem of continuity, through which any particular aspect of experience is intrinsically related, indeed determined, by an immanent "beyond" of what is given. This relation of continuity is not assumed to be simply a feature of human consciousness but is, as the calculus demonstrated, taken as a fundamental feature of the universe. Just as in constructing a circle we necessarily suggest the indefinite continuum of the space "outside" of the circle, so for Leibniz all phases of experience – including the most basic perception – involve an immanent relation to an indefinite horizon in which they are situate, a horizon of that which is not known, is possible or is tacitly assumed. Although such relations appear most readily in negative constructions, the import of continuity is not limited to the "negative" at all, but rather describes the positively and immanently given moment of experience as a "correlation" intensively connecting a particular instance of consciousness to the horizon of all of its possible relations in the world.[51]

Leibniz's work for Cassirer thus does not yield an abstract idealism, but rather his skepticism provides the basis for continually refining perception of the immanent world. Appearances that were previously overlooked – often because they were, so to speak, caught between ideality and reality – can be recognized anew. Consciousness is recognized as a complex and unfolding diversity in unity rather than reified, that is, taken as a thing; motion – since before Zeno an infamous source of confusion – is understood as pure change rather than a successive change of "places"; the infinitesimal is taken as a "real" phenomenon rather than as a ghost; and so on across the range of human experience. For this reason Cassirer describes Leibniz's philosophy as a form of "methodological seeing" (*Schauens*) in which in the strongest sense "thought is first determined in its objective meaning."[52]

This process is epitomized in what Cassirer describes as Leibniz's "deepest scientific development of the idea" in the concept of hypothesis, that is, a reflexive means through which thought fuses with intuition as a mode of questioning and of experiential, and experimental, synthesis.[53] Hypothesis allows new forms of appearance of phenomenon to be recognized in the world, but also guarantees that these are never taken as simply "given," but as codefined by idea. Although epitomized by scientific "hypothesis," this process is inherent in all experience altogether. Here subject and object, rationality and empiricism are considered coextensive or correlative, and the focus of philosophy is determining the means of judgment through which their unity appears in experience.

The monad is used not only as a means to critique the concept of the Cartesian ego, but also in fields as apparently divergent as in its use as mathematical function or to define "force" in mechanics.[54] This contrast of psychological and mathematical or physical uses of the monad at first appears to be a deeply problematic aspect of Leibniz's thought. Cassirer acknowledged that this translation of the concept of the monad through different fields appears at first to present a "riddle" in Leibniz's work, but "this riddle is solved in itself" once it is remembered that Leibniz's philosophy is based on a nominalism that holds no fundamental opposition of subject and object, rationality and empiricism.[55] Distinctions such as the ideals of mathematics, the worldly application of physics and the "inner" aspects of consciousness in relation to monads are only themselves hypotheses in the first place. In this reading, the "well-founded fiction" of the monad is a placeholder or foundational hypothesis in Leibniz's thought that allows for the reading of each particular moment of pure correlation of experience within form, and prevents any mistaken notion of subjectivity or objectivity – indeed of "substance" at all – from taking the place of pure change. Leibniz's use of the concept of the monad is not a means of reducing all phenomenon to one reality, as occurred in atomism for material phenomenon or in Descartes' concept of the ego for mental phenomenon, but is on the contrary a means for recognizing a vastly increased number of possible realities that can be defined in the universe. It is not an erasing of distinctions between fields of endeavor or realities, but the critical basis of first establishing these distinctions anew.

One of the few assumptions of Leibniz's system is that phenomenon or moments of correlation can be subordinated to broader laws understood as hypothetical judgments – the famous principle of sufficient reason – so that, as he puts it simply, "things are subordinated to laws," in the full sense that they are understood as based on processual functions. This process is equivalently present in the "laws of mechanics or of metaphysics."[56] This general set of rules will define the unity in difference found in the world through the application of calculus to mechanics, and the corollary claims of forces or "simple substances," as well as the possibility that these are from another angle "living perceptions."[57]

Leibniz's ultimate response to the Cartesian concept of the ego and subject in Cassirer's view thus encompasses the breadth of his philosophy as critical idealism: "The concepts of subject and object can, to the degree that they logically have any right to be used at all, have meaning only by demonstrating their position in regard to the fundamental relations from which we start: 'the distinction between them is first to be developed from fundamental questions [*Grundfragen*], not the reverse.'"[58] In this regard, not only does Leibniz reject Descartes' concept of the ego as subject, he completely redefines the centrality of the subject in Western philosophy altogether. In its place, he establishes a study of correlations and relations from which in certain circumstances definitions of the subject or object first occur.[59]

B. Space, time and form

Descartes' second definition of "substance," extension, is grounded in his invention of analytic geometry, the triumph of which in Cassirer's view influences nearly all of Descartes' thought. On the one hand, Cassirer notes that analytic geometry demonstrated for Descartes a new means, unknown to the ancients, through which "to fight against the given material [of the world]," while on the other it provides a "regulative criteria" for a method of thinking inspired by the coordinate system.[60] Similarly to Descartes, Leibniz's concept of space is anchored in a new mathematical innovation, that of calculus, but the logic of calculus transforms not only the concept of the substantial definition of extension, but with it that of space, perception, force and indeed reality.

For Descartes, extension is defined by the "next-to-each-other" (*Beisammen*) of identical points, which in turn form lines and planes.[61] The basic principle is the congruence of identical parts. Extension is then "substantialized" in that it is assumed to exist in and for itself, and it is further assumed, as Cassirer phrases it, that the "concept of the body is the demand for a persistent basis for all change," as Descartes famously demonstrated in his example of melting wax.[62] The basis of change is given "in" bodies, which is what properly qualifies extension as "substance," or in the medieval sense as *in se et per se esse* (being in and for itself). Descartes' geometry, and later his physics, are then attempts to further analyze and develop this fundamental thought.[63]

Cassirer sees an "inner contradiction" in Descartes' natural philosophy however, in that from the beginning it claims to exclude empirical sensibility in dealing with the extended world, yet then often makes tacit use of it, particularly in his work on physics and biology.[64] Criticized by Newton and others as a failure of overweening idealism, the problem in Cassirer's view was rather that Descartes "could not make space into a principle of determination [...] because he transformed it in the development of his system into a determined thing."[65] Rather than illuminating a given reality, as occurred with Galileo's law of falling bodies or later Newton's principles, Descartes "strayed from his own philosophical principles" of the constructive role of the method in favor of a false notion of an absolute space.[66]

In contradistinction to the extended space of Descartes or the later absolute space of Newton or Clarke, for Leibniz space and time are always relational and phenomenal. They operate as an active principle of determination of particulars. In part, this relational definition follows from the definition of both the point and of the monad. In Cassirer's view, this argument represents not only a more encompassing and powerful definition of space, but a more fruitful and "critical" use of the relation of mathematics in relation to reality.[67] By defining the monad solely as relational, Leibniz avoids from the start the assumption of material substances, much less a sole substance of extension. Space is instead entirely defined by the negative concept of the "difference" of elements, which presents a difference in unity in any particular aspect of space as a distinct form of correlation. The point for instance is a moment of infinite determination in relation to its context, not a discrete entity in itself. Cassirer summarizes Leibniz's relational definition of space as follows: "It lies in the character of space as such, that all unification, which is possible in it, at the same time is a separation. The 'next-to-each-other' [*Beisammen*] of the aspects of space are at the same time a 'difference-from-each-other' [*Auseinander*]; every part is that which the other is not and is there, where the other is not."[68] The system of space is a system of difference, the unity of which occurs only from various monadic perspectives.

This relational definition provides for both the new mathematics of topology, developed more distinctly from Leibniz's premises by Leonard Euler (1707–1783), and a definition of "space" itself that encompasses any relational scheme.[69] Descartes' vision of uniform extension based on the analytic geometry proves to be only one of many possible forms of a spatial organizing system. Space is defined solely by the difference between different aspects of a system, and can thus be defined in an infinite variety of ways without reference to any pre-given "substance." Indeed, mathematical "extension" itself proves to be one of many kinds of space. The historian of science Amos Funkenstein summarizes Leibniz's definition in a useful manner:

> When Leibniz said that there is nothing more to space and time than the notion of the difference between things he probably meant the following. Assume a possible world (U) with three "things" (a, b, c) and three properties (F1, F2, F3), in it. Assume further that a "thing" must have at least one property, and that (a) has all of them, (b) all but the last, and (c) only the first. This is the only content of spatial expression that places (a) and (b) closer to each other than to (c) and also equidistant from (c). In a world of more properties and more things the mapping could turn three-dimensional.[70]

With Leibniz, appearances are no longer understood merely by matching an intrinsic relation of subject with predicate, a "thing" with "characteristic." Rather the expression of one or more "inner conditional relations" (*Bedingungszusammenhänge*) of intensive relations redefines the mere "togetherness" (*Beieinander*) expressed in any element of the world into a dynamic relation.[71] Space is defined by these relations, and different elements of space can be open to permutation through different readings, in Leibniz's term symbolizations, of configurations of space. As an extreme case, Leibniz's model seems to defy the commonsense definition of "space" altogether: the three terms "orange," "just" and "aromatic" could be said to define a space for him. Similarly, three items with various properties could be said to constitute different "spaces" depending on how these properties are symbolized and ordered. A cherry, a tomato and red meat, for instance, might be considered "close" if symbolized as "edible," in a progressive order if symbolized as "savory" or "wide apart" if symbolized and ordered by the category "vegetable."

In Cassirer's reading the value of such a reading is again its nominalism – its reduction of all terms of definition to their bare minimum – and with this its ability to promote skepticism about accepted elements of the old system and creativity in defining forms of science and reality. That this apparently fantastic mode of defining space had real advantages can be suggested by drawing some examples from everyday experience and ideas of space, which then can be useful in understanding Leibniz's concept of function and form as it is developed by Cassirer. If one is driving according to legal rules on a modern highway and misses an exit ramp, a Cartesian definition of space would hold that one is, say, 70 meters past the ramp in absolute space. A Leibnizian definition would, however, hold that it is precisely as accurate to say that within the spatial system of the road one is as far from the ramp as the entire return circuit to the next exit and then back to your original starting place, say seven kilometers. Even taking this close analog to Cartesian space – the metric space of the planar and linear highway system – the predicates that go with certain aspects of a highway dictate that "really" one is quite far from the exit that was just missed. Space, like time, is only based on the difference between its moments within a particular framework.

Leibniz's definition of space is in turn closely connected to his definition of motion. For Leibniz, the concept of "motion" encompasses any unification of multiplicity whatsoever. "'Motion'," Cassirer writes of Leibniz, "is no longer a single empirical data, but rather a general principle, the idea of which serves to describe the constructive emergence of complexity [or compounds, *das Zusammengesetzte*] from simple elements [*dem Einfachen*]."[72] As such it will imply a transitional state not only within space, but also between any forms of becoming whatsoever. The development of an elm from an acorn, or of a word from letters, is as much "motion" as is physical movement, and each will have their optimal form of "calculus" to illuminate these transitions. In this way, Leibniz can readily consider his mathematical "calculus" as an initial and particular proof of the validity of any form of combinatorics that unifies diversity. On this basis, motion is used by Leibniz to define the characteristic feature of both consciousness and the coordination of consciousnesses in pre-established harmony.

This definition of motion allows us to further see that Leibniz's definition of space is integrally related to his definition of time. The calculus, of course, suggested their equivalence by focusing on "an infinitesimal time–space moment."[73] But this is itself because they are defined under the broadest concept of relation. As Leibniz writes, "Space and time, extension and motion are not things, but rather of modes of consideration [*modi considerandi*]."[74] Space and time are thus not only not substances for Leibniz, but not even thing-like "properties or features."[75] Rather space and time are themselves only ever inherent as "functional forms" themselves: space is "the possibility of being-next-to [*Beisammen*]," just as time is the "possibility of following in a particular way [*Nacheinander*]," as it appears in a particular form.[76]

C. Descartes' God and Leibniz's theodicy

In Descartes' system, the third major substance is the infinite and omnipotent God, whose existence is largely proved through medieval ontological arguments and the argument from perfection in the *Meditations*. The structural role of God in Descartes' system is as a regulator between the other two substances of ego and extension in that the deity serves at once as a means of 1) guaranteeing that we are not deceived by appearances, 2) linking the cogito with extension and 3) situating clear and distinct appearances within our apprehension of nature. As with Leibniz's critique of Descartes' ego and extension, here too the transition from a substantial to functional definition determines aspects of Leibniz's system that initially appear far removed from the initial theme of the deity. Indeed, Cassirer places Leibniz's theodicy in a reading of the human sciences towards the end of his survey in *Leibniz's System*, since in his reading Leibniz's theodicy is the basis for a future and secular system of knowledge that dominates the modern era.

Although Leibniz's theodicy often appears as one of the more dated aspects of his system, for Cassirer it marks the key transition to the modern worldview. Leibniz's concept of God "puts away all dualistic separations that exist between the mundane and the spiritual, the worldly and the other-worldly."[77] It unifies all of the dichotomies of the Cartesian system into a single definition of immanent temporal experience. Even the concept of the "afterlife" is characteristically transformed in this definition. Death is taken as the breakdown of a particular monadic ensemble into its constituent parts, and thus as a transmutation of energy in this world rather than a transition to another one. Even as this concept appears quaint if not bizarre – the central monad of the soul persists – it represents a distinctly modern rejection of the notion of a "transcendental" world of heaven for a fully immanent definition of nature.[78]

Leibniz's definition of God in the *Monadology* at first appears straightforward: God is essentially situated by Leibniz as "monad-in-chief" within the universe, the one monad that has full circumspection over the entire system. Since monads by definition "have no windows," however, this immediately implies that we will never have direct, much less intuitive, knowledge of God. God instead becomes the cipher for a never fully known universe and indeed *is* this universe in its possible full circumspection. Following upon Nicholas of Cusa's negative theology, which for Cassirer serves as the background to Leibniz in *The Knowledge Problem* and elsewhere, Leibniz's theodicy establishes that the divine can only be known negatively in relation to the development of humanity and the world. Conversely, humanity is only ever known through its various conceptions of reality and the divinity in a project that is inherently incomplete.

Cassirer's starting point for understanding Leibniz's theodicy is the theory of knowledge. The broadest characteristic of Leibniz's work is its contrast of discursive and intuitive knowledge, that is, the familiar distinction of knowledge through signs and symbols, and that which is given immediately.[79] The opposition for Cassirer is developed and refined by Kant, who defined it as the opposition of *intellectus ectypus*, "which needs images," and *intellectus archetypus*, which does not – a major focus of Cassirer's Leibnizian interpretation of the late Kant.[80] For Leibniz, true intuitive knowledge is ultimately only the domain of God; its principal conceptual role is to clarify the limitations of human experience. "For human beings," as Cassirer writes in later notes on Leibniz, "'knowledge' is given in no other way but in this form, only through the power and function of 'signs.'"[81] Similarly, for Leibniz the definition of "objects" is linked to the notion that they are the means by which phenomena appear for "limited minds."[82]

The definition of truth within discursive knowledge for Cassirer forms nothing less than "the basis of Leibniz's system," and was the foundation for his definition of the natural and social sciences.[83] The critical insight of Leibniz's concept of truth is that in order for a statement to be true, it is not necessary that the predicate "perfectly and without remainder resolve into the subject, but only that a general rule of development can be perceived so that we can grasp with certainty that the difference between the two becomes ever smaller."[84] In the immediacy of experience we sense this "general rule of development" in perceiving any movement *towards* a true statement, whether that the sun will rise tomorrow or that a lynx will soon catch a hare, and this movement itself is for Leibniz the evidence of the harmony of the world at some level. It is in Leibniz's definition of truth that the "scientific" basis of his system extends across all of its facets, from mathematics to epistemology and aesthetics. "It was only first through geometrical knowledge and the analysis of infinity," Leibniz writes, "that the light was given to me to see that also ideas are resolvable infinitely."[85] This is, of course, not to say that the resolution of ideas is a concept of mathematical physics, but rather that a new definition of truth itself develops in Leibniz's logic as it applies to both fields.

It is in the context of this definition of truth that we can grasp Cassirer's otherwise incomprehensible claim that in the phenomenon of motion we find the "real kernel and factual ground" for the idea of pre-established harmony. In the observation of a present state we always make a judgment that presupposes future trajectory, as much in physical motion as in the resolution of a subject and a predicate tend towards a particular resolution.[86] The problem of physical motion is merely the most intuitively approachable aspect of a general philosophy – and indeed definition of motion itself – that holds that the "present contains within itself the past, and is pregnant with the future," and which finds the "rule form" for such changes as the basis of experience.[87] Introducing the necessary transition from Leibniz's natural philosophy and metaphysics to his idea of the human sciences, Cassirer writes that in this regard "the same principle that lies at the basis of material and pure laws of movement is revealed as equally important for spiritual culture."[88]

The discursive definition of truth establishes the concept of pre-established harmony as a regulative concept: not only is any particular aspect of truth based on "a general rule of development" perceptible through limitation towards a goal, so are the interrelations of all truths. Leibniz held that monads share a common reality as itself a limit point of all monadic interactions. This reality could only ever be given or understood by the divine monad; for all others, it is always approximated through form and symbol. Thus Cassirer writes that one and the same reality could be

> grasped by various subjects under the most varied of material symbols, but it remains nonetheless identical with itself in so far as in all of these differences of symbols a particular invariable rule is followed. Individuals have in this their own worlds, when we consider the content of their representations [*Vorstellungen*]. Indeed they have their own "truth"; their unity with the whole will nonetheless be established through the form of connection that establishes these representations. In this manner "idea" and "sign," "reason" and "speech" are correlatively related to one another.[89]

Invariants within particular forms define various aspects of meaning and truth, which they themselves demonstrate as a "lessening of difference" in relation to one another, and thus suggest the limit horizon of an ultimate reality. Notably, however, the form of correlation between monads is not an identical "particular invariable rule" in itself, but the "connection" between these representations as a limit concept. As we noted earlier, Cassirer's broadest summary of Leibniz's work describes it as "a system of forces, of free-functioning [*freiwaltenden Tätigkeiten*] activities [that are] the foundation of all reality [...]. But just in this, that these forces were grasped as forms, demonstrates that they carry in themselves a rule of inner connection and self-limitations."[90] This concept in turn formed the germ of Cassirer's own "invariant theory of experience," which will largely be based on the later idea of group theory, as has been demonstrated by Karl-Norbert Ihmig.[91]

Pre-established harmony is understood foremost through the eighteenth-century definition of purposiveness, albeit a purposiveness whose ultimate goal is unknown; indeed, Cassirer notes that "purposive" was originally a translation of Leibniz's use of harmony.[92] In this regard, the concept of God is needed by Leibniz as the terminus of all possible true statements, or as the guarantee of the correlation of all monadic patterns with one another.⁹³ Cassirer argues, however, that Leibniz never makes the "dogmatic assertion" that the universe has a settled determination in God's omniscient pre-established harmony that would be accessible to humans.⁹⁴ From a Leibnizian perspective, only God as the "monad-in-chief" would know if pre-established harmony existed in the absolute sense, and would know it through an intuitive knowledge beyond human perception.

The definition of God as the "central" monad instead establishes the possibility of "a necessary concordance of all directions of consciousness to a common outcome and goal" and establishes a methodological principle, evidenced through lesser forms of purposiveness in everyday life, for all of the sciences.⁹⁵ Here too, Cassirer's interpretation inverts the standard transcendental and immanent definitions of the divine, so that he holds that Leibniz's optimism did not stem from his notion of God, but rather his idea of God *was* precisely the concept of optimism. God is the promise of the unlimited unfolding of meaning in the world. It would be difficult to overstate the import of Cassirer's nominalist reading of Leibniz's God: "Leibniz's optimism develops, not from his idea of God, as it much more is this idea of God itself, which holds its inherent content."⁹⁶

In Cassirer's reading, pre-established harmony is in this sense the culmination of Leibniz's philosophy. It defines the theodicy as based on a secular and immanent foundation, "not [as] a coordination among absolute substances, but a coordination between means of observation and the means of judgment [*Betrachtungs- und Beurteilungsweisen*] of the spirit [*Geistes*]."⁹⁷ In this regard the concept of pre-established harmony is the holistic application of the principle of sufficient reason to the relation of phenomenon in the world. If the world demonstrates any coherence at all it already shows the minimal cohesion necessary for it to form "one" world, and is harmonious since any particular element or function must have some further basis. The present instant of experience is defined in its immanence only by its place in a series linking past to future – that is, as motion – and to this degree immediacy already folds into the larger issue of "purposiveness."

Rather than a transcendent God that guarantees the coherence of soul and extension by fiat, as in Descartes or Malebranche, Leibniz's immanent definition of God is part of a new definition of experience that presents a "unity of causal and teleological understanding, of nature and of history."⁹⁸ All of experience is open to further exploration as the promise of ever more subtle processes of unity-in-difference. All of being is part of a process, and the divine is the promise that being can continually develop and interpret itself differently. The unity of "causal and teleological understanding" and "nature and history" also provides the necessary link in Cassirer's narrative between Leibniz's natural philosophy and his development of the human and social sciences: the unfolding of monadic entelechies necessarily becomes self-reflective and historical.

The concept of the pre-established harmony, despite some metaphysical overstatement, allowed Leibniz to bracket certain assumptions of Western philosophy and develop the opening of a "critical" approach to the experience of nature and history. This critical approach will, in Cassirer's reading, be completed in Kant's later concept of purposiveness, Herder's concept of multiple historical worlds and Hegel's concept of the dialectic.

Pre-established harmony is for Cassirer the focus on the problem of the "purposive" that is later refined with Kant: "A totality is called 'purposive,'" writes Kant, "when in it there exists a structure such that every part not only stands adjacent to the next but its special import is dependent on the other [...] in which each member possesses a characteristic function; but all of these functions agree with one another."[99]

Cassirer again interprets Leibniz's speculative philosophy as in fact a path towards productive metaphysical skepticism: the individual and the sciences are to search for ever new connections in the world, even as they assume that those connections already discovered could be reformulated into new patterns in the future. Nonetheless, Leibniz does suggest some interpretations of what the pre-established harmony of the universe might be in itself. In a late fragment, Leibniz suggests that the organization of the universe may function in the "manner of a plant or animal reaching towards a point of maturity" through various forms of consciousness, albeit one that will never regress or be aged.[100] As Cassirer notes, this vision of pre-established harmony as a sort of "collective organism" nicely combines Leibniz's notion of history and nature.[101] As a practicing historian, however, even Leibniz was well aware that this cosmic development was not perceptible in the cruel mechanizations of human history. As Cassirer puts it, the concept of pre-established harmony could never be absolute since "we are necessarily trapped in a subjective inability and narrowness of perception."[102]

Despite the quietism of many of Leibniz's own assumptions, by emphasizing that reason is the final arbitrator of all social definitions, Leibniz establishes, in Cassirer's view, the motive for the Enlightenment critique of "concrete social relations."[103] "The words '*sapere aude*,' 'Have the courage to follow your own reason,' that Kant saw as the motto of the enlightenment," Cassirer writes, "is here for the first time raised to the level of a full, unlimited truth."[104] In Cassirer's reading of Leibniz, the world of truths no longer "depends" on God as an outer instance, for if it did these truths would no longer be self-standing. Rather God is the limit point of all of these truths. Again, the medieval traditions of a transcendent God is inverted into a fully immanent divinity; human truths take part in, and indeed are, divine truths. "The 'Reason,' which we find expressed and embodied in every eternal truth," writes Cassirer, "is not a product or result of the divine existence. It is an expression of that which in the pure idea of God itself is thought as and what first constitutes the content of this idea."[105] In direct opposition to Voltaire's famous satire of Leibniz's optimism as based on a glib acceptance of the world, Cassirer holds that Leibniz's characteristic difference from figures such as Spinoza is his emphasis on the necessity of conflict and struggle between consciousnesses and forces in the world, between what is known and not-known.[106] God is in this reading the promise that our attempts to act well will play into a redemptive future, no matter how obscurely we see it in the present. "The deepest idea of 'belief' for Leibniz is found in this fundamental belief: that every true act [*wahrhaftes Tun*], over all limitations of outer being and of the moment, will find in the end its goal and proper determination [*Bestimmung*]." In this, "the religious ideal flows together with the cultural ideal that Leibniz deploys directly into one theme."[107]

Ironically, then, it is precisely in Leibniz's theology, often taken as what is most antiquated in his thought, that the "foundational feeling of the modern world" is developed in Cassirer's view, since it establishes the principle that there is no higher instance of truth than those revealed in reason itself – even as this itself is an infinite process, so that reason can never assume that it is complete.[108] In strong contrast to medieval and early modern ideas of Christianity, Leibniz's reading of religion became, as Cassirer writes in one of his few moments of rhapsodic prose, "a new love for the world and for finite being [*Dasein*], which now must be said 'yes' to with all of its limitations."[109] The meaning of this "yes" and its relation to limitation is in Cassirer's reading of Leibniz quite specific, and it will pertain to Cassirer's own work at least through his discourse on "infinitude" in the Davos debates twenty-five years later: it is the promise that there is always an infinity of qualitatively different transformations through which the world could be understood in the future and is understood by other monads in the present. For this reason our perception of present experience (and certainly of our selves or objects) is always provisional since it could always be proven to be rearticulated in a different form. In this manner the meaning of immanent finite reality is amplified, since – in the manner of a game such as chess or go – it always infinite even in its finitude, yet here the nature of game itself is capable of change.[110]

Leibniz's Theory of Reality

If Leibniz's definition of truth as a limit concept is for Cassirer the methodological foundation of his system and theodicy, this definition further depends upon an equally new conception of an immanent, but open ended, reality. Cassirer argued that the "definition of the worth of reality" forms the ultimate basis of Leibniz's system and the foundation of his "critique of ontology" in its medieval and Cartesian forms.[111] In a final section of the book outlining the historical development of Leibniz's entire system, Cassirer claims that the key concept uniting the evolution of Leibniz's philosophy is "the question concerning the reality of the concrete and individual," and the means through which it leads to "ever new categories of objects and ever more concrete possibilities of development."[112] The basis of this problem is Leibniz's conceptualization of reality itself. Leibniz's theory of pre-established harmony is ultimately grounded in its first condition, the fact that there is no "one" settled or substantial reality to which monads relate. In *Leibniz's System* Cassirer summarizes Leibniz's position by writing, "reality […] is no outer material something [*dingliches Etwas*], which occurrence set up as a foundational cause."[113] Nor is it a simple sensual relation provided in the present. Reality for Leibniz is rather a "conceptual setting" (*begriffliche fixierung*) that defines the present as a particular perspective on the evanescent relation to past and future, part to whole.[114]

The heart of Leibniz's idealism is summarized by Cassirer in the idea that "the unknown X, which we seek under the name of Reality, is determined [*bestimmt*] through a system of ideal conditional relations."[115] Yet, this apparent subjectivity is countered by the fact that the given is also always on some level "reality." Monadic consciousness always is presented with "a" reality, and there is none greater or "beyond it," even if different constellations of conditional relations could define it otherwise. "The Real" (*das Reale*) of any particular moment is the ideal limit form that governs a more diffuse "reality" as a unity, and is determined for Leibniz not as a "sensual-individual present" but through the "conceptual fusing in which the present state carries in itself its rules of production and its rules of continuation."[116]

The balance between the particular moment of the "Real" and the general condition "reality" is defined for Leibniz, as it will be later for Cohen, by the model of calculus. Calculus defines "the relation of the element to the structure [*Gebilde*] from which its continuity is defined," and will do so through the model of "the relation of the differential to the integral."[117] This principal is of universal value and application: "The idea that first led to the concept of the differential can be applied with methodological continuity in a manner that stops before no content of thought. As the condition [*Voraussetzung*] of every object it is not limited by any particular object before us."[118] In this case, the definition of any object or experience as real occurs within a context of an infinite set of conditions, so that neither the particular judgment of what is real nor the broader definition of reality need to be taken as copies of an outside reality. Both are rather part of one immanent reality. Indeed, despite the initial apparent abstraction of Leibniz's approach, it resolves into a new renewed emphasis on "immediate" experience and the "felt" sense of the world, as will be developed in the aesthetic reception of Leibniz by figures such as Alexander Gottlieb Baumgarten and later described in the last two chapters of Cassirer's *Philosophy of the Enlightenment*.[119]

In Cassirer's reading, Leibniz's philosophy occupies an interesting and underutilized position in the history of thought, one that culminates in the concept of the symbol. Leibniz's system is not subjective, since monads do not know themselves as objects (Kant's empirical egos) any better than any other aspect of the cosmos, and as a unity of apperception (Kant's transcendental ego) they are unfathomable other than as the continuity of experience in the world. Nor however does Leibniz's system reveal pure settled "objective" conditions, since these could never be directly known, although experience does reveal real existential relations through relations of continuity. That there is nonetheless increasing coherence in any judgment also means, however, that this is not pure relativism: some perspectives explain progressively more than others, so there is what might now be considered a probabilistic or emergent standard of truth. Similarly, Leibniz's argument is "realist" in a particular sense, since all perspectives and structures converge on a common horizon of reality, albeit one that is capable of surprising permutation and is itself a limit case defined either by an infinitely distant future or the promise of an ever unfolding divine order. Understood in this manner, Leibniz can be considered an initial source of what has been described as Cassirer's "structural realism," even as Cassirer follows Cohen in largely removing this position from the metaphysical context of Leibniz's monadology.

Leibniz's Perspectivalism and the Concept of the Symbol

Leibniz's thought in Cassirer's view is fundamentally historical, since all processes develop and can only be judged as more or less productive in time, and all processes interact with one another in a manner that encompasses all social and natural phenomenon. On the one hand, Leibniz holds to the felicitous convergence of signs in relation to truth. For instance, in response to the famous Molyneux problem – the question of whether someone born blind could, if surgery restored his or her sight, recognize spatial forms on the basis of their previous sense based only on touch – Leibniz holds that the translation of forms would readily work – the modes of formal recognition from touch or from calculating distance through temporal cues converge with those of vision, even as they are formally different.[120] In this manner, the "harmony" of Leibniz's larger system is not merely between particular events, but also this relation of "indicated" and "indicator."[121] This unity, however, is not the result of mere similarity, but is always rooted in the deeper conceptual connections linking any two signs. "Thus a perspective projection expresses its appropriate geometrical figure," Cassirer summarizes, "an algebraic equation expresses a definite figure, a drawn model a machine; not as if there existed between them any sort of factual likeness or similarity, but in the sense that the relations of the one structure correspond to the other in a definite conceptual fashion."[122]

On the other hand, the concept of pre-established harmony also entails the existence of the sharpest possible contrasts of meaning from some particular perspectives, even as they converge on one reality. Leibniz holds, for instance, that in an important sense the cosmological theories of Ptolemy and Copernicus – already in his era the model of radical perspective shift – have from a certain perspective equal validity. "It is fully the same," Cassirer notes of Leibniz's argument, "if one chooses to portray cosmic movement through a coordination point based on the sun or the earth."[123] The principle again here is that any projection of reality, any living reality of the monad, is to some extent valid in itself as long as it can provide an invariant rule. Considering the earth flat can be considered a highly productive explanatory form for a number of uses, such as mapping. From a second vantage, however, both within each system and in choosing between them, there are standards of truth. The Copernican system allows for a progressive – but never necessarily definitive – reduction of the difference between subject and predicate, between theory and findings, even as Cassirer will add that this greater accuracy also opens the possibility of new problems – and new theories – even in its success.

At the foundation of this complex problem is Leibniz's concept of the symbol. The concept of "symbol" plays an increasingly important role in Leibniz's understanding of both the problem of truth and his concepts of subjectivity and history. Symbol in its basic form is defined through the relation of part to whole epitomized by the calculus. In Leibniz's early work, Cassirer notes, the term symbol is used in a manner similar to sign, and it is limited to representation of aspects of reality that suggest "a direction of thought" linking phenomenon. The infinite quality of the symbol implicit in the analogy to calculus is important, however, because it suggests that each relation of sign and signified always occurs against a third term of the plenum of reality, so that all symbols are in some sense related to one another. Just as a small changes of letters might in some cases yield an entirely different meaning to a word or message (cat > bat) but in others might retain a similar or approximate meaning to a communication (cat > kat), symbols can appear similar or highly different depending on the frame of meaning in which they are interpreted. Leibniz's early work describes symbolic knowledge as "blind" knowledge, since there is in every case a necessary representational "gap" of sign and signified as opposed to the ideal of divine intuitive knowledge.[124]

However, Cassirer argues that in his later work Leibniz becomes interested in how some monads have greater circumspection of reality than others, so that some monads grasp the composition of "lesser" monads, "*les composes symbolisent les simples.*"[125] In this regard, the concept of symbol becomes the basis of Leibniz's understanding of all forms of correlation in the world, such as those of spatiality and time, as well as those particular to the representation of "I" or self. This use of the term "symbol" has profound implications for Leibniz's understanding of individuality, biology and life that intimate many of the later developments of Cassirer's own philosophy.[126] An animate being such as an animal or human can be understood as a complex society of monads, with one "central" monad acting as the basis of will in relation to all of the others, but this relation will no longer simply be analogous to that of signs or indications. Thus, describing the central role of biology and psychology as a symbolic relation for Leibniz, Cassirer writes

> The relation of body and soul represents the prototype and model for a purely symbolic relation, which cannot be converted into a relation between things or into a causal relation. Here there is originally neither an inside nor an outside nor a before and after, neither an agent nor an effect; here we have a combination which does not have to be composed of separate elements but which is in a primary sense a meaningful whole which interprets itself, which separates into a duality of factors in order to interpret itself in them.[127]

The relation of correlation between unity and diversity is here described as "symbolic," and this relation has its own validity previous to any form of mechanical explanation, mimesis or causal explanation. In place of Descartes' mind/body problem, Leibniz thus finds that the two represent a singular – and indeed singularly ineluctable – fusion. For Cassirer's own philosophy, this initial insight proves profoundly important as a precursor for his placing primary importance on the symbolic form of expression, which entails both "physiognomic immediacy" and a felt, rather than thought, sense of the world.[128]

Although the unity of mind and body can "interpret itself" through certain dualities, these dualities are not themselves primary phenomena. In this regard, symbol becomes the centerpiece of Leibniz's late philosophy, since it is a means of claiming that some forms of correlation cannot be broken down into the "universal calculus of indications" he dreamed of in his youth. The dualisms developed by Descartes, as well, Cassirer argues, as the dualisms that later followed the reception of Kant between nature and culture, or nature and history, are all secondary and artificial divisions of primary phenomena. The basis of these divisions can only be understood and critiqued by finding their common root in the logic of phenomenon, which in Cassirer's view first occurred through the concept of the monad and philosophy of motion developed in Leibniz's unitary philosophy.[129] Whereas for Descartes or Spinoza the world is essentially "substance," for Leibniz it is in Cassirer's view transformed into "life" and thus continuous transformation within an indefinite context that is yet capable of provisional and hypothetical form.[130]

Human consciousness for Leibniz is itself, as Cassirer puts it, "not a part, but a symbol of the universe."[131] Consciousness is an irreducible unity in diversity of all of the monadic relations from a particular perspective, but one that cannot be analyzed into parts, including being separated as ego from the world, without losing its initial meaning. In this regard, symbol renders what the Enlightenment called "confused" or "dark" knowledge into a pellucid phenomenon of experience. "Each soul," Leibniz writes, "knows the infinite, knows all, but confusedly; as in walking on the seashore and hearing the great noise which it makes, I hear the particular sounds of each wave, of which the total sound is composed, but without distinguishing them."[132] In Cassirer's view, the late use of the term symbol is central to Leibniz's revision of the human and social sciences: the "subject" is no longer substance, but is the correlation or even derivative (the "living mirror") of all of the contextual events, ideas and perceptions that are its context. The individual is never defined positively, and is indeed never strictly defined "individually" at all, but only in concert with the whole of its environment.

The Individual in the Context of the Human Sciences

Leibniz's work and metaphysics culminates in a study of what Cassirer describes as "The Concept of the Individual in the Context of the Human Sciences [*Geisteswissenschaften*]."[133] For Leibniz, the most complex form of reality accessible to humanity is the individual person or soul itself, which is at once determined and understands itself through the human sciences in four principle fields: history, law, aesthetics and the theodicy. In the "drive to unity and harmony, the highest and most valuable object," Cassirer writes of Leibniz, "is found in the formulation of the human world."[134] The false assumptions and substantialism of Cartesian, Newtonian and Lockean philosophy obscured, in Leibniz's view, even the proper starting point of these fields, and for Cassirer they have, despite partial inroads by German classicism, limited the potentials of the human sciences to the present. The "scientific foundations" of Leibniz's understanding of natural philosophy in the relational structure epitomized by calculus provides the framework of an open process philosophy capable of grounding the human sciences. In this manner, the "scientific foundation" of Cassirer's title is not the calculus itself, but the logic at the base of calculus as applied in very different ways to the natural and social sciences. As Cassirer noted in his "self-review" of the book in *Kant-Studien* in 1904, it was in fact in its import for the human and social sciences that Leibniz's philosophy was crucial to the modern world.[135] Even in an era when Leibniz's metageometry and combinatorics were taking on renewed relevance in physics and mathematics, Cassirer saw a greater importance in what Cohen had described as a metapolitics, a field that could now be demonstrated in its historical foundation in Leibniz's work.

A. *"The Subject of Ethics and the Concept of History"*

"Consciousness is not a part, but a symbol of the universe," Cassirer writes of Leibniz. "[This] new concept of subjectivity cultivates [*erzeugt*] the concept of history."[136] In Cassirer's view this idea grounds both the modern discipline of history and the gradual transformation of all of the sciences through the problem of history.[137] Consciousness as "symbol" is the living reflection of the infinite plenum of experience through one particular perspective. The historical individual is not defined by the "substance" of the

Cartesian cogito but the "one in many" of the content of this cogito as it is experienced in the world, a process of motion that is fundamentally temporal and historical. Thus, the historical individual is by definition not separate from its environment, but is the embodiment of the manifold of this environment and its tensions.

History is the primary category of Leibniz's philosophy; it is the complement on a higher level of complexity to the temporal processes revealed in the metaphysics of the monadology and in the methodology of calculus. Cassirer summarizes the strong definition of this insight for every aspect of Leibniz's system in writing: "The individual moment has no Being in itself that can be removed from the entirety of the series [*Reihe*] of continued existence."[138] Such development is not necessarily direct or smooth, however, but occurs for human perception through surprising turns and in distinct stages, so that "reality does not allow itself be conceived as a persistent embodiment of joined elements side-by-side [*nebeneinander*], but rather is knowable only as a successive unfolding through a series of individual stages."[139] Leibniz's definition of temporality and being is the foundation of his well-known claim that the "present contains within itself the past, and is pregnant with the future."[140] The present has no substantial being except for this relation to past and future; it is flowing or, as Cassirer puts it simply, "life." For history, the Leibnizian principle of limit relation, or infinitesimal relation, between past and future established the individual as negatively defined by patterns of relations both on the broad level of historical trends and in each specific instant of aesthetic experience of the present. In this regard, Leibniz's work in Cassirer's view is both at the center of the Enlightenment project of finding the ideal "rules" of present experience and the Romantic project of the experience of the present as an infinite and "felt" relation, an argument that Cassirer returns to at length in his later *Philosophy of the Enlightenment*.[141]

The eccentric nature of history, the fact that the present is defined negatively in relation to past and future, and that consciousness is defined negatively in relation to world, leads to an intrinsic ethical aspect to history. The realm of "being" (*Sein*) is always interlinked with the realm of ethics, of what "should" be (*Sollen*), both in the collective future and in the collective striving to understand the past. "It is not a question," Cassirer writes of Leibniz's ethics, "of the reality of the 'should' taking on meaning among a completed 'being,' for the being that is here under consideration is never closed but rather develops as a collective process of the work and co-work of individuals."[142] The relations of anticipation and the continuum link the individual present with both society and a field of potential so that experience is fundamentally open ended and open to reflexive influence on every level.[143] Unlike in most interpretations of Kant, there is at root no distinction of "Sein" and "Sollen" for Leibniz, even as of course these fields can be usefully delimited by different sciences, such as physics or law. This relation means that there is an intrinsic relation of history and ethics: in defining the temporal past on any level, whether through action, reflection or historiography, we are already shaping possible future actions.[144] Nor is this only true for the human and social sciences: in studying past transformations of natural scientific "truth" and claims to being, the awareness of present epistemological limitation and potential future transformation is illuminated. Characteristically, despite this emphasis on ethics, Cassirer never discusses the more proximate limitations of Leibniz's own monarchical histories and the less ideal developments of his ideas, even as he undoubtedly had a thorough knowledge of both, particularly in his role as an editor of Leibniz's work.[145]

The normative role of history in Leibniz's project is both as a practical guide for transformations of historical form and as reminder of the radical transformation of human verities – ideas of subject, object, time and fate – over time. The goal is on one hand to create the scientific and institutional basis in which this fundamental motility of "human" becomes progressively instantiated in society through rules, habits and ideals. On the other, history in itself, however, is the "originary image" (*Ur-bild*) of human development presented as an ever developing ideal goal of "curiosity" and "diversity" in which earlier definitions of reality are surpassed and redefined.[146] Just as the progressive development of non-Euclidean geometry provides ever more encompassing definitions of its "object," so ideally law, politics and social institutions provide ever more comprehensive ideas of humanity – an idea that as we have seen was developed by Cohen's metapolitics. History forms a practical means through which we orient specific human practices towards the maximal ideal and potential of a future humanity, and is thus closely linked with the theodicy of future development.

B. "Law and Society"

Cassirer writes that the "last and highest form" of the unfolding of reality for monads occurs through the "transformation from science to action [*Handeln*]," and this is both the highest point of Leibniz's metaphysics and the basis of the key science of law.[147] The sciences are important because they intrinsically establish the foundation of possible practical procedures, mores and policies throughout society, a view that was arguably far more plausible in the cameralism of Leibniz's Germany (or the technocracy of the Wilhelmine period) than it may appear today. This translation from science to activity in turn finds its highest plane in the transformation of "empirical communal living" (*empirischen Gemeinschaftslebens*), for it is the improvement and continuous criticism of this life that the widest frame of knowledge is developed.[148] For Leibniz, the various forms of this "empirical communal life" formed the "connection [*Vermittlung*] between general ethical rules and the reality of history," and thus between concepts of should/*sollen* relations and those of being/*sein* relations.[149] Positive law is the foundation for this relation and for the transformation of empirical communal life. As such it "returns on the one hand just as much to principle questions as it, on the other hand, represents in the multiplicity of its development the transformation of historical particularities."[150] Positive law establishes a developing matrix of ethical rules and habits in society that links "principle questions" to "historical particularities," ideal forms to everyday experience. As such, law is the central political aspect of Leibniz's philosophy.

The foundation of this central role of law lies in Leibniz's logic as first epitomized by calculus. It is not simply that law is *translated* into action by institutions, it is rather that a "should" statement in law – say, that all people should be treated as equal – establishes a range of possibilities and a structure of perception that becomes the basis of any action in society, and then can be multiply instantiated and interpreted through different institutions. The relation of the particular determination of a statement of law relates to the plenum of its possibilities similarly to the relation of an integral to the differential. The effect is not only to inflect norms of society through changing what society should be, but also – in keeping with the continuum of *Sollen* and *Sein* Cassirer finds throughout Leibniz's work – society's view of the reality or "being" it assumes as the background to any action.

Leibniz's ideal of individuality develops from this same relation to the productive value of law. In this regard, Cassirer writes that it can be seen that "the modern concept of personality" is itself a product of "positive law," in that only this allows for the negative definition of individuality – namely the modern abstract and universal subject in Leibniz's particular interpretation – in which the individual can be recognized as infinite.[151] In this manner, law structures from the outset the forms of modern life, a position that as we have seen is further developed by Cohen. Much as religious law and prohibitions structure a community's perceptions of reality, for instance in restrictions on diet or clothing that entail multiple contexts of respect or concern, so secular law structures the entirety of social reality. In this regard, Cassirer sees Leibniz's concept of law as fundamentally different than liberal definitions of law and property such as Locke's. "From the negative proposition," Cassirer writes, "which only defends against injury and attacks on the 'foreign sphere' of others, law develops to the objective concurrence [*objektiven Übereinstimmung*] and relational development of individual goals."[152] Law as science literally structures the prospective reality of individuals and their communal affiliations.

Positive law as a science has as its most important component, analogous to principles in mathematics, the development of natural rights in Leibniz's particular sense. Unlike other Western definitions of rights, for Leibniz rights are neither given by nature nor arbitrarily defined by contract or force of an outside agent, whether God or man. Rather, rights are invariants that progressively reveal human potential as a never fully known form of nature; they are a developmental form of social *a priori* that unfolds through relations discovered in experience. For Leibniz, following Grotius, the particular meaning of "natural right" is those rights that allow "a community of reasoning beings" to shape ever more complex forms of society.[153] They thus act as "social virtues," or virtues "that sustain and promote natural society," with natural society defined as any form of progressively more integrated and intelligent mode of sociation, ranging from marriage and family through the concept of the republic.[154] In this regard, to take a simple example, the social custom of "forming lines" would for Leibniz be an example of "natural law": it arguably creates the possibility for more complex, efficient or productive further forms of community action, and thus reveals a new aspect of the "nature" of human community. In this regard, as Leibniz importantly phrases it, law "makes details themselves calculable [*von den Tatsachen selbst Rechenschaft zu geben*] and regulates them in advance" in that it creates the apprehension and expectation of forms of behavior.[155]

Through law individuals are instantiated as infinitely varied parts of a diverse humanity defined through numerous overlapping forms of symbolization, sociation and perception. Thus in place of the multiple modes of liberty and privilege in the medieval period, modern law and constitution provide a more basic set of rules that also provides the potential for ever greater levels of individualization and harmonization.

The individual is never considered a "known" entity in Leibniz's definition of law, but is defined only negatively and in a manner that is open to future development through the progressive unfolding of forms of law.

As with Cassirer's reading of Leibniz's understanding of history and ethics, his liberal recasting of Leibniz's theory of law is a strong interpretation that leaves out several aspects of Leibniz's work. Thus Leibniz's glaring retention of the Roman law categories of knight and laborer is glossed over, even though it clearly stands in contrast with Cassirer's egalitarian reading.[156] Recasting Leibniz's basic argument is valuable as a general redefinition of the German intellectual canon for modernity and the future, and thus apparently for Cassirer the more archaic aspects can be de-emphasized.

C. "Aesthetics"

If ethics and law are the highest categories of Leibniz's understanding of the human sciences, the fields determined by aesthetics form the broadest umbrella linking these sciences to all other aspects of Leibniz's system. Leibniz's philosophy, particularly in its influence on German and Swiss aestheticians, established the basic path through which classical German thought confronted the problem of the individual's ability to form impressions of complex whole, and thus establishes the foundational epistemology of understanding the individual in all fields.[157] Aesthetics in Leibniz's definition is therefore not principally or only concerned with art. Indeed, Cassirer claims that "Leibniz's system contains no individual and self-sufficient theory of art."[158] Rather, Leibniz's system develops the foundation of the German tradition of aesthetics and psychology as a general reading of conscious processes as it relates to the horizon of relations in the world, a field that hinges on the philosophical problem of the determination (*Bestimmung*) of particulars. Although developed from the perspective of human consciousness, the same processes are presumed to have potential application to other monadic forms, although this more esoteric aspect of Leibniz's monadology is not central to Cassirer's treatment.

The problem of aesthetics was somewhat misleadingly described as the "irrationality" problem in the twentieth century, or the problem of "confused perceptions" before that.[159] Leibniz's aesthetics was initially received in German thought as concerning "obscure" or "dark" (*dunkel*) knowledge since it defined the immediacy of experience in relation to a plenum of meaning that cannot be ultimately known in itself other than through affect."[160] As Cassirer noted, "confused" perception is defined as such only in the strictly etymological sense of the term, since "in all aesthetic intuition a confluence, a fusing together of elements takes place, and that we cannot isolate the individual elements from the totality of the intuition."[161] The goal in resolving the "irrationality" problem was to do justice to the range of human experience, not only that which was logical or rational in the classic sense, but, in Baumgarten's words, that which included "the full scope of intuitively perceived reality."[162]

Leibniz's successors, notably Christian Wolff, consistently sought to make "confused" or "dark" knowledge secondary to rational thought.[163] In Cassirer's reading this is a distortion of Leibniz's original premises, for the aesthetic definition of the monad establishes an equivalent place to both sentiment (*Gefühl*) and experience (*Erfahrung*) as a whole. For Leibniz, the definition of thought is very broad and defines "the activity that represents a multitude in itself as a unity," with this unity understood as the mystery of pure correlation or motion.[164] Consciousness is primarily a "tendency" or "movement" of this unity in difference, that is, a striving beyond itself.[165] In Leibniz's terminology, the unification of a multiplicity occurs through perceptions, which are essentially "conscious" thoughts of relations we grasp, and apperception, which describes the horizon of our "unconscious" or unreflexive thoughts about the infinity surrounding us. Whereas for Descartes the "confused" thoughts that we have not yet ordered are meaningless, for Leibniz they are in fact our most meaningful general orientation towards the world – the adumbrations, as it were, of an infinity of broader relations in which we find ourselves.

Aesthetics in Cassirer's reading provides a link between Leibniz's Enlightenment and Romantic reception, as well as a means for redefining this connection by demonstrating the necessary relation of Leibniz's aesthetics to his epistemology, ethics and theodicy.[166] With Leibniz, Cassirer argues, feeling and emotion are elevated to a primary role in Western philosophy, and a role moreover that is at the center of understanding the monad's place within the plenum of reality. Aesthetics reveals that the synthesis of experience occurs as a lived texture of meaning through which we orient ourselves to the particular problem of meaning and our general setting.

The model of unity in difference found in aesthetics again establishes the intrinsic fusion of *Sein* and *Sollen* in all of experience, and is of particular import for the other three aspects of the human sciences – history, law and theodicy. The monad does not simply reflect a given reality in time, in relation to present limitations, or future possibilities, but is necessarily inflecting determinations of what "is" with possible relations of what could or should be. In this manner aesthetic experience is the basis of Cassirer's claim that Leibniz's definition of experience is "not a coordination among absolute substances, but a coordination between means of the observing and the means of judgment [*Betrachtungs- und Beurteilungsweisen*] of the spirit [*Geistes*]."[167] It thus forms the basis of "a unity of causal and teleological clarification, and of nature and history."[168]

Cassirer concludes his reading of aesthetics in *Leibniz's System* by noting that for Leibniz the model of unity in multiplicity "by no means" finds its most ready analogy in mathematics and the calculus, even as this model is of course pertinent. Rather, it is in the problem of the organism and of life, "in the progress from ideas of mechanism to those of organism," that it finds its clearest development, and this problem demonstrates that themes from nature and culture are unified in the model of aesthetics. Leibniz tellingly describes natural scientists as "poets of the real."[169] Cassirer concludes by noting that this theme will be developed directly by Goethe – a progress he develops in full in *Freedom and Form* – who in the concept of harmony and the idea of the "objective-beautiful" (*Objektiv-Schöne*) is "in inner agreement with the fundamental and unifying vision of Leibniz's philosophy."[170]

D. "Theodicy" and the future of humanity

Whereas for Pascal the "double infinity" of the infinitely great and the infinitesimal are used to "deny the human self," Cassirer writes, for Leibniz they are integrated into the individual monad: the monad becomes a "living reflection" of the universe and reveals a continuum at once infinitely subtle and infinitely vast.[171] More importantly, however, the notion of infinity is introduced both into history and into the concept of the future in the theodicy: the course of experience is capable of drastic transformations, and can pivot on apparently minor interpolations. Leibniz's pivotal idea in the theodicy that this is the "best of all possible worlds" at first appears as a "crass anthropomorphism."[172] Cassirer claims that despite this appearance the concept actually serves to open a host of new historical possibilities. Most importantly, it provides the basis for avoiding a mechanistic or deterministic understanding of the world, since the best of all possible worlds entails the maximal amount of meaning within the minimum of means. From the limited perspective of the human, this is a guarantee that the world, even as it displays reason, will develop in endlessly surprising turns, and cannot be reduced to the present condition of what is obvious or given. In this regard, the theodicy is the final guarantee that the human sciences could develop the maximally open definition of humanity itself. The individual is related to the future as an ethical problem in the sense of Kant's idea of a cosmopolitan history: both history and present action have to be oriented towards a redemptive ideal of humanity based on a present "setting" of forces determined by the past. In a broader sense, the theodicy in Cassirer's reading further suggests the conditionality of the will of God within what is revealed as the best of all possible worlds, since God is defined not by caprice or power, but as the correlate of the most perfect and harmonic world. This concept also for Cassirer opens the possibility for the conditionality of logical and ethical concepts of God within a rational universe.[173]

Leibniz's work establishes the constructive power of law and the sciences to inflect individual development within a universe in which humanity and nature are understood as mutable yet rule based. In this regard, the critique and development of the sciences becomes the basis of ethics and the foundation of any present politics. The subject becomes secondary to the constellation of forces in which it is shaped and through which it can be understood. "Autonomy of will," as Cassirer later summarizes it in his description of Leibniz in *Freedom and Form*, "no longer relates to a unity that is given in the subject," but "to that which is sought and defined in the social and in knowledge itself."[174] His emphasis is not surprising, since the same goal was the culmination of the Marburg school's project as developed by Cohen, particularly Cohen's attempt to move the Enlightenment project from one of autonomy, or independence, to one of what he defines as *autotelie* (Greek: *autós*, self and *télos*, goal) or the freedom of collective or individual development within a set of limitations focused on a future goal.[175] Cassirer finds each of the key aspects of Leibniz's understanding of the social and human sciences to promote this project: first, in a definition of history that reveals permutations of even the most obvious features of reality, subjectivity and objectivity; secondly, a definition of law as a constructive force that links being and norms, present and future, to create new definitions of personality, association and state or political union; thirdly, a definition of aesthetics as the template for an open-ended model of perception, thought and experience; and finally, a definition of the theodicy as the guarantee of a correlatively open-ended determination of humanity and the world.

Cassirer's *Problem of Knowledge* series (1906–1940) applied this Leibnizian and Marburgian approach to a philosophy and methodology of society and history by demonstrating that "the gradual transformation of definitions of self as well as of definitions of objects is accomplished by no means solely through closed philosophical systems, but rather through the manifold experiments and endeavors of research, as well as through the entirety of the intellectual culture."[176] For this form of critique to have traction in modern society, however, a modern critique of the natural sciences had to be carried out, analogous in some ways to Leibniz's critique of Descartes' substantialism. It was to this project that Cassirer turned in *Substance and Function*, where he hoped to demonstrate that the natural sciences in the last several decades had themselves already developed the most thorough critique of substantive thought and general model of "science" then available, and with this the key development of Leibniz's basic principles.

Notes

1 L 431. Leibniz is the pivotal figure for periodizing modernity itself: "It is only with Leibniz that the older medieval system is replaced," Cassirer writes in *Freedom and Form*, "and the questions of the relation of science brought up, and with it the entire spiritual focus of the age changed" (FF 22). See also: L 482; ECW1 433; SF 91ff.
2 Sieg, Ulrich. *Aufstieg und Niedergang des Marburger Neukantianismus: Die Geschichte einer philosophischen Schulgemeinschaft* (Würzburg: Königshausen & Neumann, 1994), 324.
3 Meyer, Thomas. *Ernst Cassirer* (Hamburg: Ellert and Richter, 2007), 49. As Meyer notes, Cassirer's *Knowledge Problem* functions in this regard as a "position paper" for the Marburg school in the modern era, a role performed by Natorp's *Platos Ideenlehre* in the ancient period.
4 EL 24.
5 Sieg, 330.
6 It is telling in this regard that reviews of Cassirer's own works through the First World War often refer to him not as a neo-Kantian, but as a Leibniz scholar. See for instance, Felix Bàlyi's review of Cassirer's *Das Erkenntnisproblem* in the supplement to the Munich *Allgemeine Zeitung* 192 (1906): 342–3 where Cassirer's "thorough and original" Leibniz research is said to establish the foundation for his rise as one of the "great contemporary writers in philosophy."
7 Cassirer, Ernst. *Leibniz: Hauptschriften zur Grundlegung der Philosophie* (Hamburg: Felix Meiner Verlag, 1904, 1912); Leibniz, G. W. *Hauptschriften zur Grundlegung der Philosophie*, vols 1 and 2, ed. Ernst Cassirer, trans. A. Buchenau (Hamburg: Felix Meiner Verlag, 1966 [1904]). The books originally appeared as part of Meiner's popular Philosophical Library series.
8 Cassirer, Ernst. "Newton and Leibniz," *Philosophical Review* 52 (1944): 366–91.
9 Ibid., 389.
10 Ibid., 366. See also, Cassirer, Ernst. "Leibniz" in *Encyclopedia of the Social Sciences*, vol. 9 (New York: Macmillan, 1933), 400–402.
11 This developmental and even playful set of transitions in Leibniz's philosophy is most clearly suggested by Adams, R. M. *Leibniz: Determinist, Theist, Idealist* (New York: Oxford University Press, 1994).
12 EK2 126–91; ECW3 101ff.
13 EK1 1–202, 311, 375ff., EK2 126ff.; ECW3 101ff.
14 On the relation with Bayle and skepticism, particularly in terms of the development of antinomies, see EK1 592.

15 L 298.
16 Reill, P. Hans. *The German Enlightenment and the Rise of Historicism* (Berkeley: University of California Press, 1975), 6ff.
17 PE 275ff.
18 ECW17 218.
19 EL 8ff.; SF 232.
20 ETR 391.
21 In particular, see Schmitz-Rigal, Christiane. *Die Kunst offenen Wissens: Ernst Cassirers Epistemologie und Deutung der modernen Physik* (Hamburg: Meiner, 2002). See also her "Ernst Cassirer: Open Constitution by Functional A Priori and Symbolical Structuring," in *Constituting Objectivity: Transcendental Perspectives on Modern Physics*, ed. Michel Bitbol, Pierre Kerszberg and Jean Petitiot, *The Western Ontario Series in Philosophy of Science*, vol. 74 (New York: Springer Verlag, 2009); "Science and Art: Physics as Symbolic Formation," *Synthese* 179 (2009): 21–41; and "Die Kunst der Wissenschaft" *Philosophia Naturalis* 40, 2 (2003): 255–91. Schmitz-Rigal's understanding of Cassirer's "open knowledge" has interesting historical similarities to Heinrich Levy's concept of Cohen's foundational ideas as an "open-system" that is further developed by Cassirer (ibid. 35).
22 FF 52.
23 L 198.
24 L VII.
25 L VII.
26 L XI.
27 LC 230–33, FF 43. Windelband, Wilhelm. *A History of Philosophy with Special Reference to the Formation and Development of its Problems and Conceptions* (New York: Macmillan, 1926), 464.
28 Cassirer was, of course, by no means alone in stipulating that Leibniz was the keystone for understanding German thought. This view was common to most German philosophical histories at the turn of the century. Wilhelm Windelband's influential philosophy of history is typical of the early twentieth-century reading of Leibniz's importance, describing him as the "many-sided founder of German philosophy" (Windelband, 443). The renewal of interest in Leibniz is usually associated with the nearly simultaneous publication of three new texts on Leibniz by Bertrand Russell in England, Louis Couturat in France and Cassirer in Germany. Cassirer reviews Couturat and Russell's books at the end of *Leibniz's System* (471–500). Partly for nationalistic reasons, partly for affinities within existing philosophical systems, this initial interest carried a persistent resonance in Germany in a manner it did not in the other countries.
29 L 419.
30 L 239, L 419.
31 L 421.
32 K 426.
33 L 379ff.
34 Cassirer, Ernst. *Descartes' Kritik der mathematischen und naturwissenschaftlichen Erkenntnis*, published as an introductory addendum to *Leibniz's System* (3–104).
35 L 108.
36 Ibid.
37 Funkenstein, Amos. *Theology and the Scientific Imagination* (Princeton: Princeton University Press, 1986), 185.
38 L 317.
39 FF 37.
40 PE 228.
41 L 154.
42 L 186.
43 L 188.

44 L X.
45 L 317.
46 This reception serves as the centerpiece of both Cassirer's own reading of Kant and his biographical description of Kant's development in his *Kant's Life and Thought* (1918). K 98ff.; EK2 611ff.
47 SF 303.
48 K 195.
49 K 195.
50 FF 425ff.
51 On infinite or limitative judgments, see Cohen, I 82; for the example of *Geist*, 82ff., SP1 90ff.; Wolfson, Harry A. "Infinite and Privative Judgements in Aristotle, Averroes, and Kant," *Philosophy and Phenomenological Research* 8, 2 (1947): 173–87; Gordin, Jakob. *Untersuchungen zur Theorie des unendlichen Urteils* (Berlin: Academie-Verlag, 1929); Funkenstein, *Perceptions of Jewish History* (Berkeley: University of California Press, 1993), 271–84.
52 L 110.
53 L 110.
54 L X.
55 L 316.
56 L 187–8.
57 The transition between these claims is addressed by Adams, *Leibniz*, 326–33 and in L 403; ECW1 361.
58 L 316.
59 L 317.
60 L 97.
61 L 141.
62 L 32.
63 Cassirer directly makes use of Lange's argument at L 306.
64 L 44.
65 L 45.
66 L 46.
67 L 46.
68 FF 43.
69 On Euler and his relation to Leibniz, see EK2 472ff., 501ff.
70 Funkenstein, *Perceptions of Jewish History*, 108.
71 FF 28.
72 EK2 128.
73 L 369.
74 EK2 155.
75 L 228.
76 L 229.
77 FF 57.
78 L 373.
79 L 110.
80 On the centrality of Leibniz for Kant, and on Kant's critique of Leibniz's monadology, see EK2 730ff.; ECW 3 611ff.; and K 99ff. See as well Cassirer's notes to symbols in "The Concept of the Symbol" (PSF4 229), citing Kant, *The Critique of Judgment*, paragraph 77.
81 PSF4 229.
82 Adams, *Leibniz*, 219, citing Leibniz, Gerhardt edition, VII: 563.
83 FF 27.
84 EK2 149.
85 EK2 148.

86 L XI.
87 FF 35.
88 L 380.
89 FF 294–5.
90 FF 52.
91 Ihmig, Karl-Norbert. *Cassirers Invariantentheorie der Erfahrung und seine Rezeption des "Erlanger Programs"* (Hamburg: Felix Meiner, 1997). Cassirer used the group theory of Felix Klein to define a series of "invariants" between any permutations of sets, and these invariants will form the only definition of a "truth" within a particular system or, in an important parallel, perception. This continuity between the exact sciences, perception and the social sciences was most clearly carried through in Cassirer's *Objectives and Paths of Reality-Knowledge*, a text never published in large part due to Cassirer's exile (ZW 83–119). Later, Cassirer used Einstein's theory of relativity to develop a similar theory of "covariance" that underlined his own theory of symbols and produced a similar argument: despite the diversity of theories about the world and forms through which reality is apprehended, as well as probability that today's truths will be superceded, there is nonetheless both a basis for "truth" and a demand for its plurality (SF/ETR 391ff.). Thus, Cassirer writes, from the "critical view, the object is no absolute model to which our sensuous presentations more or less correspond as copies, but it is a concept, from which reference to presentations have synthetic unity." This concept of the theory of relativity no longer presents truth in the form of a picture "but as a physical theory, in the form of equations and systems of equations, which are covariant with reference to arbitrary substitutions" (SF/ETR 393). For a further reading, see: Ryckman, A. "Einstein, Cassirer, and General Covariance – Then and Now," *Science in Context* 12 (1999/4): 585–61.
92 L 279.
93 Introduction to HS; ECW9 527.
94 L 398.
95 LS 400; FF 32.
96 FF 58.
97 L 422.
98 L 422.
99 K 287.
100 "Universum est ad instar plantae aut animalis hactenus ut ad maturitatem tendat. Sed hoc interest, quod nunquam ad summum pervenit maturitatis gradum, nunquam etiam regreditur aut senescit" (L 444, quoting from Leibniz, *Die Leibniz-Handschriften der königlichen offentlichem Bibliothek zu Hannover*, ed. E. Bodemann (Hannover: Hahn, 1895), 120).
101 L 398.
102 L 399.
103 L 407.
104 FF 30.
105 FF 30.
106 L 394ff.
107 FF 57.
108 "In this manner the idea that appears to link Leibniz most clearly with theological dogma, in truth is the starting point of the modern scientific style of thought" (L 374; FF 21).
109 FF 64.
110 Leibniz writes about games in "Annotatio de quibusdam Ludis; inprimis de Ludo quodam Sinico, differentiaque Scachici & Latrundulorum, & novo genere Ludi Navalis" (A note on certain games, especially a certain Chinese game, and on the difference between chess and latrunculus and on a new kind of naval game), trans. Richard J. Pulskamp (Cincinnati, OH: Department of Mathematics & Computer Science, Xavier University, 2009), from *Miscellanea Berolinensia* (1710), 22–6.

111 Cassirer's statements here are broadly analogous to the role of reality he notes at the beginning of his Leibniz text (L 111; ECW 1 102). Similarly, his introduction to Leibniz's *Hauptschriften*; note the importance of the key theme of the limit in relation to objects and metaphysics, and how this will determine "knowledge of reality" (*Wirklichkeitserkenntnis*) (HS 11–12; ECW 9 527–8).
112 L 433.
113 L 262.
114 L 257.
115 L 254.
116 L 255.
117 L 154.
118 In mathematics, it explains how the reality of line connects to surface, and ultimately provides a new basis for *analysis situs* of any phenomenon; in mechanics it leads to a similar development in real world applications leading to "higher dimensional" analysis; and most importantly, in dynamics – as particularly outlined in Leibniz's "A Species of Dynamics" – it leads to the awareness of the relation of "impulse" to a quantity of motion, and then in turn to how acceleration, as a second order phenomenon can be described from an instance of velocity (L 170, 189).
119 PE 338ff.
120 Leibniz addresses the Molyneux problem in its original formulation in the *New Essays on Human Understanding:* "Suppose a Man born blind, and now adult, and taught by his touch to distinguish between a Cube, and Sphere of the same metal, and nighly of the same bigness, so as to tell, when he felt one and t'other, which is the Cube, which the Sphere. Suppose then the Cube and Sphere placed on a Table, and the Blind Man to be made to see. Quære, Whether by his sight, before he touched them, he could now distin-guish, and tell, which is the Globe, which the Cube." In *New Essays on Human Understanding*, ed. Peter Remnant and Jonathan Bennett (Cambridge: Cambridge University Press, 1981), 137. On the Molyneux problem, see: "Molyneux's Problem," Stanford Encyclopedia of Philosophy, 5 July 2011. Online: http://plato.stanford.edu/entries/molyneux-problem/ (accessed 4 February 2013).
121 L 416.
122 ETR 392, quoting Leibniz, G. W. F. *Die Philosophischen Schriften von G. W. Leibniz*, edited by C. J. Gerhardt, 8 vols (Berlin: 1875–90), vol. 2, 7, 167, 233; vol. 7, 44, 263ff.
123 "Zur Phoronomie und Dynamik" in HS 109; ECW9 530.
124 Leibniz, G. W. F. "Betrachtungen über die Erkenntnis, die Wahrheit und die Ideen (1684)" in *Hauptschriften zur Grundlegung der Philosophie*, vol. 1, ed. Ernst Cassirer (Hamburg: Felix Meiner Verlag, 1966), 24–5.
125 L 418.
126 "The organism contains in itself the guarantee of infinity and inexhaustibility in itself," since there is no possibility of explaining or presenting it through a finite set of conditions, as there is a machine (L 404; ECW1 360). The model for this unity in Cassirer's earlier reading was the mathematical concept of the series, which describes a unity in multiplicity which is both inherent to the parts and only existent through them, whereas in Cassirer's later reading it is the immediacy of this relation that is emphasized: "Wie das Gesetz der mathematischen Reihe zu seiner Darstellung die Ausführung in die Mehrheit der Glider verlangt, so kann sich das formale Gesetz der Entwicklung nur darstellen, indem es sich in der Hervorbringung des Phänomens des organischen Körpers und seiner Veränderung bethätigt" (L 408; ECW1 364). On this topic see Ihmig, *Cassirers Invariantentheorie*. Unlike the later example, however, the presence of immediacy is given in the body immediately in a manner incapable of analysis, and is for not only analogous but identical with perception and consciousness itself: "The unity of the function of presentation cannot be separated from the manifold of the presented contents" (L 408; ECW1 364). The relation of body and soul, like the relation of immediacy in perception,

is a unity of the particular with the cosmos, but this relationship is primarily and first grasped as thoroughgoingly "alive." Similarly, the function of expression is both the simplest relation of one in many and many in one, as well as that which all others depend on.

127 L 9, 407–8.
128 Cassirer's turn to expressive relations as beyond "the concept of thing and the concept of causality" is usually considered to be a late turn in his work as he approaches the irrational and myth (PSF3 94). It appears to have a more proximate foundation in Leibniz's idea of the symbolization of the body.
129 L 364, 380–81.
130 FF 65.
131 L 428.
132 Leibniz, G. W. F. *Leibniz: Selections*, ed. Philip P. Wiener (New York: Scribners, 1951), 530.
133 This section is the culmination of the systematic part of *Leibniz's System*. The book as a whole ends with a long study of the historical genesis of Leibniz's system (L 435).
134 L 402.
135 Cassirer, Ernst. "Selbstanzeige," *Kant-Studien* 7 (1902): 376; ECW9 439.
136 L 428.
137 L 316–17.
138 L 156.
139 L 378.
140 FF 35.
141 L 415/411; PE 312–61.
142 L 432.
143 "The new concept of the present is here made particularly clear [...]. The present, which we are used to grasping as sensory reception, logically is much more a condition through the act of anticipation" (L 296; ECW1 265).
144 The general outlines of Cassirer's reading of the centrality of Leibniz for German historiography, developed at length in the *Philosophy of the Enlightenment* (1931), has been developed and confirmed in later studies such as Reill's *The German Enlightenment*, 214–15, 218.
145 Leibniz's world history, for instance, was certainly developed by a number of authors into a racial system, and one of his early works define non-Christians as "semi-beasts." Nonetheless, as Peter Fenves has argued, Leibniz's monadology is largely inimical to racism on much the same grounds as Cassirer outlines. See Fenves, Peter. "Imagining an Inundation of Australians; or, Leibniz on the Principles of Reason and Grace," in *Race and Racism in Modern Philosophy*, ed. Andrew Walls (Ithaca, NY: Cornell University Press, 2005), 73ff. Franklin Perkins has noted that Leibniz is one of the few early philosophers to demonstrate an interest in other cultures, particularly China, as intellectually equal to that of Europe. See Perkins, Franklin. "Leibniz and His Correspondence with the Jesuits in China," in *Leibniz and his Correspondents*, ed. Paul Lodge (Cambridge: Cambridge University Press, 2004), 159–60.
146 L 401
147 FF 56.
148 L 403.
149 L 405.
150 Ibid.
151 L 394.
152 L 409.
153 L 403.
154 Ibid.
155 L 406.
156 Cassirer argues that Leibniz's reading is an improvement over Roman law, since the laborer is no longer considered as an "object." Rather the relation is now considered as one based

on more or less understanding, and the master must "justify his power based on intellectual dominance" (L 454). Although Cassirer sees this turn of argument as ultimately opening the path towards a democratization of relations, in itself it is clearly not one – and if anything has an uncomfortable resonance with the position of the German Mandarin's relation to power.
157 EK2 566ff.
158 L 411.
159 The most widely recognized study of this problem is still Alfred Baeumler's *Das Irrationalitätsproblem in der Ästhetik und Logik des 18. Jahrhunderts bis zur Kritik der Urteilskraft* (Berlin: Wissenschaftliche Buchgesellschaft, 1981 [1923]). The stakes of the debate over this topic become particularly clear if we look ahead to Cassirer's long-standing debate with Baeumler, arguably the leading scholar of the irrationality problem in the 1920s. By the early 1930s he was infamously also an adamant supporter – indeed the principal philosophical advocate and later director of the university system – of the Nazi cause. Directed in part against Cassirer's *Freedom and Form*, Baeumler's *The Irrationality Problem in the Aesthetics and Logic of the 18th Century through the Critique of Judgment* (1923) remains a standard work on the meaning of Enlightenment aesthetics. See Baeumler's polemics against Cassirer in Baeumler, Alfred. *Das Irrationalitätsproblem in der Ästhetik und Logik des 18. Jahrhunderts bis zur Kritik der Urteilskraft* [intended first volume of: *Kants Kritik der Urteilskraft: Ihre Geschichte und Systematik*] (Darmstadt: Wissenschaftliche Buchgesellschaft, 1975), 36ff., 57, 244–51; 274–307, 354. Cassirer states his position against Baeumler in 1931 in terms of the development we have followed in Leibniz on the model of calculus as follows: "The aesthetic 'unity in multiplicity' of classical aesthetic theory is modelled after this mathematic unity in multiplicity. It is a mistake to think that the principle of unity in multiplicity is contrary to the spirit of classicism and that the opposite pole of classicism is most effectively expressed in this principle. In the realm of art the spirit of classicism is not interested in the negation of multiplicity, but in shaping it, controlling it, and restricting it" (PE 289 and n. 10). For a brief overview of Baeumler's pivotal place in the history of aesthetic theory see Zammito J. H. *The Genesis of Kant's Critique of Judgement* (Chicago: University of Chicago Press, 1992)., 19–20.
160 L 412.
161 PE 347.
162 PE 348.
163 ECW17 218.
164 L 460 citing Leibniz, G. W. F., in "Cogitatio est activa quaedam repraesentatio multorum simul facta in re per se una," Gerhardt, III: 376.
165 L 414.
166 Ibid.
167 L 422.
168 Ibid.
169 L 423.
170 L 422.
171 L 419.
172 L 431.
173 L 432.
174 FF 160.
175 Cohen, Hermann. *Kants Begründung der Ethik, nebst ihren Anwendungen auf Recht, Religion und Geschichte* (Hildesheim: Olms, 2011), 237ff.; Cassirer, "Kant," in *Encyclopedia of the Social Sciences* (1932), 539b. The concept is extensively developed in Natorp's social philosophy in *Vorlesung über praktische Philosophie* (1925).
176 EK1 8.

Chapter Five

SCIENCE AND HISTORY IN CASSIRER'S *SUBSTANCE AND FUNCTION* (1910)

Cassirer's *Substance and Function* (1910) translates the key theoretical arguments of *Leibniz's System* and Cohen's philosophy into a modern scientific idiom, weaving the basic themes through a survey of the modern development in logic, mathematics, geometry, physics, chemistry and psychology. Cohen praised the book for its development of the Marburg critique through an immanent discussion of the contemporary sciences, a project that, he noted in a letter to Cassirer, his own work had never accomplished.[1] The central argument of *Substance and Function* is that the contemporary primacy of the "*logic of the mathematical concept of the function* [...] is not confined to mathematics alone [...] but extends over into the field of the *knowledge of nature*; for the concept of the function constitutes the general schema and model according to which the modern concept of nature has been modeled in its progressive historical development."[2] The text thus "represents a single problem," the functionalist definition of part and whole, "which has expanded from a fixed center, drawing ever wider and more concrete fields into its circle" to present a new phenomenology of experience and theory of perception.[3] *Substance and Function* provides a clear definition of the meaning of function and form in the "system of the sciences" that will then be available as a model for all sciences, and as a means of critique of their limitations, thus completing Cohen's vision of a general critique of science in society.[4]

"Only the function fulfilled by the universal in connection with the particular," Cassirer writes, can serve as a true basis of modern epistemology and a modern philosophy.[5] This definition culminates in the theory of representation and judgment as symbolic in Leibniz's sense, as we saw in our earlier survey of the conclusions of *Substance and Function*, in which "each member of experience possesses a symbolic character, in so far as the law of the whole, which includes the totality of members, is posited and intended in it. The particular appears as the differential, that is, not fully determined and intelligible without reference to its integral."[6] In *Substance and Function*, however, Cassirer develops a more comprehensive model than Leibniz's – as well as of Cohen's – use of calculus alone to explain this relation of part and whole, a model that is largely based on the modern theory of groups. As in Leibniz, mathematical theory is used as both an analogy and a demonstration of a general process that obtains in perception and experience, a point that Cassirer returned to throughout his career, notably in a 1938 essay, "The Concept of Group and the Theory of Perception," as well as a related 1945 lecture, "Reflections on the Concept of Group and the Theory of Perception."[7]

The connection between *Substance and Function* and *Leibniz's System* is explicitly made in relation to logic and mathematics, but implicitly found throughout the wider argument.

Cassirer holds that the contemporary mathematical themes at the heart of his analysis are themselves ultimately Leibnizian: "There is a continuous development beginning with Leibniz [...] of a general science of quality, up to modern projective geometry and the theory of groups."[8] Similarly, he writes that the "evolution of modern mathematics has approached the ideal, which Leibniz established for it" and which culminates not in a "general science of magnitude, but of form."[9] This *Kombinatorik* or "characteristic" of form is the basis of a functional reform of all of the sciences, which by the end of the text encompasses psychology and, by implication, the social sciences.

To reduce Cassirer's argument to claiming that the "functionalism" of science only culminates in the mathematical modeling of physics or involves "abstraction" from phenomenon (rather than, as he argues in *Substance and Function*, a new form of abstraction that reveals phenomenon) mistakes the nature of his philosophy. The functionalist "correlation" of part to whole entails a logic that is, as Cassirer first demonstrated with Leibniz, of pervasive importance across the sciences and for all domains of philosophy. Cassirer describes this general study as equally determinative of all spheres of knowledge: "Psychology and criticism of knowledge, the problem of consciousness and the problem of reality, all take part in this process. For in fundamental problems there are no absolute divisions and limits; every transformation of the genuinely 'formal' concept produces a new interpretation of the whole field that is characterized and ordered by it."[10] The basic argument of *Substance and Function* is further developed by the historical study of *Freedom and Form*, where it is extended to the critique knowledge and practice in the social sciences. Ultimately the "circle of interpretation" introduced in *Substance and Function* expands to form the basis for Cassirer's developing understanding of language, art and myth in his philosophy of symbolic forms.[11] Without first understanding Cassirer's argument in *Substance and Function* it is not possible to fully understand these later works, as Cassirer himself indicates by frequently directing the reader back to his foundational argument.

Substance and Function was hailed by Cohen as the "fulfillment" of his *System of Philosophy* and with it of the Marburg project he was promoting, yet as we noted it paradoxically constituted the key break in the unity of the two philosophers.[12] The basis of this break was epitomized by the problem of the infinitesimal and demonstrates its centrality for both philosophers. As we noted earlier, Cohen read *Substance and Function* as continuing his late philosophy in "in its main outlines."[13] It did so both in the "general method of the leading thought and in the prevalence of the meaning of the infinitesimal."[14] Given that Cassirer only occasionally indirectly discusses the infinitesimal theme this claim at first appears odd, but Cohen properly sees that Cassirer's "logic of the mathematical concept of the function" and reading of group theory is a direct development of this theme.

Within this broader harmony, however, Cohen writes that their "unity is endangered" by Cassirer's significant shift of emphasis to the role of pure relations as the basis of reality. This can only be, Cohen writes, to the degree these are functions, which find their "root of ideal reality" in the infinitesimal element, that is, in the manner the infinitesimal suggests both continuity and reality within relations. For Cohen, this further suggests a unity among all uses of the infinitesimal, linking its physical, psychological and biological uses into a single system. Cassirer, on the other hand, uses the model of the infinitesimal solely as a tool for understanding relations in themselves, and with this develops a historical reading of the problem of unity of knowledge. As with his reading of Leibniz's use of the infinitesimal, it is for him a means of illuminating improper substantial assumptions and defining correlation. The problem of reality, as Cohen notes, is for Cassirer resolved into "antinomy" that is not further grounded in the infinitesimal. "This," Cohen writes, "is the crossroads."[15] In Cassirer's path, a new philosophy of pure relations is developed that has as its principle unity its historical development, and even this suggests different definitions of evidence and truth (in the Leibnizian sense) for different fields.[16]

It is a fundamental irony of intellectual history that Cassirer, Cohen's chosen successor and the undoubted authority of the Marburg school in relation to science, most clearly saw the severe limitations of Cohen's literal reading of calculus's relation to philosophy and with this, in attempting to decant its philosophical importance into a new form, radically altered it. In doing so, Cassirer broke not only with Cohen's philosophy of science, but – in a manner with broad historical implications – with the concept of the linear development of the sciences as it had formed in the nineteenth century. Far from philosophy developing organically and consecutively through scientific conceptualization and knowledge, as Cohen assumed, Cassirer – perhaps the only figure at the time who could have actually attempted such an exegesis – came to emphasize what he would later describe as the "polydimensionality" of human experience.[17] Later, in the last volume of *The Knowledge Problem*, for instance, Cassirer could write, "There are now as many single forms of theory of knowledge [*erkenntnistheoretische Einzelrichtungen*] as there are different scientific disciplines and interests."[18] In place of the systemic unity of culture and science that Cohen promoted, Cassirer explicated different trajectories of "reality-knowledge" (*Wirklichkeitserkenntnis*), to use the term from the title of his 1936 overview of *The Knowledge Problem*, as they branched and wove through different fields. The consequence was a new understanding of both history and science that still hinged on understanding Cohen's concept of the infinitesimal, but reframed its meaning.

Due to its general and formal use, for Cassirer – unlike Cohen – the basic metaphors and analogy of calculus and group theory can be understood philosophically without having to be extensively drawn out through the technical details of these fields. They function as analogies that suggest "the characteristics," in the Leibnizian sense of a general set of rules and patterns, "of the paths and progress of experience."[19] Indeed, despite Cassirer's own magisterial surveys of these developments, from his *Substance and Function* through *Determinism and Indeterminism in Modern Physics*, he insists on a fairly simplified and schematic use of both as metaphor in a manner that would be accessible to any university student, using a basic definition of integral calculus that is no more complex than that used by Kant in the *Critique of Practical Reason*, and a definition of group theory that he could explain to an undergraduate philosophy club in a few paragraphs. He was nonetheless aware of the often daunting formalism of his argument, and would largely assume it as background after *Substance and Function* was published.

Only when placed in the context of the functionalism developed in Leibniz and Cohen's work does Cassirer's argument become fully intelligible. Such a reading will also suggest why the reception of the book would far exceed even its considerable influence in the philosophy of science, leading to the uniquely pervasive role in the social and human sciences that Hans Blumenberg and others have described.[20] Typical of this influence, for instance, is Cassirer's reception in law, suggested by Siegfried Marck's use of Cassirer's work and title as the basis of his 1925 *The Substance Idea and Function Idea in the Philosophy of Law*.[21]

Cassirer begins *Substance and Function* with a survey of several contemporary fields that demonstrate the functionalist turn he champions, of which we can briefly sketch his reading of logic, mathematics and chemistry before developing the broader argument of the second half of the book. Entitled "The System of Relational Concepts and the Problem of Reality," the second half of the book suggests the key theme of Cassirer's argument: the issue of how a system of relations or functions relates to reality, as a general problem of meaning in the sciences, as a specific epistemological problem and a moment in the phenomenology of experience.[22] At the center of this study, is the "advance from purely logical considerations to the conception of the knowledge of reality," and the problem of reality – as earlier in his study of Leibniz – will prove to be the keystone to Cassirer's project.[23]

Logic, the Science of Mathematics and the Transformation of the Concept of Number

Substance and Function aims to complete the development of modernism initiated by Leibniz, which Cassirer depicts as still held in check by a dominant reading of logic and being stretching to the foundations of Western philosophy in Aristotle. Until the beginning of the last century, Cassirer writes, whenever "the question as to the relation of thought and being, of knowledge and reality, has been raised, it has been dominated from the first by certain logical presuppositions, by certain views about the nature of the concept and judgment."[24] These presuppositions, he claims, have been overturned by the developments of the sciences in the previous century. The logic of Aristotle, Cassirer notes, "had not had to retrace a single step" since the medieval period even if it had "not also advanced a single step." In the modern period, however, it has been thrown into question both by the "historical development of science itself and [...] the systematic presentation of its content by the great scientists."[25] Looking at the recent history of the exact sciences of "arithmetic, geometry, chemistry and physics," Cassirer argues that each is dominated by theories of the manifold in which the intensive relation of part to whole creates a revolutionary transformation of transcendental logic, which in turn demands a new philosophical starting point for understanding the development of any science. "The work of centuries in the formulation of fundamental doctrines," he writes with uncharacteristic boldness, "seems more and more to crumble away; while on the other hand, great new groups of problems, resulting from the mathematical theory of the manifold, now press to the foreground."[26] In light of this change, "criticism now begins to be applied to those very doctrines that have persisted unchanged historically in the face of profound changes in the ideal of knowledge."[27]

The foundation of this criticism will be the functional theory and intensive logic developed in the Marburg school critique. The "theory of the manifold" proves analogous to Cohen's own study of intensive logic and the infinitesimal problem, which Cassirer writes "gains new confirmation" in developments in mathematics since Leibniz's calculus, but is now generalized and transformed through "the more inclusive" problems of the "analysis of relations," particularly in set and group theory.[28] While the "mathematical theory of the manifold" is developed from a similar starting point as Cohen's theory of the infinitesimal, it has a broader and indeed different application. The theory of the manifold "appears as the common goal toward which various logical problems [...] tend and through which they receive their ideal unity."[29] Characteristically, despite the extensive thematic links between the two philosophers, Cassirer refers to Cohen only three times in the text, and avoids nearly all use of Marburg terminology. In tracing the historical development of logic and the sciences themselves, Cassirer develops an immanent critique based on contemporary science.

Cassirer describes *Substance and Function* as a "critique of knowledge" (*Erkenntniskritik*) that will investigate the meaning of "concept" as it has developed in the modern sciences and as it affects "all the problems and solutions of nature."[30] Cohen's original use of the term "critique of knowledge" in the *Infinitesimal Problem* described the search for pure forms of transcendental logic within the activities of science; indeed, he claims that "the critique of knowledge means the same thing as transcendental logic."[31] Cassirer advances a similar project in attempting to find new modes of transcendental logic within the development of the sciences in the previous century. But far from remaining solely on the level of logic, "the new form that is beginning to take shape is also a form for a new content," and this new content has broad implications for "the problem of consciousness and the problem of reality."[32]

Cassirer begins his account by critiquing the traditional "logic of abstraction" initiated by Aristotle, and the related categories of knowledge dominated by the Aristotelian "thing," and the species/genus distinction.[33] Aristotle's logic, Cassirer writes, is closely linked to the concept of substance in its assumption that "only in a fixed thing-like substratum, which must first be given, can the logical and grammatical varieties of being in general find their ground and real application."[34] Aristotle's initial premise is to explain the world through categorization, a process modeled on the species concept of biology:

> The biological species signifies both the end towards which the living individual strives and the immanent force by which its evolution is guided. The logical doctrine of the construction of the concept and of definition can [for Aristotle] only be built up with reference to these fundamental relations of the real. Determination of the concept according to the next higher genus and its specific difference reproduces the process by which the real substance successively unfolds itself in its special form of being. Thus it is this basic conception of *substance* to which the purely logical theories of Aristotle constantly have reference.[35]

With this initial set of precepts, however, "the category of relation is forced into a dependent and subordinate position" to that of substance.[36] For this reason, in Cassirer's view, Western thought following Aristotle is dominated by parallel substantive ideas of the category of "thing and its attributes," a "whole and its parts" and a substantive logic of "subordination and superordination." Each species is based on abstraction in the sense of a simple negation of a particular substantive quality to determine broader species types.³⁷ The general model of development is based "merely on the principle of similarity," and stipulates as well that each species can stand under only one genus and at only one level of specific difference – any scientific translation between the genera of species constitutes a category mistake, in a phrase Cassirer uses throughout his career, a μεταβασισ εισ αλλογενοσ, a *metabasis eis allogenos* (transition to another genus).³⁸

The flaw of the Aristotelian vision of logic and being is its foundation on a particular form of negation. "If we merely follow the traditional rule for passing from the particular to the universal, we reach the paradoxical result that thought, in so far as it mounts from lower to higher and more inclusive concepts, moves in mere negations. The essential act here presupposed is that we drop certain determinations, which we had hitherto held; that we abstract from them and exclude them from consideration as irrelevant."³⁹ Beginning with Leibniz, and with increased tempo in the late nineteenth century, the metaphysics of substance was in Cassirer's view gradually replaced in the practice of science by a logic of relation or function. As we have seen with Leibniz and Cohen, these new forms of logic were understood as introducing forms of negation – and in Cassirer's view a new definition of abstraction – that are not merely subtractive but rather productive and determinative, a role Cohen found epitomized by the transcendental logic at the base of infinite or limitative judgments.

The foundation of understanding these forms of transcendental logic will be the function. "In opposition to the general logic of the generic concept, which, as we saw, represents the point of view and influence of the concept of substance," Cassirer writes, "there now appears the logic of the mathematical concept of the function."⁴⁰ This transition was itself first heralded by Leibniz's studies of nonsyllogistic logic, particularly as it developed from the Hamburg logician Joachim Jungius (1587–1657), a development that Cassirer later describes as presaging the development of logic in the late nineteenth century of "Pierce, Schröder, and Russell."⁴¹ The problem with the form of traditional logic is not that it is inaccurate, but that it is partial and unable to critique its own assumptions. "Logic, in its traditional form, is founded on the thought of identity, and seeks to reduce all types of connection and inference ultimately to identity."⁴² In the logic of the "manifold," Cassirer will instead claim to develop a logic that in the manner of Leibniz understands unity as difference.

Cassirer borrows a "drastic example" from Rudolf Hermann Lotze of the limits of abstraction based on similarities in traditional logic: "If we group cherries and meat together under the attributes red, juicy and edible, we do not thereby attain a valid logical concept but a meaningless combination of words [...]. Thus it becomes clear that the general formal rule in itself does not suffice; that on the contrary, there is always a tacit reference to another intellectual criterion to supplement it."⁴³ A new approach becomes apparent if we focus first on this "tacit reference" within an assumption of a plenum of meaning and attempt to develop a logic that recognizes and incorporates this horizon of meaning. Drawing on both Lotze and Benno Erdmann, one of Cassirer's graduate teachers, Cassirer suggests such a logic as essentially a variant on that suggested by Cohen's philosophy of origin and definition of limitative judgment.

The first insight of such a logic is to note that any particular logical distinction always contains a tacit intensive relation to the field in which it occurs. The traditional logic of abstraction, as Lotze had noted, already demonstrates its weakness when we notice how it translates into psychological terms, in that it is "never satisfied to advance to the universal concept by neglecting the particular properties without retaining an equivalent for them."[44] When we develop the concept of "metal" using gold, silver, copper and lead, we theoretically need to remove from this abstract concept all of the particular properties of these metals. We do not understand the concept of "metal" purely on its negative terms, as "neither of this or that [particular] hardness and resisting power." In practice, however, we tacitly add "the positive thought" that metal is "of *some* degree of hardness, density or luster" in a manner that is not explicitly recognized in the traditional logic of abstraction. Far from being a "psychological" hold over, this is for Cassirer a necessary feature of the transcendental logic of the function, and one that is equally necessary for both logic and phenomenology.

The process at work here resembles the functional attribution of variables, even as it is not traditionally recognized as such: "We represent this systematic totality when we substitute for constant particular 'marks,' variable terms, such as stand for the total group of possible values which the different 'marks' can assume. Thus it becomes evident that the falling aside of the particular determinations is only in appearance a purely negative one. In truth, what seems to be canceled in this way is maintained in another form and under a different logical category."[45] The key element of the form of negation suggested by the functional variable is its relation with the problem of continuity. The concept of function always contains a positive, constructive and open-ended quality epitomized by the active use of the "variable." In a mathematical variable, the nonspecialist can only imagine that the variable "works" when supplied with a specific content, but in Cassirer's transcendental logic this is not the case. The form of negation implied by the variable "maintains" a positive form and defines a positive plenum of meaning based on what a particular value "is not." Cassirer's operative model here appears to be Cohen's definition of limitative judgment, but he completely avoids this reference or terminology, and instead develops the argument solely using other contemporary sources.

Such functional variables can overlap in an object, say a particular metal, in such a way that they progressively determine its place within this differential logic. Similarly, Cassirer writes, "We can abstract from the particular color only if we retain the total series of colors in general as a fundamental schema, with respect to which we consider the concept determined which we are forming."[46] This tacit dependence of the determination of any phenomenon on a next order "form" that exists under a different, and usually unstated, category demands a transition from earlier formal logics to a new transcendental logic of the concept.

Cassirer's central argument is that the functional variable that defines a series of elements operates as a "point of view" to first make them visible, but is always of a necessarily different but intrinsic related logical order to the contents themselves. "The similarity of certain elements," Cassirer writes, "can only be spoken of significantly when a certain 'point of view' has been established from which the elements can be designated as like or unlike."[47] The sensuous presentation of a group of four elements is not sufficient, for instance, to define the concept of "four," for instance, which can only occur if the determination "four [...] is given by its place in an ideal and therefore timelessly valid whole of relations."⁴⁸ Whereas both Aristotelian logic and empiricism take it for granted that we can gain a sense of the "similarity" of contents as a "self-evident psychological fact," there is for Cassirer always a further process of functional definition enabling this judgment. There is, Cassirer writes, "a characteristic contrast between the member of the series and the form of the series," in which one always has a different logical form than the other – a difference that Cassirer claims is an "irreducible fact."⁴⁹ For this reason, "the content of the concept cannot be dissolved into the elements of its extensions, because the two do not lie on the same plane but belong in principle to different dimensions."⁵⁰ The "concept" as serial form is a generating principle, and it always demands a search for a different mode of logic than is present in any one element.

By series, Cassirer means a group of elements in which a "generating principle enables us to connect the individual members into a functional whole."⁵¹ On its most basic level, one aspect of Cassirer's definition of "form" is itself simply the generative rule for a particular series. A mathematical series (such as 1, 2, 3, 4...) is defined by the generating principle of each element, in this case the simple addition of one element to the previous. There is a crucial difference, and an indissoluble link, Cassirer writes, between the "member of a series and the form of the series": a member has no single meaning apart from its place in the series and conversely "there is no danger of hypostasizing the pure concept," that is the form, or "of giving it an independent reality with particular things" since it is determined by its members in its particular rule-form.⁵²

In this manner the functional definition of form avoids the Aristotelian definition of both "species concept" and "thing." The relation of part to whole here is analogous to the manner in which Cassirer noted the particular series of colors being inherent in the observation of any one of them, or of the perception of a point inherently and tacitly relating to positionality and the "relation of spatial 'coexistence.'" Here it can be seen that "form" is not a thing or a lens between subject and object, but an ordering principle for the particular. This principle, however, is not to be understood as simply an ideal imposition of the mind or as an empirical continuity in the world, but rather as the phenomenological foundation of any experience precisely in its appearance as "reality."

The logical nature of the elements of the series can be considered to have the logical form of variables in the sense suggested by Cassirer. A form or series is solely defined through its internal relations, and for this reason its meaning can change with any transformation of an element. Thus if we remove "3" from the series above (to have 1, 2, 4...), the meaning of the whole is completely altered. Importantly, it could now have several forms as a series, such as a series of numbers defined by the doubling of the previous number or the completely different series found in squaring of the previous number. Just as in gestalt psychology we might see figures with "multistability," for instance a Müller cube that appears to "shift" backwards or forwards in space, so here the "variable" elements define the whole form differently. In this regard, Cassirer writes in reference to the gestalt theory of "form-quality" initiated by Ehrenfels' "On the Qualities of Form [*Über Gestaltqualitäten*]" (1890), it is not the "mere coexistence of parts" that defines such structures, but "the reciprocal action they exert on each other."⁵³ Such action is already dynamically at work in any relations of series form in which variable elements relate both to an immediate setting and, negatively, to the entire range of possible permutations of experience.

Cassirer writes that this simple feature already suggests the general insight of the problem of series: "A series of contents in its conceptual ordering may be arranged according to the most divergent points of view; but only provided that the guiding point of view itself is maintained unaltered in its qualitative peculiarity."[54] Traditional logic tacitly relied on a principle of similarity that is here overcome: "The real weakness of the theory of abstraction is apparent in the one-sidedness of its selection, from the wealth of possible principles of logical order, of merely the principle of similarity."[55] This is evident in the simple model of the series itself, so that "side by side with series of similars in whose individual members a common element uniformly recurs, we may place series in which between each member and the succeeding member there prevails a certain degree of difference," such as "equality and inequality [...] spatial and temporal relations, or causal dependence."[56]

The application of series occurs in all concept formation and experience. In this regard the entire Aristotelian definition of logical abstraction is overturned, since "abstraction is no longer uniform and undifferentiated to a given content."[57] True abstraction is indeed in this definition most similar to "intention" as it had recently been defined in the new field of phenomenology, notably through the influence of Edmund Husserl's *Logical Investigations*.[58] Thus, citing Husserl, Cassirer claims that abstraction entails a "directed 'intention' [...] the intelligent accomplishment of the most diversified and mutual independent acts of thought, each of which involves a particular sort of meaning of the content, a special direction of objective reference."[59] Indeed, reversing its sensualist definition, Cassirer claims in this particular sense that "abstraction" is a principal feature not only of thought but of all perception and empirical experience, as well as of all transcendental logic, it is "the logical prius," since all series will always have a latent meaning and this new sense of abstraction defines this meaning. Content is now defined by what something "*means* in the system of knowledge; and thus, its meaning develops out of the various logical 'acts' which can be attached to the content [and] which imprint [*aufprägen*] upon it different objectively directed 'intentions.'"[60]

Although presented initially as a logical theme, the concept of functional variables thus proves equally applicable to a "pure phenomenology," and indeed for Cassirer both fields are rooted in the same transcendental logic and theory of judgment. The functional concept exists before any division of subject and object, so that Cassirer can claim it is equally valid in each realm. "Thus the circle of our subject is complete, since we are led, from the side of the 'subjective' analysis, from the pure phenomenology of consciousness, to the same fundamental distinction [*grundlegende Unterscheidung*]," he concludes, "the validity of which has been shown in the 'objective' logical investigation."[61] The concept of function performs this universal role, and provides a logic of experience that addresses "a deeper phenomenology of the pure thought processes" that is in some ways a harbinger of later intellectual movements such as the logical development of a calculus of indication by G. S. Brown, the development of a "logic of biology" by Cassirer's student Adolf Meyer-Abich or the later development (in part from the synthesis of these two fields) of systems theory.[62]

Substance and Function as a whole contends that all sciences could be understood using the overlapping system of generative series concepts. The "system of sciences," Cassirer writes, presents us with "the totality and order of pure 'serial forms'" at a given time, and a survey of some of their key moments will be the topic of the text.[63] The goal of *Substance and Function* will be to demonstrate and refine the meaning of such "generative principles" and variables in the sciences, and the surprising transformations among them.

Crucially, this project is based on a simple but fundamental assumption: that we first have a world, and indeed this particular world, which is for Cassirer a general basis of the analysis of experience, so that "the law of experience only 'results' from the particular cases because it is already tacitly assumed in them."[64] Or as Cassirer writes, "The advance from the individual to the whole, involved here, is possible because the reference to the whole is from the first not excluded but retained, and only needs to be brought separately into conceptual prominence."[65] The world is not given simply as "a" world or in single phenomenon, that is in things, and nor can it be understood through these means. It is rather an unfolding problem of generative relations between "serial forms." Any process of analysis is not a rational abstraction or destruction of the given, but rather a turn towards an immanence that is constantly open to dehiscence and further development. Although this "universal law" might appear to conclude with Kantian intuition of space and time or categories, in Cassirer's reading these too are to be defined purely in terms of difference and capable of further rearticulation – as the progress of geometry and physics in the nineteenth century had demonstrated. A transcendental logic that begins with the form of productive negation stipulated by Cassirer's model of functional variables aims to explain both the continuities of this process and its possible transformations.

On the first page of the book, Cassirer describes the general theme of *Substance and Function* as having developed from "investigations concerning the concepts of the series and of the limit," with the latter presumably beginning from Cohen's work on the infinitesimal and the former his own innovation. The question of how this concept of the "series" overlaps with the related concept of the "limit" in the sciences is never fully resolved as a single model in the text, and perhaps intentionally so.[66] Cassirer writes that the "specific results of his [mathematical study of the series and limit] could not be included in the general exposition of this book."[67] In general, however, the law of production of the series is related to the concept of a limit in relation to a particular element, particularly once this model is applied to perception and judgment. This is ultimately found in Cassirer's central definition of judgment and representation, as noted above, where "each particular member of experience possesses a symbolic character, in so far as the law of the whole, which includes the totality of members, is posted and intended in it. The particular appears as a differential, that is not fully determined and intelligible without reference to the integral."[68] Even in basic logic, this same relation occurs, as Cassirer suggests by summarizing his argument in reference to the contemporary logician, Hermann Lotze: "But precisely to the extent that the concept is freed of all thing-like being, its peculiar functional character is revealed. Fixed properties are replaced by universal rules that permit us to survey a total series of possible determinations at a single glance."[69]

The manner in which a total series is grasped in its "possible determinations at a single glance" is ultimately a limit concept, one that is at work in all conscious functioning. In this regard, Cassirer writes, judgment is not particular to higher order logical definitions, but proves true of any perception whatsoever: "The process of perception is not to be separated from that of judgment."[70] Unlike the use of series and group in the work of his teacher Benno Erdmann, or the phenomenology of Husserl, Cassirer claims (to lapse into technical language) he posits no distinction between "first and second order" objects, matter and form, hyletic and noetic factors.[71] Against these dualisms, Cassirer argues that all objectives of intention are defined solely by relational form. In this manner any particular "objective" of thought fits within a "systematic totality [*Inbegriff*] to which those marks belong as a special determinations."[72] Cassirer further demonstrates this theme in the problem of number in mathematics, through the example of Dedekind's "cut" that we discussed earlier, as well as group theory and geometry.

Series and Group

What Cassirer describes as the "great new groups of problems" addressed by *Substance and Function* focuses on the "mathematics of the manifold" found in "the series and group."[73] First introduced in relation to logic, and then described in the overview of mathematics, these become the basis of Cassirer's theory of process, change and history in the second half of the book. *Substance and Function* begins with a study of the "serial forms" that define the "system of the sciences," and this definition of form is shaped first by the figure of series, and then the more inclusive figure of the group.[74]

Cassirer elaborates his idea of series in *Substance and Function* through the analogy of mathematical group theory, which is again applied as equally a logical and phenomenological schema. In the simplest sense, a group can be considered a collection of interconnected series or operations. Cassirer defines a group as follows: "A totality of operations forms a group, when with any two operations their combination is also found in the group, so that the successive application of different transformations belonging to the totality leads only to the operations originally contained in it."[75] The importance of this model is that "here, in a manner of speaking, the dualism of 'element' and 'operation' is annulled: the operation itself has become an element."[76] In this manner Cassirer can claim that he has developed a completely immanent definition of form. The model of the series effectively eliminated any dualism between part and whole, and thus "avoided any danger of hypostasizing the pure concept, of giving it an independent reality with the particular things."[77] The parts defined the whole and, conversely, the generative order of the whole, of form, in turn defined each part. Yet the whole also revealed a different qualitative meaning even in this unity: it "belongs in principle to a different dimension" that cannot be reduced to "simple sensuous intuition" even as it shapes this intuition.[78] With the move to group theory, even the dualism of operation and element is overcome, and with it a new awareness is provided of both interconnectedness and different orders of meaning. As Cassirer later puts it, the application of group theory to experience is "nothing other than an exact expression of what we mean by a 'self-contained' sphere, or system, of operations."[79] In this regard, Cassirer's *Substance and Function* forms one of the founding texts of the later intellectual movement of "systems theory" both in engineering and in sociology.

Cassirer links his approach, and form of negation, with the work of his teacher Benno Erdmann, who had developed a form of logic and definition of perception tied to group theory in mathematics.[80] The group of indications by which we unify an object create an internally consistent pattern within an order that in turn suggests its place in a whole. Through the idea of continuity Cassirer had advanced in relation to the series, we can imagine how even a "self-contained" system of operations also relates to the plenum of meaning as a whole. The operations that define our experience of a certain form of metal, iron for instance, principally apply in relation to the basic understanding of "metal" as a concept, but could also be reinscribed into a number of larger constellations of operations and groups, such as "electrical conductors" or "tools of warfare." What appears within a group "as an element may prove, otherwise considered, to be a very complex system [...]. Thus there arises a complex whole of intermeshed [*ineinandergriefende*] syntheses, which stand in a certain mutual relation of superordination and subordination."[81] Cassirer's writing will use the terms series and group somewhat erratically, presumably due to the relative rarity at the time of the second term in particular, with series usually used more often for temporal developments and group used for complex phenomenon in the present, but also with series sometimes considered as a subset of group.

The Invariant Theory of Truth and the Problem of Reality

The importance of group theory becomes apparent in relation to invariants within a group, and thus to Cassirer's "invariant theory of truth."[82] Although present as "an ideal" in Leibniz, Cassirer writes, the theory of invariants was most clearly defined in Felix Klein's *Erlanger Program* (1882) in geometry.[83] Invariants define which elements or functions remain consistent under a particular transformation in mathematics, geometry or indeed, in the context of Cassirer's structural genesis as developed so far, any phenomena. The mathematics of group theory approaches Leibniz's "original vision" of invariance, in which "permanence is not related to the duration of things and their properties, but signifies the relative independence of certain members of a functional correlation, which prove in comparison with others to be independent moments."[84]

The historical inception of the theory of invariance is found in projective geometry, the geometry developed in the Renaissance to explain the optical projection of three-dimensional forms on a flat plane in a camera lucida or in perspective drawing. Projective geometry, as noted earlier, explains how invariants emerge, for instance, when the regular space of Euclidean geometry is "projected" against a line at infinity, as occurs in perspective painting. Far from being a curiosity, the forms of invariants developed in projective geometry legitimized the varieties of non-Euclidean geometry, beginning with the work of Girard Desargues (1591–1661). A number of surprising but valid invariants were found to define different projective approaches.[85] Thus under the principle of duality in projective geometry, points and lines prove to be fully interchangeable in all theorems.[86] The mathematician Felix Klein (1849–1925) used the principle of invariants to redefine all of geometry as the study of its different forms of invariants: "Geometry

[…] is distinguished by the fact that only such properties of space are called geometrical as remain unchanged in a certain group of operations."[87] Moreover, "each special form of geometry is then coordinated with a definite group of transformations […]. The conceptions of constancy and change are seen to be mutually conditioned by each other."[88]

For Cassirer, form is defined as the set of invariants in a group not changing under a set of transformations. "All experience," he writes in *Substance and Function*, "is directed to gaining certain 'invariant' relations, and first in these it reaches its conclusion."[89] Notably, in its simplest sense an invariant form shares features with Ernst Mach and Christian von Ehrenfels' definition of gestalt. The same melody within two keys, for instance, or the same words written in two colors, are invariants despite, strictly speaking, different "elements."[90] As with the model of the series, in more complex situations different sets of invariants might coordinate "the same" manifold differently. With the group defined through a combination of operations, Cassirer is thus able to find a basic model for a multivariate logic that, as he suggests at the beginning of his text, would contrast with classical logic.

In analogy to projective geometry and later non-Euclidean geometry, this very broad definition of form and concept allows for more fundamental and surprising transformations than possible under the Aristotelian definition. Thus any particular "object" for Cassirer is defined primarily through its place within a system of invariants rather than, as for instance in Locke, its simple location in time and space. A cue ball, for example, might be understood as a geometric example of a sphere, a chemical example of a certain form of polymer plastic, an example in physics of an object for the study of shading in art, or an object that is only intelligible within the rules and further equipment of a game of pool. Importantly, in each case, the "form" and "matter" of the object are defined differently against a various set of invariants. In the theory of invariants we find the culmination of Cassirer's argument that "matter *is* only with reference to form, while form, on the other hand, is valid only in relation to matter."[91] For Cassirer, the only "full" definition of the object "cue ball" was not found by abstracting it from all of these definitions to its place as a "thing" in time and space, but through the full, and unlimited, definition of its possible functional relations.

As mentioned earlier, for Cassirer the Kantian *a priori* of the categories and the intuition of time and space are themselves defined as ultimate invariant functions, but as such they are always in theory capable of being redefined or recontextualized as a new form of invariant.[92] Invariants are considered *a priori* only when they appear to be a necessary precondition for any experience whatsoever. "Only those ultimate logical invariants can be called *a priori*, which lie at the basis of any determination of a connection according to natural law."[93] Thus for instance: "According to the critical theory of knowledge, space and not color is '*a priori*,' because only space forms an invariant in every physical construction."[94] "Space" is not defined in the sense of Newton's absolute space, however, but functionally as the totality of topological transformations. "Space as such," as Cassirer later formulates it, "conceived as the mere possibility of coexistence, has no definitive and unambiguous form but stands equally open to the most diverse forms of formation.[95] Space thus appears to be a true universal invariant, but not in the sense of the absolute space of Kant's intuition of space, but in the Leibnizian definition of a pure system of relations.

Similarly, "laws" of nature are not presumed to be permanent, but through a theory of invariants prove to be merely more stable forms of "rules" of nature. "The 'fundamental laws' of natural science seem at first to represent the final 'form' of all empirical processes," Cassirer writes, "but regarded from another point of view, they serve only as the material for further consideration. In the further process of knowledge, these 'constants of the second level' are resolved into variables."[96] A "general" form to experience can only be defined as a convergent sense of forms in the Leibnizian sense, but this could only exist as a possible series defined towards an unlimited future.[97]

Chemistry and the Functional Definition of the Atom

Chemistry is of particular epistemological importance for Cassirer because of its "intermediary position" between an ideal and empirical science. "Chemistry seems to begin with the purely empirical description of the particular substances and their composition," Cassirer writes, "but the further it advances, the more it also tends toward constructive concepts."[98] For this reason, the field will also be exemplary in summarizing the argument of the first half of Cassirer's text concerning the constructed nature of all scientific knowledge, that is, the constructed nature of "substance" from "function," as well as the role of series and group in forming such definitions. On the one hand, a chemical element or compound literally is an example of a series when it is categorized, respectively, in the periodic table or in relation to related compounds. Thus, for instance, when there appears "a new chemical compound in a series of known compounds of similar constitution" it is immediately "fixed in its definition character as a member of a series."[99] Yet the meaning of the atom, to take an example Cassirer develops at length, also exists in a number of other series and as an example of an invariant that changes in the course of its history.

The starting place for Cassirer's argument is again Leibniz, for whom space and time are not given absolutes with corresponding objects "in" them, but modes of determination. Cassirer's work is again explicitly predicated on the assumption that Cartesian and Newtonian definitions of space entail an unwarranted set of assumptions that Leibniz's definition avoids. "For [Leibniz]," Cassirer writes at the beginning of his survey of chemistry, "both concepts [space and time] were only another expression of the thorough-going spatial and temporal determinateness, which we demand for all being and process."[100] There is no such "thing" as space, in this reading, only the particular "objectives" of a particular formal determination of spatial relations. The original atomic idea of Leucippus and Democritus has to be qualified against this Leibnizian definition from the start, since atoms cannot immediately be assumed to be self-evidently "things" in a given "space." Extension and bodies are both the product of a common fusion of theory and data, schema and sensation, which has historically changed in each era of understanding the atom. "We inscribe the data of experience in our constructive schema, and thus gain a picture of physical reality; but this picture always remains a plan, not a copy, and is thus always capable of change."[101]

It is on this basis, removed from Cartesian and Newtonian assumptions about space, that a comprehensive understanding of the modern development of chemistry and physics for Cassirer begins. The same element of "construction" found in mathematics, and the same hybrid concept of ideality and reality, Cassirer argues, proves to be present in a new form in the empirical sciences. The old idea of element that developed until Lavoisier was "as a generic property belonging to all the members of a definite group and determining their perceptible type."[102] That is, from certain "striking sensuous qualities" the element was hypostatized as a general form, so that historically, for instance, sulfur was considered the basis of combustibility.[103] Gradually the goal of finding quantitative proportions between elements in their combination was added to this conception, leading to the law of the definite proportions for their admixture and combination. With Dalton, these proportions were converted into an "equivalence number" for each of the elements, and with this a relative atomic weight.[104] Through the work of numerous researchers, this gradually led to a unitary table of atomic weights, developed through such innovations as in Avogadro's understanding of the relation of molecules to volume, Dulong-Petit's understanding of heat determination, and Mitscherlich's idea of the isomorphism of crystal form suggesting equivalent and parallel connections of atoms.[105]

The various moments of this process demonstrate a related "logical direction of thought" sustaining the concept of the atom even as it emerged among its competing interpretations.[106] Atomic elements were not of course truly "given" in themselves as objects of physical research – that is individual atoms were not directly "seen" or even directly manipulated without some form of mediation – but are hypothesized based on ideal forms. The initial empirical research on numerical relations in atoms was, for instance, codefined by the "Pythagorean doctrine of the 'harmony' of the universe" combined with the sensuous experience in experiment.[107] These ideals were in turn developed through a further ideal "latent assumption" that any particular experiment will "remain valid for all times and places," which forms the basis of the science of chemistry.[108]

The atom is thus first and foremost a *conceptual* tool through which experience is understood. It acts as a functional variable for a series of observations and correlations of evidence. In this regard the concept of the atom is characteristic of any scientific concept: "It becomes clear that the atom is never the given starting-point, but always only the goal of our scientific statements."[109] Even as each researcher often had different specific assumptions about the atom, the general concept of the atom itself operates to combine all of these systems of research into a broader series. "The complicated relations between certain systems are not expressed by our comparing each system individually with all the others," Cassirer writes, "but by putting them all in relation to one and the same identical term."[110] The relation of this term, in this case "atom," to its context becomes one that on closer inspection extensively defines the "content" of this term, even as new functional definitions of this content enable new fields of research. In this regard, Cassirer writes, "The atom functions here as the conceived unitary center of a system of coordinates, in which we conceive all assertions concerning the various groups of chemical properties arranged [...]. In truth we are concerned not so much with relating the diverse series to the atom, as rather with relating them reciprocally to each other through the mediation of the concept of the atom."[111]

Although the atom is regarded as "substance" as a practical definition in order for empirical research to proceed, in truth this is a useful and "real" fiction based on a concept that then proves to be of ever greater functionality to the point of a complete change of definition of "substance" itself.[112] As we move from the concept of the atom as the smallest thing, to that of an object with orbits, and then to an electron- or quantum-based entity, its "content" absolutely changes even as its function as an ideal limit case remains the basis of research. "This function," Cassirer summarizes, "remains as a permanent characteristic of the concept of the atom, although its content may completely change; thus *e.g.* the atom of matter becomes the atom of electricity, the electron."[113] As Cassirer later summarizes it in his *Determinism and Indeterminism in Modern Physics* (1936): "All branches of science take part in this consolidation of the atomic concept, a consolidation, however, that is the exact opposite of its immediate hypostasis, its immediate conversion into an object [...]. And whenever, by means of the atomic concept, a new field of experience is opened up, the atom itself gains, as it were, a different and new appearance."[114]

The "concepts" of chemistry overlap as variable functions to define an object in a manner loosely analogous to the concepts of mathematics, but these "new appearances" again by no means entail a reduction of chemistry to mathematics or even pure functional relations. Rather the "concepts" and functions of chemistry both reveal and construct reality in the empirical sciences in a new manner, just as a similar process of correlation of functional variable and series will provide new forms of construction for law or the social sciences. The mind "reaches out" to develop the concept of the atom from ideal determinates or functional variables, which are always realized in the intuition of the hypothesized object as mediated by instrumentation, experiment, etc. The "object" of consideration, the atom, as with the mathematical object, can be defined as a "being-different-than," but this is now not a pure mathematical relation, but rather defines its place in a relational field of ideas and practices, concepts and intuitions.

In this regard, Cassirer's history of the atom of chemistry returns to the specific meaning of the "critique of knowledge" first set forth by Cohen, which develops a history of the relation of part and whole, the perceived and its context, as a history of the ever protean fusion of concept and intuition, or as he describes it elsewhere, "thought and being."[115] Cassirer summarizes one aspect of this process when he writes that the "atom of chemistry is 'Idea,'" in the strict meaning Kant gave this term – in so far as it possesses "a most admirable and indispensably necessary regulative use, in directing the understanding to a certain aim, towards which all the lines of its rules converge and which, though it is only an idea [*focus imaginarius*], that is, a point from which, as lying completely outside the limits of possible experience, the concepts of the understanding do not in reality proceed, serves nevertheless to impart to them the greatest unity and the greatest extension."[116] The atom is in this case functions as the limit of the whole "totality and order of 'pure series forms'" within the development of the sciences. The atom is also a limit in a different sense than Idea, in that it is presented phenomenologically in experiment as the limit of perception, allowing us to pull together various empirical and technical elements of a diverse series of phenomena into one coherent scheme. It is experienced as a unity-in-difference through its relation to the manifold of difference and structure in which it develops.

Cassirer's approach to the empirical sciences in *Substance and Function* is in this regard consonant with one of Cohen's most controversial conclusions, epitomized by his statement that "the stars are not given in the heavens, but rather we indicate these objects as given in the science of astronomy."[117] Cohen's statement was widely attacked as implying a radical constructivism, indeed one misunderstood to suggest the arrogance that the stars themselves were "created" by astronomy.[118] Cassirer's study of the history of chemistry implies the same relation for the microscopic world as Cohen's quote for the macroscopic. Just as the concept of the atom has radically changed with new definitions from the electron to quantum electrodynamics, so we can imagine the meaning and experience of our intuition of "stars" similarly changing with radical transformations in the functional network of concepts in which they are defined. That there is a reality to the "stars" is not questioned by Cohen's comment, or condescendingly subordinated to human science. Rather the opposite: Cohen intends to leave room for future realities of the sciences that are not yet grasped, and to ensure that each science is considered only ever partial. That there is *a* reality to the stars or atoms is not questioned; that we have access to *the* absolute reality of them is. The same argument, as Cassirer will note in *Freedom and Form*, carries over to the social and human sciences as well where the idea of "person" is itself fundamentally defined by procedures of law and politics, even as this does not so much constrain as first realize a notion of freedom.

An extreme example from the world of science fiction can clarify this central aspect of Cohen and Cassirer's work. If, following the outlines of Olaf Stapledon's classic 1937 novel *Star Maker*, we were in the future to discover that in fact the extremely high-temperature environment of our sun and other stars somehow contained or indeed were themselves intelligent beings, we would have to fully redefine the meaning of the term "star," and we would do so within a new development of astronomy (presumably an entirely new branch of astrobiology). Far from arrogantly presuming that the sciences "create" reality, Cohen's insight is designed to protect us from the arrogance of thinking we know, or experience, a completed reality in the present.

"The System of Relational Concepts and the Problem of Reality"

In the second half of *Substance and Function*, Cassirer describes the general implications of his study of the exact sciences for his philosophy, clarifying both the specific figure of the "mathematical manifold" and situating this model as the basis of a theory of history and reality. Cassirer's conclusions are similar to those of his study on Leibniz, and develop his strongest argument for how the individual (understood in the most universal sense to include a spectrum of objects or subjects, things or persons) is revealed through forms of knowledge, even as collectively these forms of knowledge are open to reflexive transformation. Turning to the evidence of both the exact sciences and contemporary empirical psychology, Cassirer again argues against the opposition of subject and object, both of which are resolved into transformations in the general field of knowledge. "If we consult immediate experience unmixed with reflection, the opposition of 'subjective' and 'objective' is shown to be wholly foreign to it."[119] To the degree these terms are useful at all, it is only as part of continuum determined by the rate of change of experience:

> We find connections that hold their ground through all further experimental testing and through apparently contrary instances, and remain steadfast in the flux of experience while others dissolve and vanish. It is the former, that we call "objective" in a pregnant sense, while we designate the latter by the term "subjective." [...] The result of thus deriving the distinction between the subjective and objective is that it has merely relative significance. For there are no absolutely changeable elements of experience at any stage of knowledge we have reached, any more than there are absolutely constant elements."[120]

The variable relation of "the subject and object of knowledge," understood as one of fixed and fluid relations, is for Cassirer a fundamental relation that constitutes each era, and can be historically reconstituted, as he demonstrates in *The Knowledge Problem*. Indeed, "this one opposition conceals all others within it and can progressively develop them."[121] Cassirer argues that in avoiding the subject/object opposition, an entire set of dichotomies of knowledge is avoided: "Thus the question is no longer what absolute separation underlies the opposition of "inner" and "outer," the "presentation" and the "object"; the question merely is from what standpoints and by what necessity does knowledge itself reach these divisions [...]. The same content of experience can be called subject and objective, as it is conceived relatively to different logical points of reference."[122]

"Knowledge itself" is not absolute, but is defined in its simplest sense as the system of ordering signs, hypotheses and practices in a given era through patterns of invariants. The basis of the structure of knowledge is in itself historical and relative, even as it provides access to hypothetical "invariants."[123] These include "ultimate invariants" that approach – but only as limit ideas – the absolute invariants of Kantian categories and intuitions of time and space, but within the larger model of group theory these are now understood as always theoretically capable of reinscription into a more encompassing form.[124]

As with *Leibniz's System*, the overall argument of the second half of the book is with a "copy theory" of knowledge, in which "the world" is somehow copied into knowledge, truth or the "mind." Since for Cassirer any particular sign or object was itself related to the whole, a new model of knowledge is necessary that avoids this fundamental opposition:

> The particular element, which serves as a sign, is indeed not materially similar to the totality that is signified – for relations constituting the totality cannot be fully expressed and "copied" by a particular formation – but a thoroughgoing logical community subsists between them, in so far as both belong in principal to the same system of explanation. The actual similarity is changed into a conceptual correlation; the two levels of being become different but necessarily complementary points of view for considering the system of experience.[125]

Rather than a binary relation of appearance/reality, sign/signified, each sign now relates tacitly to the entire system of meaning or experience.

The deepest continuity of the historical exegesis in *Leibniz's System* and the systematic argument of *Substance and Function*, however, is in Cassirer's definition of truth and reality. Cassirer writes that the signal innovation of *Substance and Function* is his approach in solving what he describes as the fundamental problem of the modern era: the "confusion of truth and reality."[126] By defining "truth" as a copy of a given reality, and thus assuming they are the same, modern science fails to understand both its own limitations and potential for creativity. The converse argument, for the subjective or solipsistic projection of truth and reality onto the world, is based on the same false definitions of both terms, only now turned against its own claim to absolutism.

Following his readings of both Leibniz and Cohen, Cassirer instead describes the various sciences as approaching particular versions of the truth through increasing patterns of internal consistency in a manner similar to Leibniz's definition of truth. As we saw, this held that for a statement to be true, it is not necessary that the predicate "perfectly and without remainder resolve into the subject, but only that a general rule of development can be perceived so that we can grasp with certainty that the difference between the two becomes ever smaller."[127] Cast through a new structural model based on group theory to explain broader truth claims, Cassirer uses this definition to redefine the meaning and transformation of the particular sciences. Truth is defined not by the "accuracy" of the copy, but by the relative movement of signs towards greater comprehensiveness and the relative reduction of obscurity: "The abolition of an absolute standard in no way involves the abolition of differences in value between the various theories. [...] The system and convergence of the series takes the place of an external standard of reality. Both system and convergence can be mediated and established, analogously to arithmetic, entirely by comparison of the serial members and general rule, which they follow in their progress.[128]

The difference between competing theories and values within these theories can be considered analogous to a converging series. In this manner, Cassirer recasts Leibniz's theory of truth into the more complex model of the series and group theory. In *Substance and Function* the electron theory of the atom over time proves to be more effective than the orbital theory, to both more accurately predict and describe phenomenon and to better lead to new theories. Within the electron theory one technique of analysis can prove more accurate than another. It is not necessary to claim access to an "absolute" reality in order to claim one series is more effective than another for one purpose, although nor is there any absolute given preference for one series over another.

The problem of the "knowledge of reality," Cassirer states at the beginning of *Substance and Function*, is in turn the key to redefining the "original opposition of thought and being" and with it nearly every classic opposition in earlier philosophies.[129] Following again on both Leibniz's and Cohen's definitions of reality, Cassirer holds that reality is not a thing but a limit concept which cannot be defined by any one science or field of endeavor. This appears to be what Cassirer means when he claims that "the system and convergence of the series takes the place of an external standard of reality."[130] Within the invariant model of group theory, there is no external reference, but rather forms of invariance that appear through the "convergence of the series." Reality in itself is defined as a temporal process of metamorphosis and development, defined neither through "subjective" construction nor as an "object" outside phenomenon.

Indeed, for Cassirer the entire understanding of reality based on an absolute or given spatial framework has been a fatal mistake for modern philosophy. So severe has this mistake been, Cassirer writes, "that all questions connected with the concept of the real are considered solved, as soon as the question of the reality of the 'external world' has been finally decided."[131] Instead, following Leibniz, a *provisional* sense of reality is determined within the logic of each science by the convergence of series surrounding a certain problem as they define a temporal order, so that: "If we proceed from the particular content of experience at a particular moment of time, there are given in it [...] certain *lines of direction* [...]. As science grasps the whole of these demands [...] it progressively gains the concept of the real."[132] Reality is itself an ideal functional concept that, as in Leibniz and Cohen, mediates between a particular experience of the Real and an absolute possible horizon of reality.

Using the structural concept of group theory, Cassirer is able to describe how a given science or field is capable of nearly complete permutations within its definition of reality even as it progressively pursues specific truths. Cassirer considers the critique of the relation of truth and reality as the basis both for an attack on the tacit ontology of modern science and for a productive new definition of the phenomenology of experience. Cassirer's theory of reality is the basis of his critique of the contemporary ontologies of Hartmann and Heidegger, as well as of contemporary life philosophy.[133] Even though in *Substance and Function* the problem of reality is largely confined to the natural sciences, it forms the capstone of both Cassirer's phenomenology and theory of liberalism as based on multiple forms of "reality" in *Freedom and Form*.

Historical Transformation and Paradigm in Cassirer's Theory of Structure

Cassirer's use of group theory combines a model of structuralism with a means of describing the genesis of new problems, one that will be applicable both on the level of individual experience and historical change. The importance of Cassirer's innovation in this regard can be measured by placing his theory in the context of the theory of structuralism on the one hand, and the practice of the history of science on the other. As in the linguistic structuralism of Ferdinand de Saussure and the Slavic schools of linguistics, or the anthropological structuralism of Claude Lévi-Strauss, the model of the group and series defines any particular element in terms of its relation to a structural whole. A comparison of Cassirer's theory in *Substance and Function* with these forms of structuralism provides a helpful summary of his system. In a manner more comprehensive than most forms of structuralism, Cassirer holds that the element is first created and defined through structure. This is the value of the mathematical models of series, limit and group, which demonstrate that any element is always defined by a series-form, and further that the operation through which this occurs can in turn be understood as part of a further group of operations.

A comparative advantage of Cassirer's theory over later forms of structuralism is that the system of series or groups for understanding any element or operation can be arbitrarily chosen, so that the system is perspectival: an element can "mean" something radically different in different groups, or within the evolving structure of one group. Since elements are defined purely negatively, by their place within a group, they have no positive value in themselves but rather indicate a place, much like Dedekind's definition of number or Cohen's definition of the infinitesimal. Unlike most varieties of structuralism, the system of signs taken to be at work is not fully arbitrary, but is rather motivated by earlier conditions: systems have inertia. There are real invariants in this system, but unlike other varieties of neo-Kantianism, these invariants are always themselves taken as capable of reinscription within a new system, as notably occurs for instance to aspects of the Kantian categories in relation to quantum theory.[134]

Unlike nearly all later forms of structuralism, however, Cassirer's emphasis on the interrelation of series and group provides a new understanding of structural genesis, that is, the question of how new phenomena occur within the structure. Any change within an operation in the system of groups potentially causes an overall change to the form of the whole. This is not always the case, however, since it is possible to have some changes to a structure that cause no noticeable changes in values up to a certain point, much as in chemistry below certain tolerances of titration a solution remains the "same," while above it the solution completely changes.

Cassirer's model resembles Saussure's famous analogy of a chess game to the synchronic and diachronic development of language, in which every "turn" changes the structure of the whole.[135] For Saussure, however, this model of a set of operations is an occasional metaphor: his definition of the structure and his use of linguistic analogy has no equivalent means of explaining functional patterns such as exists in both the rules of chess and in a particular game in the manner of Cassirer's use of group theory. Indeed, it is notable that since his system is initially based on the difference of purely arbitrary phonemes, it is arguably based on an identical series of fully random substitutions of a form that Cassirer's logic explicitly critiques. Cassirer's uses of the variable as the basis of functionalism instead entails both a difference within each element – that of continuity – and the ability for each functional element, whether of series or group, to be a part of a fully different operation.

Using the mathematical analogies of series and group, Cassirer in short arguably develops a more sophisticated first sketch of structuralism in his 1910 essay than Saussure did in his lectures of roughly the same period of 1906–1911. It is not surprising for this reason that some notable thinkers influenced by structuralism, such as Pierre Bourdieu and Roman Jakobson, would consider Cassirer's work as foundational. The most important consequence of this variety of structural genesis occurs in Cassirer's description of the history and philosophy of science in *Substance and Function*. Cassirer describes his basic model for a dialogic history of science using the example of Newton:

> Such principles as, for example, those on which Newton founds his mechanics, do not need to be taken as absolutely unchanging dogmas; they can rather be regarded as the temporarily simplest intellectual "hypothesis," by which we establish the unity of experience. We do not relinquish the content of these hypotheses, as long as any less sweeping variation, concerning a deduced element, can re-establish the harmony between theory and experience. But if this way has been closed, criticism

> is directed back to the presuppositions themselves and to the demand for their reshaping. Here it is the "functional form" itself, that changes into another; but this transition never means that the fundamental form absolutely disappears, and another absolutely new form arises in its place. The new form must contain the answer to the questions, proposed within the older form; this one feature establishes a logical connection between them, and points to a common form of judgment, to which both are subjected.[136]

Again, the form of knowledge in a particular system can radically change through a transition in any element of it, just as within a group any change in an operation in the group can change the meaning of the whole. This leads to a system that is similar to the "punctuated equilibrium" of evolutionary biology, that is, a system in which models can be relatively stable for long periods of time and suddenly change. In this way, Cassirer writes, "the instrumental character of scientific concepts" is not contested, but each operation is placed in a "projected unity" that is capable of change.[137] The model for this project unity is that of group theory, in that every operation is defined by one coherent group of interrelations. Cassirer is presumably suggesting, for instance, that Newton's laws of gravitation explain the "group" of operations defining the relation of gravitation among the planets, a group that is later reinscribed as a special case within Einstein's general theory of relativity. Ultimately, it is only the historical permutations of a model as a whole that can be taken as "true." Thus Cassirer concludes, "No single astronomic system, the Copernican as little as the Ptolemaic, can be taken as the expression of the 'true' cosmic order, but only the whole of these systems as they unfold continuously according to a definite connection."[138] Truth resides in the development of invariants within a particular system; the reality of one "cosmic order" is only given as the never given totality, extending into the future, of these systems.

"These concepts," Cassirer continues, "are valid, not in that they copy a fixed, given being, but in so far as they contain a plan for possible constructions of unity, which must be progressively verified in practice."[139] In this manner, entire systems of facts can be absorbed from one system into another yet have their meaning utterly changed, such as the observations of Brahe, which "enter into the system of Kepler, although they are connected and conceived in a new way."[140] The basic concept of paradigm shift as defined later by Ludwik Fleck and Thomas Kuhn is in part suggested in this passage, a connection Kuhn later highlighted and recent observers such as Michael Friedman have confirmed.[141] Cassirer does not, of course, address Fleck and Kuhn's questions of the biases of knowledge or sociological groups in a similar manner.

So complete can the transition between forms be, however, that there is no positive element connecting the two other than the continuity of the history of problems, or even *a* problem, that they are addressing. In this regard, the most radical feature of Cassirer's scheme is that "the new form must contain the answer to the questions, proposed within the older form; this one feature establishes a logical connection between them."[142] The logical series itself is thus governed by continuity between elements, but because of the possibility for a complete transformation of a group through a change in any one of its elements – through, for instance, a new answer to a question as the "one feature" necessary for logical connection – this entails a potential for a nearly complete transformation of meaning between historical epochs or problems. Cassirer's approach has an important correlation with the philosopher Hans Blumenberg's theory of epistemic change based on question and answer positions, although Blumenberg interestingly reverses the directionality of the problem: each age receives from its predecessors "answer positions" to which it forms new questions.[143]

The developmental structures of each of the natural sciences themselves each have to be understood in the context of other forms of meaning. Indeed, not only is "the 'individual' reality" or moment of experience open to change and reinscription within each science, but it also has to be understood on a level "beyond the circle of scientific concepts and methods" since the "individual of natural science" exhausts "neither the individual of aesthetic consideration nor the ethical personalities, which are the subjects of history."[144] Again, "individual" is used in the same very broad sense as in Cassirer's reading of Leibniz, and presumably derived from the concept of the monad: the individual is any particular determined by the broader structure of a field or science, whether subjective or objective. "The individual, as an infinitely distant point," Cassirer writes, "determines the direction of knowledge," even as it can never be known in itself.[145] For this reason it cannot be totalized by science, aesthetics, ethics or any other field. There is never an absolute definition of an individual outside of a particular form and a particular definition of knowledge.

Thus for instance, the chemical "compound" is defined as "the real scientific expression of the empirical reality of the body" not simply because it demonstrates the body's composition, but because of the manner in which it operates as "an individual 'thing' or particular event [that] stands within the totality of real and possible experiences."[146] An individual compound is a relational determination correlating to all features of the concept of that compound: its relation to other substances, its qualitative appearance to human sense, and its historical use and discovery. In this manner the definition of the compound is both strictly defined and open to permutation, since "it inserts it into various typical series, and thus refers to the totality of such structures."[147] On the one hand, this is simply a return to the initial argument of the definition of element by series, but on the other in its reference to the "totality" of these structures it returns to Cohen's understanding of the "Real" and "reality" since this totality is by definition infinite, even as it has very clear local rules.

By arguing that the true individual can only be defined in the relations of the sciences themselves, Cassirer displaces the problem of the individual into a teleological problem, which is now defined as a limit ideal to science as well as to history, aesthetics and ethics. In this regard, *Substance and Function* also reinscribes the earlier argument of the secular meaning of Leibniz's theodicy as a teleological development of reality. Since reality is always a developmental problem occurring through multiple fields, the particular "Real" itself is grounded on an overdetermined history and teleology. Cassirer thus concludes the first section of the book by noting that by placing science within the history of aesthetic and ethical form, a new "teleological order of the real" is created: "It is a new teleological order of the real, that is added to the mere quantitative order, and in which the individual first gains its full meaning. Logically speaking, the individual is taken up and shaped by different forms of relation. The conflict of the 'universal' and the 'particular' resolves into a system of complementary conditions, such as only in their totality and cooperation can grasp the problem of the real."[148]

The determination of the "real" necessarily goes beyond its particular resolution in a single science, since it rests on the infinite "totality and cooperation" of all "complementary conditions" on which it depends. The determination of a particular compound, for instance, rests on an infinite series of assumptions, say, about qualitative features such as color or the universality of the chemical system itself, that by definition exceed the science of chemistry.

In this context, the particular unity of truth within each science, as Cassirer puts it in *Substance and Function*, is "not one of principle, it is merely classificatory."[149] Similarly, Cassirer notes, "The establishment of facts occurs in the different special sciences under very different conditions."[150] Progressive definitions of truth exist within each science, but the concept of reality only exists as a limit idea shaped by the interrelation of all fields of human endeavor. For Cassirer, it becomes questionable whether this procedure ultimately leads to one science in the modern period at all, or rather it breaks into different, potentially incommensurable fields – a conclusion that Cassirer reached by the time of the third volume of this *Symbolic Forms* series and the forth volume of his *Knowledge Problem* text in 1940.[151]

Here, Cohen's description of the ideal context of the relation of the particular "Real" to the wider "reality" is newly defined. The particular moment of the Real for Cassirer is defined, as it was by Cohen, as both the limit concept of all possible realities and an ideal focus within experience. In a manner similar to Cohen's relation of his projected four volumes of *The System of Philosophy* – covering essentially epistemology, ethics, aesthetics and, as their keystone, psychology – or Cassirer's own overview in *Leibniz's System*, so here Cassirer creates a necessary relation of experience to each of the other fields of endeavor, and claims that interrelation of these fields is always incomplete. The individual is defined negatively in relation to the structure of each field, but also by the difference among each of the fields. Reality taken as a whole necessarily moves from epistemological to aesthetic and ethical definitions, and with this into a temporal problem relating to the general teleology of humanity.

The theoretical importance of this multiplication of frames for determining truth is suggested by Cassirer's repeated use of it at the conclusion of nearly all of his major works on natural science, starting with *Leibniz's System* itself. Thus in *Einstein's Theory of Relativity* (1921), he concludes the book by noting that in fact art, history, music and many of the sciences have different realities, and it is crucial that none of them be "hypostasized as the whole."[152] Even space and time are defined relatively not only in the sense of Leibniz, but through every major field in which they are of importance, such as history, painting or music: "What space and time truly *are* in the philosophical sense," Cassirer writes, "would be determined if we succeeded in surveying completely this wealth of nuances of intellectual meaning" of all of the fields, a process which is implied to be infinite.[153] Similarly, *Determinism and Indeterminism in Modern Physics*, a work on quantum physics, concludes by noting, "Modern physics had to abandon the hope of exhaustively presenting the whole of natural happening by means of a single strictly determined system of symbols."[154] This lack of completeness is ultimately true for all fields of science. In this regard science is again part of a larger unlimited human experience, so that when "we seek to realize the full concept of reality [it] requires the cooperation of all functions of the spirit [*Geistes*] and can only be reached through all of them together."[155]

In each of these cases, Cassirer is highlighting that the definition of the real within any specific science always has to foremost recognize that it is neither self-contained nor a closed totality, but rather opens out onto an unknowable horizon of potential. The real and reality are tools by which particular truths are gained in the sciences, but they are never given in themselves other than as ideal limit concepts. Read against his earlier works and Cohen's theory of the Real, however, the concept of reality has a pivotal systematic importance within his philosophy. The ideal function of reality as a limit problem is both inherent in all perception and crucial for understanding the interaction of larger forms or systems of knowledge. Whereas Cohen assumed the architecture of reality to unfold from a single plan, however, and this plan had as its key method the revelation of the concept in the infinitesimal, Cassirer's model of the group and series introduces a far more variegated model of change. The difference of the two is emblematic in some ways of a generational difference suggesting a break from an earlier linear, unified and progressive definition of historical change to one that is multivariate, disparate and capable of sharp reversals or transformations.

The Problem of Reality in the Contemporary Philosophy of Science

Cassirer directly develops his theory of reality in relation to the philosophy of science in *Substance and Function* through a discussion of the "modern criticism of knowledge" epitomized by Émile Meyerson's *Identity and Reality*.[156] Meyerson's work puts forth the contradiction, in Cassirer's reading, that "in order to understand reality by our mathematical concepts it is concluded that we must destroy it in its real nature, i.e. in its multiplicity and changeableness."[157] Reality, in this view, "withstands the efforts of thought and sets up definite limits, that it cannot transcend. […] It is only by such opposition from reality, that being itself does not disappear in the perfection of knowledge."[158] In opposition to this reading of an "abstraction" from reality in the classical sense, Cassirer argued that his model of history and science overcame what he describes as the "antinomy" of reality and ideality. The opposition of ideal structures and reality is part of a historical antinomy that has been recognized since the debates over nominalism and realism in the medieval period, and has persisted to the modern era.[159] The opposition of the terms ideal and real, and the arguments that derived from them (such as the opposition of the "true" world from the real and given world), were an antinomy in the sense that the original opposition was in fact premised on false suppositions of "substance."

The problem Meyerson confronts at first appears consistent with the larger narrative of the first half of Cassirer's text, where each of its characteristic sciences is placed in a development model in which conception relates to reality in progressively more comprehensive and fruitful functional manner. The "construction concepts of chemistry," for instance, place the "relation of the universal to the particular" in a new light, because the more we look at them "the sharper the separation becomes between the system of our concepts and the system of the real."[160] In chemistry, the problem raised by the relation of reality to concept that has been an "antinomy" since at least the time of Aristotle at first appears most clearly developed. "For all 'reality,'" Cassirer argues, "is offered to us in individual shape and form, and thus in a vast manifold of particular features, while all concept, according to its function, turns aside from this concrete totality of particular features."[161] The more developed chemical ideas have become, say, in Lavoisier's theory of the conservation of mass, the more the qualitative features of the elements appear to become secondary before increasingly more comprehensive modes of understanding, say, in numerical measurement. "The further the chemical construction of concepts proceeds, the more sharply the particulars can be distinguished."[162] For many commentators, this appeared to suggest that chemistry was a progressive "abstraction" from reality, that is, a field that used the "conceptual pyramid" of the classical logic of negation to categorize and delimit meaning, even as it moved away from the qualitative diversity of impressions.[163]

The key contemporary philosopher to fall prey to the antinomies of reality in Cassirer's reading was Heinrich Rickert, who was one of the principal figures of the Southwest or Baden school of neo-Kantianism. In Rickert's words, "The gap between concepts and individuals, which is produced by natural science, is thus a gap between concepts and reality in general."[164] Cassirer holds this antinomy to be false since science does not function to "name" or map reality in the sense of the commonsense understanding of regular words or a copy theory of knowledge. Rather it provides rules for the generation of meaning, it provides "a plan and not a copy."[165] These rules prove their usefulness in relation to each other and through an internal standard of consistency in which explanation and operation grow increasingly more inclusive and effective. More importantly, not only is this not a "gap," but concepts *illuminate* and reveal new aspects of reality. In short, reality is not "out there" but is defined historically and internally, so to speak, in relation to the unfolding of conception and experience itself.

To advance this argument, Cassirer refines Meyerson's opposition of a unitary science of "real nature" in its "multiplicity and changeableness" by noting "this explanation itself needs a limitation," which is that in neither the case of science nor any other approach to "reality" are we dealing with a given substantial whole. Rather in both science and everyday life we find only "functional orders and correlations." "These functional correlations," Cassirer writes, "do not exclude the element of diversity and change, but are determined only in it and by it."[166] In the case of natural science, Cassirer writes, "It is not the manifold [of experience] as such that is canceled, but only one of another dimension; the mathematical manifold is substituted in scientific explanation for the sensuous manifold."[167] The sensuous manifold itself is in Cassirer's reading already determined by form, as empirical gestalt psychology had already argued against concepts such as Mach's "elements" of sensation. Cassirer later notes of this problem in the *Philosophy of Symbolic Forms*, where it finds a more comprehensive response: "The most important thing, Goethe has said, is to recognize that all fact is in itself theory. If this applied to any fact at all, it is to the fact of simple sensation."[168]

The implication is that there might be many "dimensions" of the manifold of experience that could be compared and there is in any case no single "reality" against which they can be defined. Science can be understood not only as relating to unity and similarity, but also with equal primacy to discontinuity and difference. Cassirer avoids combining the two models of "series" or "group" and "limit" that he claimed initiated his project until the end of *Substance and Function*, when they are loosely taken as the figures, defined as "function and structure" for the "interpenetration" that "determines the complete concept of knowledge."[169] It is here that the opposition of "concept" and "reality" that has been presented as an antinomy throughout the text is resolved:

> Knowledge realizes itself only in a succession of logical acts, in a series that must be run through successively, so that we may become aware of the rule of its progress. But if this series is to be grasped as a unity, as an expression of an identical reality, which is defined the more exactly the further we advance, then we must conceive of the series as converging toward an ideal limit. This limit "is" and exists in definite determinateness, although for us it is not attainable save by means of the particular members of the series and their change according to law.[170]

Essentially, the various structures of knowledge themselves can provisionally be assumed to form a series, one which demonstrates developmental invariants so that "we may become aware of the rule of its progress" as a "possible order and a formal serial concept."[171] Function, on the other hand, is defined through the concept of the limit, a moment within structure that "exists in definite determinateness" both in its coherence in the moment and in its overall orientation towards the future synthesis of different forms of knowledge. Function, however, unifies the problem of ideality and reality only as a limit problem oriented towards the future. Indeed, for Cassirer it is only the historical movement of structure that lends it an ultimate unity. It is only historicism and its ability to point towards a unifying feature, Cassirer later summarizes at the end of his *Knowledge Problem* series, that resolves the "danger of extreme fragmentation" in knowledge.[172] Rather than being a "metaphysical unity" it reveals that "diversity is [...] no longer the antithesis to Being but its correlative."[173] In this manner, "though [historical science] could offer no unified and common *solution* to the problem [of the unity of knowledge,] they did work at a common *task*."[174] It is in this fraught sense that the relation of structure and function thus indeed lends "all knowledge a static and dynamic motive."[175]

Thus the overall architecture of Cassirer's system meets, to put it somewhat paradoxically, in the difference of two or more systems of difference. Nor can these differences be reduced to a particular form of judgment, as Cohen had assumed was the case of the infinitesimal forming the "root of ideal reality." Cassirer is well aware, as he notes, that the "modern form of logic [...] has shown with increasing distinctness that it is impossible to reduce the diverse forms of judgment to a single type of identity."[176] Nominal reality cannot be considered "one thing," and should not be assumed to be static; indeed, to be consistent it must be considered difference itself, indeed that which cannot be placed in any pattern of relation so cannot be known. Or as he puts it: "We are not concerned here with an opposition between the 'concept' and absolute 'reality,' but with a distinction wholly within the system of concepts" – which is to say, the concepts ordered by serial, group and limit forms.[177] Concept is, as we have seen, itself in Cassirer's definition not separate from material, and entails no separation of ideality from reality, subject from object, or empiricism from idealism. There is simply an immanent domain of reality, but as with Cohen's work this domain is provisionally unified for perceiving subjects through the ideal concept of the "Real."

A wider meaning of the relation of limit and series, structure and function, becomes clearer if we look at how this theme is revealed not only in historic shifts in patterns of knowledge, but in individual perception. There is a "double meaning," Cassirer writes, between how the problem of reality is understood in its purely "logical meaning" and how "this whole is *realized* in knowing individuals."[178] Following from Cohen's distinction of the limit concept of the "Real" given in experience and the broader "reality" to which it relates, Cassirer's *Substance and Function* develops a "phenomenology" of experience to demonstrate how the tenuous unity-in-difference of knowledge is consolidated for the individual. As with Cohen, reality proves itself to be an ideal function, which links the certainly given – but only amorphously known – plenum of all possible realities to the particular "hypothesis" of the immediately given Real. Thus, discussing the problem of the "psychological" appearance of real phenomenon, Cassirer writes, "The question here can never be how we go from the parts to the whole, but how we go from the whole to the parts. [...] Only the total result itself is 'real' in the sense of experience and of psychological process, while its individual components have only the value of hypothetical assumptions."[179] Such hypothetical assumptions apply in the particular relation of part and whole, and are described by Cassirer as appearing within a series of invariants that are described as being related to Ehrenfels' concept of gestalt forms and modeled on the problem of group theory.[180] All processes of perception are, Cassirer writes, "not to be separated from judgment," since it is only in "acts of judgment that the particular content is grasped as a member of a certain order and is thereby first fixed in itself."[181] Such a process always entails the wider judgment of reality, however, since this suggests how any particular moment of judgment is orientated within the whole of experience.

In developing this "double meaning" of how reality is actualized by "knowing individuals," Cassirer is led to some radical conclusions based on the initial logical argument of the determination of the reality of any individual through its structure. As we have seen, sociologists such as Elias, Bourdieu and Luhmann noted that *Substance and Function* constitutes one of the first and most consistent definitions of the priority of structure or system over individual in describing society. Less recognized was that the model of group theory at the base of this model in itself radically changes definitions of both the social individual and historical change. Not only is the empirical ego, to put it in Kantian terms, defined as open to permutation as any other object, but the transcendental ego is redefined in a radical minimalist fashion:

> The images in the presentation of the individual [...] always have a certain inner form of connection with each other no matter how variegated and diverse in their succession; and without this inner connection, they could not be grasped as contents of the same consciousness. They all stand in at least a temporal connection, in a definite relation of earlier and later; and this one feature suffices to give them a fundamental common character through all diversity of individual form. No matter

how much the particular elements may differ from each other in their material content, they must nevertheless agree in those determinations, on which the serial form rests, in which they all participate.[182]

Cassirer here suggests a very minimal interpretation of Kant's transcendental unity of apperception in line with his theory of series and group: all contents of consciousness are negatively defined as related with each other, yet the *only* form of continuity necessary for consciousness is a "temporal connection, in a definite relation of earlier and later." As with Leibniz, the ego of any form is here largely abandoned in favor of an argument for the temporal continuity of experience, but this experience is itself open to radical permutation. This pure relation of temporal continuity allows a "maximal" definition of personal and structural change in history as suggested by group theory. The "temporal connection" that links forms is, if we place it in the context of Cassirer's earlier discussion, further qualified by the fact that the mode of temporal continuity can presumably be an "answer" position to an earlier "question," rather than a direct transmission of original meaning. Cassirer's "history of problems" (*Problemgeschichte*) implies that there would minimally always be the trace of the "answer" positions for earlier questions, but the questions themselves may have been forgotten. In this regard, Cassirer's description of historical change again finds a similarity to a later theory of historical change, namely that developed in Michel Foucault's claim in *The Order of Things*, which suggests a radical shift between eras of knowledge, or what he describes as *epistemes*, that can lead to a complete and sudden shift in the meaning of an era and its definitions of self and object.[183]

It is in the context of Cassirer's study of a "concrete totality of *productive* functions" that Cohen's heated disagreement with the tenets of *Substance and Function* is now clear. Cohen attacked Cassirer for failing to recognize the infinitesimal as the "ideal root of reality" and claimed that Cassirer mistakenly described the problem of reality as part of an "antinomy" with ideality.[184] Whereas Cohen's concept of the infinitesimal provided a unified definition of reality in its progressive unfolding of knowledge across all fields, and assumed the common anchor of reason and the development of humanity across all fields, Cassirer's ceases to have any such guarantees. The consequences of this turn are as marked in the natural sciences as they are in the social and human sciences, in the realm of logic as they are in the ethical problem of how "knowing individuals" relate in the world. For Cassirer, the only basis of unity in the human project is history itself in conjunction with the speculative vision of a redemptive future, and thus is recast in terms of a Kantian concept of ethics as "cosmopolitan history" or a Leibnizian concept of perfection as revealing the maximal reality through the minimal, or most elegant, means. It is in this regard telling that the final volume of *The Knowledge Problem* concludes in Cassirer's own era by noting that historicism will both insure that "metaphysics in its old dogmatic form could never rise again" and provide a "consolidating and unifying power […] in this moment of extreme fragmentation."[185]

By the time of *Substance and Function*, in other words, history does not merely productively recast the Marburg school project, as Cassirer had noted since the first volume of *The Knowledge Problem*, it rather now becomes the principle source for the unification and coherence of this project itself across all of the sciences. The proper historical transformation of each science allows for a productive development of various truths; the recognition of the historical concrescence and mutual delimitations of the sciences, arts and ethics intimates the limit condition of reality. Yet there is no guarantee of any of this happening, and indeed there is no pre-existing "ethical subject" or definition of the unity of knowledge to promote this project.

It is not surprising in this regard that Cohen would find *Substance and Function* a devastating revision of his ideas, for it changes a linear and progressive notion of the development of science into what appears as a fragile philosophy of difference with only local rules of order. It is notable that in his response to Cassirer's work, Cohen immediately casts it in terms of his student Anton Gawronsky's recent dissertation on the *Judgment of Reality*, which develops a view of this problem far closer to his own.[186] Cassirer's logical and phenomenological philosophy in *Substance and Function* is in this regard the necessary foundation for his later argument for the political use of myth, for it explains how it is possible for new political "techniques" of myth to completely remove ethical reflection or individual subjectivity from society, since these are not pre-given by human nature, but are rather the product of form and history. Cassirer's apparently old-fashioned defense of the need for philosophy to turn towards the understanding and maintenance of the development of "civilization" meant the need to cultivate the basis of different critical modes of subjectivity and objectivity through the study of forms of knowledge and activity.

It is only in the aptly titled *Freedom and Form* that Cassirer recast the perceived weakness of his system in terms of Cohen's absolute claims into what he considers the strength of a positive definition of an ethics of liberal politics and the related critical development of science and form. The foundation of this turn in Cassirer's thought had already been developed in the first volume of *The Knowledge Problem*, a text that presented a new model of the necessary relation of philosophy and history, and had already departed from key tenets of Cohen's work. Before addressing this development, however, we can conclude by demonstrating the continuity of the core set of ideas established in *Substance and Function* with Cassirer's later projects as it related to the problem of group theory, truth and reality.

The Concept of the Group and Perception Revisited

The importance of Cassirer's early model of group theory is evident in a set of articles revisiting his theory two decades later, culminating in "The Concept of Group and the Theory of Perception" (1938), as well as a lecture in the last year of his life, "Reflections on the Concept of Group and the Theory of Perception" (1945). Here Cassirer develops his argument concerning the relation of group theory to perception and meaning more concisely in relation to both the mathematics of Felix Klein and gestalt theory, but its basic form had not changed. Developments in empirical and gestalt psychology since 1910, Cassirer argues, confirm his earlier argument that the group of mathematical theory can form the basis for a fruitful understanding of the transformation of structures (such as those of language, cultural forms, etc.) that affect the perception and judgment of the present moment. "So far as I know no modern philosopher has discussed and analyzed the concept and used it for the solution of epistemological problems," Cassirer writes, "As for myself, I am convinced that […] such a discussion is highly desirable and even necessary."[187]

Cassirer begins his 1938 essay by noting that "the perceptual image," as much as the geometrical figure, "involves that reference to certain possible groups of transformation. It changes when we refer to a different group and determine the 'invariants' accordingly."[188] In short, all perception is judgment, and judgment can best be described by the pattern of invariants in groups.[189] Cassirer uses the now familiar gestalt examples of "figure-ground" illusions, as well as the principle "perceptual constancy," that is the perception of an object as constant under changing conditions, to develop his argument further.

Within this argument, the theme of the limit idea of form reappears, linking the model of "series and limit" as Cassirer had announced in the preface to *Substance and Function*. "All perception is confined to the 'more or less,'" first defined by Plato and refined by Kant, Cassirer writes.[190] Similarly, the "phenomenon tends towards the idea but never reaches it and necessarily falls short of it."[191] In this regard, Cassirer's vision of the invariants can be placed within the general architecture of Cohen's reception of Kant's system, since each invariant is a "monogram" of imagination in the sense of Kant's schema, "a schema towards which the particular sense experiences are orientated."[192] The phenomena of "towards which" is thus the function of the infinitesimal as it occurs within the Kantian schema, while the pattern of invariants within a group functions as the "structure" within which this schema appears.

Importantly, the conclusions of the second half of *Substance and Function* are also confirmed in Cassirer's 1938 article: the relation of subject and object, internal and external experience, as well as form and matter are all revealed as equivalent creations of the underlying development of functional form, defined by invariants. The "subject," for instance, is defined as never directly given in perception or direct intuition, but rather only by continuity within particular forms of transition. "The 'I' only comes into being," Cassirer writes in a key passage, "in the 'focal point' in which it succeeds in grasping itself in the mirror of its own utterance [*Äußerung*]." In this regard: "Every utterance […] is a new art of hearing [*Hörens*] […] and this mode of hearing leads to a form of 'obedience' [*Gehorsams*]" to language.[193] This point is particularly true within the functional forms of language, but would have the same function in relation to any form, such as art, law or politics. Invariant forms reveal both the objective of meanings in the world, as well as the particular subset of these defined as "subjective." It is only in the "symbolic character" of the relation of the part to the whole that there is a continuity of experience of the self. Cassirer demonstrated the political application of this idea in a continuous argument from *Leibniz's System* through *Freedom and Form*: the very fragility of the "self" demands that structures – such as political constitutions, laws and a critical system of the arts – are in place to insure and protect its most productive form through time.

Cassirer's model of series or group and limit is the theoretical basis for the historical narrative of forms he presents in *The Knowledge Problem* and develops in *Freedom and Form*, in which subjective and objective positions, along with nearly every other facet of experience, are capable of permutation or redefinition. The political import of Cassirer's balance between constructivism and realism in his theory of form is consistent throughout: only through changing the "rule-forms" of society, in any of the sciences or their practices, is actual human freedom both demonstrated and instantiated. Conversely, the rules and structures of "civility" are of paramount importance, because only through these structures can groups and individuals attain relative freedom. In this regard, Cassirer's *Substance and Function* was part of a common text with his next book, *Freedom and Form*, which began to extend its structures into a critical theory of society.

Notes

1 Cohen to Cassirer, 24 August 1910, St Moritz. Ernst Cassirer Nachlass, Beinecke Library, Yale University, gen. ms. 355, box no. 2, folder 45; Schilpp 1949, 22.
2 SF 3, 21. Cassirer's italics.
3 SF 232–3.
4 SF 26.
5 SF 232.
6 SF 300.
7 CG, as well as a lecture in the last year of his life, "Reflections on the Concept of Group and the Theory of Perception" (1945), SMC 271–93.
8 SF 201.
9 SF 43, 91ff.
10 SF 26.
11 PSF1 69.
12 Cohen to Cassirer, 24 August 1910, 21.
13 Ibid.
14 Ibid.
15 Ibid.
16 SF 232; EK4 17.
17 PSF 3 13.
18 EK4 17 (German edition; EK4 24).
19 Cassirer, Ernst. "Dualität der Principien." Ernst Cassirer Nachlass, Beinecke Library, Yale University, box 39, folder 750.
20 Blumenberg, Hans. *Wirklichkeit in denen Wir Leben* (Stuttgart: Philipp Reclam, 1981), 164–5.
21 Marck, Siegfried. Foreword to *Substanz- und Funktionsbegriff in der Rechtsphilosophie* (Tübingen: J. C. B. Mohr, 1925); on legal personification, 23–5, 83–147; on von Gierke 92–104.
22 SF 3 309.
23 SF iv.
24 Ibid.
25 Ibid.
26 SF 3.
27 SF 4.
28 SF 99–100.
29 Ibid.
30 SF iv; Cohen subtitled the *Infinitesimal Method* "A Chapter in the Foundation of Knowledge Critique," the term is distinguished from the earlier "Erkenntnistheorie."
31 I 6.
32 SF 26.
33 SF 17, 344n11.
34 SF 8.
35 SF 7.
36 SF 8.

37 SF 17–18.
38 ZLK 180–81.
39 SF 18.
40 SF 21.
41 Cassirer, Ernst. "Leibniz and Jungius" (J 366); "Leibniz" entry in Encyclopædia Britannica (ECW17, 318).
42 SF 344n11.
43 SF 7.
44 SF 21.
45 SF 23. Translation modified.
46 Ibid.
47 SF 25.
48 Ibid.
49 Ibid.
50 SF 26.
51 Ibid.
52 Ibid.
53 SF 333.
54 SF 16.
55 SF 16, 343ff.
56 SF 16.
57 SF 25.
58 Ibid.
59 Ibid.
60 SF 25. Cassirer's use of "*aufprägen*" (imprint) here, can be read as the first appearance of "*prägen*" in the sense that is later developed in the idea of "symbolic prägnanz." Cassirer's later claim that symbolic prägnanz both bridges and surpasses Kantian and Husserlian definitions of intention, addressed in Chapter Two, can be read as developing from combing this concept of intention with the infinitesimal model of representation (SF 300); PSF 191–200.
61 SF 25.
62 SF 24. On Abich-Meyer see Moynahan, Gregory. "Ernst Cassirer, the Case of 'Clever Hans,' and the Politics of Theoretical Biology" *Science in Context* 12, 4 (1999): 549–74; and Amidon, Kevin S. 'Adolf Meyer-Abich, Holism and the Negotiation of Theoretical Biology," *Biological Theory* 3 (October 2008): 357–70.
63 SF iv.
64 SF 248.
65 SF 249.
66 Although Cohen and Cassirer were aware that the French philosopher Antoine Augustin Cournot (1801–1877) had demonstrated that Newton and Leibniz's Systems of calculus could be mutually grounded in functional theory, and later attempts were made to explain calculus through set theory by mapping coordinates of calculus onto sets, neither attempted to literally use this scheme to map this theory out in detail (L 497). See: Cournot, Antoine Augustin. *Traité élémentaire de la théorie des fonctions et du calcul infinitésimal* (Paris: L. Hachette, 1857), 44ff.
67 SF iii.
68 SF 300.
69 SF 23.
70 SF 341.
71 SF 25.
72 SF 22.

73 SF 3–4.
74 SF 25–6.
75 SF 89.
76 PSF3 353.
77 SF 26.
78 Ibid.
79 PSF3 353.
80 SF 24.
81 SF 267.
82 SF 235.
83 SF 91.
84 SF 91.
85 EK4 30ff.
86 Approachable presentations of this principle are presented in Maor, Eli. *To Infinity and Beyond: A Cultural History of the Infinite* (Princeton: Princeton University Press, 1997), 108ff.; and Ogilvy, C. Stanley. *Excursions in Geometry* (New York: Dover Publications, 1969), 107ff.
87 SF 89, 91.
88 SF 249.
89 SF 250.
90 SF 332.
91 SF 311.
92 SF 261ff.
93 SF 269.
94 SF 270.
95 PSF3 45.
96 SF 266.
97 SF 267.
98 SF 204.
99 SF 23
100 SF 184.
101 SF 185.
102 SF 205.
103 Ibid.
104 SF 206.
105 SF 207ff.
106 SF 204.
107 SF 206.
108 SF 243.
109 SF 208.
110 SF 196ff., 208.
111 SF 208.
112 SF 209.
113 SF 211.
114 DI 144ff.
115 I 3ff.; SF iv.
116 SF 211; CPR 672.
117 I 127.
118 Blumenberg, Hans. *The Genesis of the Copernican World*, trans. Robert M. Wallace (Cambridge, MA: MIT Press, 1987), 110.
119 SF 272.
120 SF 273; PSF3 421.

121 SF 271.
122 SF 272, 275.
123 SF 268; Ihmig, Karl-Norbert. *Cassirers Invariantentheorie der Erfahrung und seine Rezeption des "Erlanger Programs"* (Hamburg: Felix Meiner, 1997).
124 SF 131ff.
125 SF 285.
126 SF 127ff.
127 EK2 181; ECW3 149.
128 SF 321.
129 SF iv.
130 SF 321.
131 SF 287.
132 SF 286; L 110–11.
133 For more on this topic, see Gordon's reading of Cassirer's critique of Heidegger's deworlding (*Entweltlichung*) in *Continental Divide: Heidegger, Cassirer, Davos* (Cambridge, MA: Harvard University Press, 2010), 248ff. and 288.
134 Cassirer explores the problem of quantum theory in *Determinism and Indeterminism in Modern Physics: Historical and Systematic Studies of the Problem of Causality*, trans. O. Theodor Benfey (New Haven: Yale University Press, 1956 [1936]), 109–214.
135 de Saussure, Ferdinand. *Course in General Linguistics*, ed. Charles Bally and Albert Sechehaye, trans. Wade Baskin (New York: McGraw-Hill, 1966), 88ff.
136 SF 268.
137 SF 321–2.
138 SF 322.
139 Ibid.
140 SF 321.
141 Friedman, Michael. "Ernst Cassirer and Thomas Kuhn: The Neo-Kantian Tradition in the History and Philosophy of Science," in *Neo-Kantianism in Contemporary Philosophy*, ed. Rudolf A. Makkreel and Sebastian Luft (Bloomington, IN: Northwestern University Press, 2010), 177; Friedman, Michael, "Kuhn and Logical Empiricism," in *Thomas Kuhn*, ed. Thomas Nickles (New York: Cambridge University Press), 34.
142 SF 268.
143 Blumenberg, *Legitimacy*, 379ff.
144 SF 233. This schema does not produce a "dualistic opposition" between science and history, Cassirer clarifies, since both are themselves different means of generating and explaining individual events within broader patterns, and both are themselves internally variegated.
145 SF 233.
146 SF 214.
147 Ibid.
148 SF 233.
149 SF 232.
150 Ibid.
151 EK4 17.
152 SF 455.
153 SF 456.
154 DI 212.
155 DI 213.
156 Cassirer is referring particularly to Meyerson, E. *Identity and Reality*, trans. Kate Loewenberg (New York: Dover Publications, 1962), 384ff.
157 SF 324.
158 Ibid.

159 SF 221ff.
160 SF 220.
161 Ibid.
162 SF 224.
163 SF 6.
164 SF 220–21; Rickert, Heinrich. *Die Grenzen der naturwissenschaftlichen Begriffsbildung* (Leipzig: J. C. B. Mohr, 1902), 235ff.; *The Limits of Concept Formation in Natural Science: A Logical Introduction to the Historical Sciences*, ed. and trans. Guy Oakes (New York: Cambridge University Press, 1986).
165 SF 185.
166 SF 324.
167 Ibid. Translation modified.
168 PSF3 27.
169 SF 315.
170 Ibid.
171 SF 310.
172 EK4 324.
173 Ibid.
174 EK4 325.
175 SF 315.
176 SF 344. Translation modified.
177 SF 229n85.
178 SF 326.
179 SF 335.
180 SF 333n4.
181 SF 341.
182 SF 266.
183 Foucault, M. *The Order of Things: An Archaeology of the Human Sciences* (New York: Pantheon, 1973); Foucault, M. *The Archaeology of Knowledge & The Discourse on Language* (New York: Pantheon, 1972); Florence, Maurice (Michel Foucault). "Michel Foucault" in *Dictionnaire des Philosophes*, ed. Denis Huisman (Paris: Presses Universitaires de France, 1984), 941ff.
184 SF 127ff., 165, 220ff.; Cohen to Cassirer, St Moritz, 24 August 1910.
185 EK4 324–5.
186 Gawronsky, Dimitri. *Das Urteil der Realität und seine mathematischen Voraussetzungen* (Weimar: Hof-Buchdruckerei, 1910). Cohen writes that now that he has read *Substance and Function* he cannot wait to hear Cassirer's response to Gawronsky's work on the same theme. Gawronsky remained one of the Cassirer's best friends, and contributed two important pieces to Cassirer's *Festschrift*: "Ernst Cassirer: His Life and His Work" and "Cassirer's Contribution to the Epistemology of Physics," in *The Philosophy of Ernst Cassirer* (La Salle, IL: Open Court Publishing, 1949), 15.
187 SMC 273.
188 GC 16.
189 SF 341ff.
190 GC 16.
191 GC 17.
192 GC 32.
193 GC 136.

Part III

LIBERAL DEMOCRACY AND LAW

Chapter Six

LIBERALISM AND THE CONFLICT OF FORMS: *THE KNOWLEDGE PROBLEM* (1906–1940) AND *FREEDOM AND FORM* (1916)

Astute contemporary commentators such as Karl Mannheim recognized both the originality of *Freedom and Form* and its continuity with Cassirer's earlier work. Summarizing the text in 1917, Mannheim writes, "Cassirer transforms the concepts of *Freedom and Form* into functional concepts [...] with the theory recently applied to the history of exact science in his last book (*Substance and Function*)."[1] *Freedom and Form* presents an outline of a theory of liberal civilization that is the culmination of Cassirer's functional model of form, which his narrative recasts through an alternative history of German classicism. Although the political format of the text is new, however, its core argument is already initiated in *The Knowledge Problem*, particularly the lapidary introduction to the series. Cohen and others immediately recognized that the content of *The Knowledge Problem* provided a comprehensive historical foundation for the Marburg school project in an interlocking series of readings ranging from Nicholas of Cusa, Leibniz and Shaftesbury, through Humboldt, Kant and Hegel. Its most important innovation, however, was found in the form and style of the text itself, which was integral to its overall argument that the critical study of different forms of knowledge in history is the basis of an ethics not of autonomy and the individual person, but of the historical forms and sciences that create the individual, whether defined objectively or subjectively.

Both *The Knowledge Problem* and *Freedom and Form* redefine the Enlightenment goal of autonomy along Leibniz and Cohen's model, arguing that autonomy in the modern period should not be defined for an insular autonomous subject, but rather through the changing social and collective patterns of form in "concrete historical life" that first determine such subjects.[2] Thus a key development of *Freedom and Form* claims that "the contradictions between freedom and form" are harmonized through the manner in which "true freedom is directed at the development [*Erzeugung*] of rule-form [*Gesetzesform*]."[3] In discovering the means to change rule-forms in the natural, social and human sciences, as well as in art, politics and in ethical life, both texts promoted the conditions for a distinctly German liberal society. The texts thus presented a version of what Cohen had defined as metapolitics, but in Cassirer's hands this was not based on the unified development of reason, but on a liberal agonism of forms and meanings that can only be understood historically. For both, however, the goal was a continuation of Cohen's focus on *autotelie*, or the freedom of collective and individual development within a set of limitations of form, one developed as much socially as individually.[4]

In his unified definition of form, Cassirer transforms the contemporary understanding of both nature and culture, claiming that all natural forms are theoretically capable of permutation and all cultural forms have aspects of continuity that are invisible to human actors. Cassirer's argument is suggested in his reading of Kant in *Freedom and Form*, for instance, where he claims that "the *a priori* problem and the freedom problem are simply different expressions of one and the same fundamental postulation."[5] As understood within Cassirer's theory of invariants, however, this approach takes on a new meaning: forms of experience that appear *a priori* can always be reinscribed into a new framework, and this process in turn reveals new forms of continuity and reality in experience. Similarly, all aspects of social life are defined through patterns of form that are open to transformation, even in basic terms such as subjectivity or work, but also contain continuities that exceed human control. For this reason the transformation of rule-forms always contains a deeply ambiguous relation to freedom: "Every positive determination [*Bestimmung*] contains within itself, in its real historical development, at the same time something negative; every solution contains in itself equally a new binding commitment [*Bindung*]."[6] There is no "pure" form of freedom, as is sometimes taken as the goal of individual autonomy, rather every system develops certain patterns of freedom, of potential for reflexive productive growth or productivity, and at the same time introduces new constraints.

Cassirer's text suggests a new way of thinking about the methods of natural science and the humanities that completes the Marburg school's critique of science as it was to define a new "conflict of the faculties." In the process, Cassirer develops a new liberal vision of German society that places the continuous agonism of different forms and actors in society at the center of his narrative of intellectual history, effectively removing his predecessors Lange and Cohen's emphasis on a unifying role for reason or its correlative in the collective development of socialism.

Freedom and Form presents a new model of the development of historical form as a theory of liberal civilization (Cassirer indeed frequently uses the German *Zivilisation* rather than *Kultur* in the text) and thus in many ways is the culmination of Cassirer's early work. When Cassirer later concurred with Albert Schweitzer that the key failure of philosophy in the century before the Second World War was that it "philosophized about everything except civilization," he may have seen his own work from *The Knowledge Problem* to *Freedom and Form* as a first tentative corrective to this tendency.[7] The text was in any case certainly considered by him an antidote to the xenophobia of the war years, particularly that surrounding the use of German *Kultur* as a "fighting term," and formed the basis of his retort to later fatalistic approaches to civilization such as Oswald Spengler's *The Decline of the West*.[8] The boldness of Cassirer's project in *Freedom and Form* is suggested by the rare self-praise with which he introduces it, claiming that his functional approach had led him to resolve "the central problems of being [*Wesen*] and appearance [*Erscheinung*] in philosophy" in a new manner, as well as to rethink "the relation of methods of natural science and the humanities [*Geisteswissenschaften*]."[9] The true nature of German society since the time of Luther and Leibniz, Cassirer argues, is its critical relation to its own forms of sociation, nature and belief, and from this it has developed its modern definition of freedom as the ability to gradually change these forms, a definition that culminates in the modern period in Goethe, Humboldt, Fichte and Hegel. The Marburg school approach continues to serve as the thread that allows Cassirer to weave his interpretation of these figures together, but to grasp why Cassirer understood his approach to *Freedom and Form* as new we have to first outline its development in *The Knowledge Problem*.

The Knowledge Problem and the History of Form

In the introduction to *The Knowledge Problem* series, Cassirer makes it clear that the form of the text itself will be part of a philosophical argument. Historical analysis, Cassirer notes at the beginning of the text, provides a crucial demonstration and "auxiliary means of understanding" philosophical functionalism apart from the logical analysis that otherwise is the only path to comprehending it. This path of logical development, Cassirer himself admits, is otherwise extremely difficult to grasp – a reference presumably to *Substance and Function* as well as to Cohen's *System of Philosophy*.[10] In historical demonstration, however, the same complex philosophy can be presented in itself, since in it the "phantasy of the absolute disappears from the first step" in its demonstration of a succession of differing historical realities, varying subject and object positions and permutations of even the apparently most stable *a priori* forms.[11] Through the presentation of history, Cassirer continues, "the highest goal that the internal critique of principles strived for [in philosophical functionalism] is won practically without effort and in full clarity," and all aspects of the absolute and substance from metaphysics are dissolved.[12] Through the practice and examples of history, the "system of relational concepts and the problem of reality," which Cassirer described at length in the second half of *Substance and Function*, are demonstrated in motion, so to speak, rather than tediously explained. The consequences of this are themselves both philosophical, in that they present a condensed experience, as it were, of the Marburg project, and practical, in that a new style of history is developed that allows each science and field of experience to be interpreted as maximally fungible. "In seeing the presuppositions of the sciences [...] in their historical relativity and conditionality," Cassirer writes, "we open a vision into their relentless progress and ever renewed productivity."[13]

Cassirer's life-long project of *The Knowledge Problem* can be considered the centerpiece of his work, and developed the underlying premise of his philosophy in a narrative and historical form. The first two volumes were published in 1906 and 1907, and a third completed in 1920. The forth, a retrospective history of his own period, was delayed by migration and war until 1940. These works, untranslated except for the last volume, formed the conceptual basis of Cassirer's notable – and more widely read – trilogy of texts from the late 1920s, *Individual and Cosmos*, *The Platonic Renaissance* and *The Philosophy of Enlightenment*, as well as his historical studies of philosophers ranging from Descartes and Goethe to Rousseau and Newton. Cassirer's philosophical overview of *The Knowledge Problem* was written in 1931 in a text, left unpublished until recently due to his emigration, emblematically entitled *Objectives and Paths of Reality-Knowledge*. Rather than surveying the contents of *The Knowledge Problem*, which have appeared elsewhere in our narrative, we will look primarily at its form and methodology.

The principal consequences of Cassirer's historical functionalism are to reveal that "the reciprocal relation of subject and object," as well as that of "thinking and being," are historically variable, and "not only their contents, but also their meaning and function" are capable of radical transformation.[14] The basis of understanding how the "meaning and function" of historical transformation occurs – and how it nonetheless suggests underlying continuity – is explicated in *Substance and Function* through the model of group transformation. Already in *The Knowledge Problem*, however, we find a clear application of a number of these ideas. Taken as present reality, Cassirer writes, the "system of fixed concepts" that dominates an era appears irrevocable, analogous to the "iron cage" of history for Weber, and appears as "a second independent and irrevocable reality that stands against us."[15] Understood historically, however, it can be seen that even as there is no escape from this cage, its form and set of determinations can be changed in time, a process that can be demonstrated in case studies by showing how invariants can change for certain fields, as for instance, Cassirer later noted, had occurred in the theory of relativity.[16]

The general argument of Cassirer's history is that "the intuition [view – *Anschauung*] that every era possesses of nature and the reality of things is only the expression and the reflection of its knowledge-ideal."[17] As in Cohen's psychology, the basic concept here is that the present intuition is the integral in relation to the differential of the full knowledge of an era, both its reflection and actualization. In this sense "each historical epoch newly formulates the problem of the reciprocal relation of being and thinking, and establishes through its ideal of knowledge their ordering and specific positioning."[18] The skeptical relation to the problem of reality, and the need to situate it within a critical reading of the various sciences, is here cast as the central theme of the work, even as this play of various realities will be demonstrated rather than explained as it was in *Subject and Function*.

Science and philosophy for Cassirer are both "symptoms of one and the same intellectual development" that defines reality for a particular era.[19] Even to say that philosophy and science influence each other suggests that the basic point at stake is overlooked, for they are rather part of the entirety of processes in interaction at a given time.[20] Forms of knowledge should not be mistaken as mediation between subject and object, or culture and nature. They are the basis of reality production in the manner outlined in *Substance and Function*, first defining poles of subject and object, culture and nature, within forms of greater or lesser stability in a manner that can be revealed historically. Thus the relative subjective and objective poles of experience are open-ended forms that change with each era and would continue to change in the future.[21] The form of knowledge in a particular epoch determines the realities of an era so that, for instance, the medieval period varies radically from the Renaissance in definitions of subjectivity and objectivity. "In general we have to be clear, that 'subject' and 'object' are not given or obvious possessions of thought [*Besitz des Denkens*], but that every truly creative epoch must first acquire them and must spontaneously impress upon them their meaning."[22] The rules that pertain for the "knowledge-ideal" of a particular period are not entirely conscious to the historical figures of that period, and even its philosophy is "only a small historical exception and highpoint within the full assay of the era."[23]

History should focus not only on the statements of past actors, but also the practices and endeavors of a period in order to grasp transitions in their "latent state."[24] In this regard particular activities such as scientific experimentation may be on the leading edge of conceptualization, but this is equally true for arts, literature or the inspirations of a religious visionary – fields developed by Cassirer's followers, such as Erwin Panofsky's work in art history.[25] The sciences develop as much through their forms of experiment and instrumentation as through ideas, and the laws of the sciences are capable of permutation even as they point to real continuities in experience. In synthesizing this combination of ideals and practices, laws and permutations, the experience of a given era can be partially reconstructed. It is for this reason that Cassirer writes, in a manner that stands opposed to Rankean history, "The problem of 'experience' is not just given, but is one of the most difficult problems and must be derived from the practices and theories of each age, not only in what it says, but also in its activities."[26]

Cassirer's specific methodology of historical reconstruction is explicitly revisionist, and begins by establishing a unified knowledge-ideal of a certain period as a postulate. There can be no absolute proof that this ideal exists or that it is correctly discerned, but "the more this postulate is realized in the development and soundings of particular phenomena, the more it has demonstrated its rights and its 'truth.'"[27] In this regard, Cassirer's definition of truth in history is the same as that of any other science, and follows the Leibnizian model: it is not based on a direct transcription of reality, but on the progressive internal narrowing of error and the discovery of continuities, measurable only within its own modes of presentation and, correlatively, its increasing range of explanatory ability. *The Knowledge Problem* aims to establish the "conditions on which the modern idea and system of knowledge developed itself," even as these conditions were not entirely grasped by individuals of the period.[28] If approached without pre-established schemas and "as indirectly as possible," Cassirer writes, historical patterns that have their own "immanent historical logic" stand out.[29] In this regard, Cassirer's approach to history is related to Cohen's metapolitics, in that it attempts to describe new continuities and relations of historical forms that may depart substantially from earlier assumptions, while also suggesting new normative forms from which specific modes of knowledge, society and politics can develop.

Eschewing a linear definition of history, Cassirer writes that different knowledge ideals do not develop by demonstrating a "peaceful one-after-another" development (*Aneinanderreihen*) of the steady growth of ideas, but through the "sharpest dialectical contradictions" in which the "manifold of fundamental intuitions [*Grundanschauung*] come into conflict with one another."[30] Unlike a Hegelian dialectic, however, in both cases the dominant metaphor of change appears to be that of group theory in which there is not a simple succession nor even a simple narrative dialectic, but a manifold of interrelated elements in which any change in a part can lead to a change in the whole.

In place of any assumption of a common cause of history, Cassirer attempts to develop a minimal or even nominalist methodology in which the only focus is "continuity of thought in the individual phase of occurrences [*Geschehens*]; this is all that is necessary in order to create the unity of processes."[31] Any particular "developmental series" (*Entwicklungsreihe*) in history – the term is used in apparent parallel with the use of the term series in *Substance and Function* – reveals a "subject that lies at its base and which it represents and voices [*äußert*]."[32] The use of a single perspective, that of subject or theme, thus defines a series, and the same phenomena could thus fit into many different forms of series. In this manner, Cassirer attempts to develop the broadest possible definition of historical change, just as he tries to develop the broadest definition of scientific change in *Substance and Function*. *The Knowledge Problem* illuminates this transformation of rule-forms on the most fundamental level by tracing changes in the reality of series such as subject and object, time and space, individual and society, through different historical eras.[33]

The use of one series is necessary for the coherence of this narrative, but is not to be reified (*verdinglicht*) in any sense, whether in the sense of materialist histories or in the Hegelian sense of "the Idea, a progress of the 'world spirit,' or the like."[34] Cassirer's theory of history itself does have one foundational limit, which is the rule developed from group theory that their must be one element of continuity between forms. He argues, however, that this is a necessary hypothesis that "functions as all true scientific conditions" solely to make possible the field of history itself. In *The Knowledge Problem*, this element of continuity is combined with the thesis of the discontinuous eras of knowledge, which have their own unity as complex ensembles of knowledge, even as they always share one serial connection with the earlier form.

In addressing the specific domains of philosophy and science in *The Knowledge Problem*, Cassirer seeks to provide an overview of both the reflection on knowledge within a given era and the manner in which knowledge is developed empirically through the "manifold experiments and endeavors of research."[35] Nicholas of Cusa's negative theology, to take a pivotal example, is consonant with a number of experiments on the relation of the finite and infinite in the Renaissance and Baroque, which in time lead to Leibniz's development of a distinctly modern form of this problem in the calculus and in monadology.[36] The various physical and mathematical experiments of Newton, Euler, Fontenelle and Boscovich gradually lead into the ideality of the intuition of space and time as understood by Kant.[37] This in turn establishes the basis of the crisis of intuition in mathematics in the nineteenth century, and the new definitions of physics at the end of the century.[38] In each case, it is the dialog of limited perspectives that leads to a general play of knowledge itself, which possesses a creativity and potential beyond any of its participants' awareness.

Cassirer's later systematic overview of the philosophical meaning of *The Knowledge Problem* series, *Objectives and Paths of Reality-Knowledge* (1937),[39] makes evident that Cassirer intends the historical format of *The Knowledge Problem* not only as a methodological application of functionalist philosophy, but also as an introduction to the philosophy of historical experience itself. In Cohen's psychology and Leibniz's history, we will remember, the experience of history or society is symbolic in the sense that individual experience is negatively defined as the differential shaped by all of the integral of manifold experience. "Consciousness is not a part, but a symbol of the universe," as Cassirer had summarized it for Leibniz, and this "new concept of subjectivity cultivates [*erzeugt*] the concept of history."[40] The converse is also true: historical experience defines a new form of negatively determined subjectivity. Cassirer writes in *Objectives and Paths* that historical awareness creates a phenomenology of the present in which individual experience is transformed by an understanding of the mutability of subject and object, and the manner in which the present and future are inflected through the past. Here, he writes, one encounters the "truly Symbolic, where in the terms of Goethe, the particular represents the general, not as dream or shadow, but as the living-instant revelation."[41] This "living-instant revelation" is ecstatic in the literal sense that the subject of history, both as experience and in reflection, partially escapes their usual understanding of personhood, objectivity and experience, even as this constitutes a greater immersion into the forces of both present reality and historical change. Historical consciousness, understood functionally, intensifies the present and is indeed the ultimate mode of the symbolic as introduced by Leibniz and, in Cassirer's reading, further developed by Goethe – in which the differential of all experience is perceived through the integral of a particular experience.

In Cassirer's later work, this experience of history is essentially the experience of immanent infinity caused by the torsion of history; indeed this was key to Cassirer's position on "infinity" in his later comments on his debate with Heidegger at Davos.[42] History, understood symbolically, reveals the maximal functional definition of experience. History removes all substantial or commonsense orientations and replaces them with functional relations; indeed history epitomizes functionalism since, as Cassirer later puts it, "as far as the reality of history is concerned, it is impossible to take it as some rigid thing-like object."[43] In this regard history leads to the "overcoming not only of the category of the thing and thingness but also categories of 'personal' being."[44] Far from being the desiccation of being or of the experiencing individual, however, Cassirer clearly sees this "living-instant revelation" as of the utmost aesthetic, ethical and political importance. The reason for this is that it reveals individual experience as at once a collective and historical experience, as both an ecstatic moment and the starting point of critical analysis. As Cassirer states on a related point in notes on the Davos debate: "Infinite time is for Heidegger a mere fiction [...] for us it is not mere objective physical time, but a time specific to mankind – a change in the subject of temporality."[45] Unlike the Hegelian understanding of history and philosophy, for Cassirer the model of group and infinitesimal meant that there was always a multiplicity of possible transformations open to society, and that these could only be grasped through an absorption into the infinite texture of present problems catalyzed by historical understanding. To perceive the world and humanity through the perspective of infinite time, or through Goethe's sense of the symbolic is also to first perceive the myriad ways the present could be otherwise, and thus the first step toward a critical analysis of the present.

Freedom and Form as Critical History

Cassirer begins *Freedom and Form* by noting that one of his intended goals is to transform the German debates concerning national character, and the personality of the state, that dominated wartime discourse. In its narrowest sense, the book was written in response to the question that Cassirer argues was the obsession of the wartime period: how to define the "being" (*Wesen*) of the German people, and more specifically the "folk-individuality" (*Völker-individualität*) of any of the European peoples.[46] Cassirer's stated goal is to make German wartime nostalgia for an idealized past into something productive in Goethe's sense; to provide a reconfiguration of elements that would lead to a creative new synthesis on which a future reality can be built. History looks to the past to find what is "existing and continuing" in the present, as Cassirer put it, so that it can form a key "decisive turning point" in relation to the future.[47] Understood through the model of form based on group theory in *Substance and Function*, this statement implies that by accenting different series of changes in the past, a new set of permutations of form within the present can be emphasized, and (as in group theory) even moderate changes in these forms can lead to profound transformations in the future. The being of Germany is found neither as a completed substance nor through an escape into an oceanic world of neo-Romanticism, but is rather found when the reader immerses themselves in the currents of change revealed in the present through the new formal combinations of the German – and European – past.

The dominant definition of form at work in *Freedom and Form* is historically limited to the changing institutional and ideal forms of the previous four centuries of German history. It is through changing the rule-forms of institutions and ideals in society, notably the state itself, that German and European development occur. Simply stated, Cassirer develops an argument for the possibility of a liberal state, a coherent set of rights and a recognition that nature – whether of a people or in the exact sciences – is never directly given. The recognition of this mutability in each category, however, occurs in part through a growing recognition of the concept of form and its relation to reality itself. Cassirer's historical and philosophical starting point in the book is thus not surprisingly Leibniz's definition of form, which is developed directly from *Leibniz's System*. According to Cassirer's reading of Leibniz, form is defined as encompassing any organizational scheme of functions. The modern definition of form for Cassirer will begin here and culminate in Hegel, and focus on what Cassirer defines as the modern "objective spiritual form" that includes forms of "society and state, science and law."[48] In Leibniz's definition, as Cassirer had also argued in his 1902 text, form is the always provisional organization of the unknowable forces of the world.[49] Leibniz's definition of phenomenal reality, further developed through German classicism, holds that the "reality of phenomena consisted in the determination [*Bestimmtheit*] of their ordered coordination [*gesetzlichen Zusammenhangs*]," and form defines this ordered coordination in regard to phenomena ranging from aesthetics and ethics to law and natural science.[50] Again, *Substance and Function* tacitly provides the modern definition of this concept, notably in group theory, but Cassirer avoids the use of his earlier terminology and rather develops an analogous set of ideas through the material of German classicism.

The central historical contrast of *Freedom and Form* is between the medieval world's "extensive manifold of stages being organized in terms of increasing levels of perfection," and the modern world's move towards understanding reality as "an intensive manifold of form-giving functions [*gestaltender Funktionen*], which link with one another into a complete system of activities."[51] In its most recent incarnation, Cassirer's form is similar to Hegel's objective spirit, in that it is not merely an ideal set of functions but includes an ensemble of practices, mores and structures one might, for instance, find in the "form" of the judicial system. The closest contemporary analog to the all-encompassing definition of form at work here is that of sociologists, notably, as previously mentioned, Cassirer's colleague and former teacher, Georg Simmel.

Cassirer cites examples of earlier transformations of rule-forms, including the development of the religious concept of work from scholasticism through Lutheranism and early capitalism, and the transformation in the idea of natural rights from Grotius and Leibniz through Kant.[52] Changes in the scientific worldview also constitute the development of rule-forms, as occurred in the development of mathematical physics from Copernicus and Descartes to Leibniz.[53] Even apparent natural truths such as the definition of motion are capable of reinscription within a new context, such as occurred in the redefinitions of motion from Aristotle to Galileo and Newton. Similarly, apparently fixed and *a priori* aspects of philosophical experience are capable of reintegration into wider patterns of meaning, such as occurred in the shift from Descartes' emphasis on the *cogito* as substance to Leibniz's focus on the necessary manifold *contents* of this *cogito* as function.

The key author for understanding the complex modern definition of form is, surprisingly, Goethe. Cassirer sees Goethe as creatively extending Leibniz's concept of form through the intermediary of Kant. Cassirer considers Goethe's theory of form as the "middle point of all of the various directions of argument of [*Freedom and Form*], now turned in a new direction" and claims that in Goethe's "theory of the intuition of nature" (*Naturanschauung*) the problem of freedom and form finds "its culmination."[54] Even if Goethe was not a "follower" of Leibniz, Cassirer claims that Leibniz's monadology was, consciously or unconsciously, the basis from which his "world picture was built," just as it was for figures such as Lessing and Herder.[55] Cassirer then situates Goethe's theory of form as the basis of a new reading of Humboldtian liberalism, one in which Humboldt's emphasis is not simply the agonism of different individuals, but the agonism of competing definitions of form and reality. A similar harmony is brought up in the final pages of the *Freedom and Form* to suggest the manner in which the "historical recognition of objective spirit" in Hegel harmonizes the counterpoised concepts of Schelling's idea of nature and Fichte's idea of freedom – and this "objective spirit" presumably finds its modern exemplar in Cassirer's own definition of the historical study of functional forms.[56] Despite the sophistication of Cassirer's dialogic development of Schelling, Fichte and Hegel, however, it is ultimately Goethe's philosophy and theory of form that is the "ideal middle point" of the text.[57] In the present analysis, then, we will look only at Goethe's theory of form as it is developed by Humboldt, which is effectively the centerpiece of Cassirer's historical and theoretical argument for a new mode of liberalism and a new definition for the congruence of the "methods of the natural science and those of the human and social sciences."[58]

The Historical Narrative of *Freedom and Form*

Modern German history is defined by different forms and contentions for freedom that, in Cassirer's wording, "dialectically confront and demarcate each other" in three historical stages, and which can be summarized through the figures of Luther, Leibniz and Wilhelm von Humboldt.[59] The agonism of these forms provides distinct lines of development in retrospect, even as their interplay is often contingent in development. The historical dialectic of the relation of freedom and state began, in Cassirer's reading, with the Reformation's concept of freedom of religion and conscience as developed through the competition among different sects in relation to the state. In the Enlightenment, the conflict of both opinions and new social and economic forces led to a new concept of the freedom of thought and opinion, an argument first fully developed in Cassirer's reading in Leibniz's concept of the legal individual. These two movements of freedom of conscience and freedom of expression led first to the concept of a distinct sphere of state power, presumably in the Peace of Augsburg and the Treaty of Westphalia, when the separation of the state from personal religious issues partially occurs. It then led to the concept of the citizen as individual that culminates for Cassirer in the work of Leibniz and Wolff as it is then received by Western constitutional theorists, notably William Blackstone and Thomas Jefferson.[60]

The third and most important stage of the development of freedom and state occurs through the aesthetic definition of the individual and experience, the problem that was often, following Humboldt, defined as an infinite "qualitative individuality" as opposed to the "quantitative individuality" defined by the legal and demographic person. For Cassirer, the crucial starting point for this line of development is again Leibniz's monadology, but this is actualized in the problem of aesthetics in two practical issues, individual autonomy and its relation to particular forms of knowledge and sociation. Again, the locus of both problems is the individual as a radically unknowable infinity, only graspable to themselves or others through actions and forces in the world.[61] Goethe's concept of form provides the highpoint of the modern definition of form, which balances freedom against determination.

The argument culminates in Cassirer's reading of Fichte and Hegel, where he contends that Fichte's legal philosophy focuses on the concept that "the content of law does not develop from the being [*Dasein*] of various different subjects, but rather the recognition of the necessary validity of a norm of law is itself the *condition* [*Bedingung*] for the [possibility] that the 'I' recognizes outside of itself not only, for instance, empirical things, but free, self-determining [*sich selbst bestimmende*], reasonable beings."[62] The self is reciprocally defined, so that as a "free, self-determining, reasonable being" it is not for Cassirer an insular self or Cartesian *cogito*, but develops in relation to the matrix of various forms of understanding and individuals. Drawing from Hegel, Cassirer argues that the concept of "spirit" is in fact only the concatenation of different forms of "objective spirit," that is, institutions and practices, and these are guided not by a singular idea, much less the idea of the state, but the often contingent and dialogic play of forces in society.[63] This argument is then placed in rapport with Fichte, in that Cassirer argues that these forces can in turn only be guided by the ethical relation to the possible future of humanity in general.[64] "The idea of the state appears for Fichte in that, and only in that, it learns from within its own resources to question beyond itself [*noch über sich selbst hinauszufragen*]."[65] Ultimately such questioning leads to the paradox that German intellectual history is particularly useful in teaching us to permanently question the past and the conflict of forms in a manner that makes it "productive" for the future. Neither any one nation nor any one epoch can lay claim to this criticism, but "it will remain to the credit of German intellectual history," that it has demonstrated the developmental basis of this criticism as it continues into the future.[66] Thus Cassirer's argument concludes with the concept that the "essence" of the German people *is* their history, and this history is one of critical confrontation among institutional forms and with the state leading towards freedom.

The true foundation of this argument in the ideal of individual autonomy developed in the work of the philosopher, linguist and diplomat Wilhelm von Humboldt (1767–1835). Humboldt argued that the individual, and particularly the interaction of individuals, is by definition more complex than any "outer instance" of state definition, cultural form or political representation can directly grasp – a concept that forms the basis of Cassirer's definition of liberalism.[67] Unlike the liberalism of Mill, in Cassirer's reading the liberal state is defined not only by the free interaction of individuals, but also the maximum free interaction of all elements of form understood as both ideas and objective spirit, as both modes of understanding the world and practices for doing so. The starting point of this turn is Leibniz's original theory of the state as itself a functional ensemble: "What is original in Leibniz's concept of the state" is that "the elements of the state are always defined by a necessary relation with one another."[68] The state is such an "intelligible constitution" of form, but due to its nature – and the intensive understanding of part and whole in Leibniz – it is also always codefined by the concrescence other forms both within it (such as guilds, associations or sanctioned religions) and outside it (such as other states, ideals of humanity or heretical beliefs) so that it is irreducible to one "thing."

Without constant and critical attention to the developments of forms of meaning and sociation, particularly the historical definition of subject and object through which politics occurs, a true liberal state cannot exist. For this reason, both the university and a critical form of public communication are critical to Cassirer's argument. Cassirer understood the historicization of subject and object as developing the work of Jacob Burckhardt, particularly Burckhardt's *Civilization of the Renaissance in Italy* (*Die Kultur der Renaissance in Italien*) (1860). However, for Cassirer, Burckhardt "in fact portrayed only one aspect" of this development in the objective development of modernity, notably in the concept of the "technology of power" leading to the "state as a work of art."[69] What is missing from Burckhardt is a convincing account of the development of modern subjectivity. It is this project that Cassirer develops in his narratives on Cusa and Luther in *Freedom and Form*, and more extensively in the *Problem of Knowledge* and *Individual and Cosmos*, which detailed the history of the subject through their interlocking histories of religion, philosophy and science.[70] From within this approach, Cassirer can depict both subject and object definitions as correlatively defining one another: the concept of "personality" develops out of the matrix of earlier natural science and religious elements that was intimately related to the new "objective world" of mathematical physics and Machiavellian power relations. Only an understanding of society that recognizes these developments, and reflexively seeks to transform them through the sciences, art and politics, can progressively develop greater freedoms.

Goethe's Philosophy of Form and Symbol

Although appearing in the center of the book, Goethe's reading of the problem of form completes the key themes of Cassirer's text and brings each term of its title to its full definition.[71] In the structure of the essay, Goethe's definition of form provides a means of developing Leibniz's key themes while avoiding the "fiction" of the monad. Presented through a compelling narrative of Goethe's development, Cassirer can introduce both a phenomenology of experience and what proves to be a translation of key aspects of *Substance and Function*, and with it in part Cohen's insights, in a popular fashion. Goethe's definition of form then works as the catalyst for *Freedom and Form*'s broader study of form in "society and state, science and law."[72] The importance of Goethe for the philosophy of science and aesthetics, Cassirer writes, is the manner in which his definition of form provides a new answer to the problem of the relation of the general and the particular, which in turn define form as the manifold of appearances as it relates to a particular reality. Goethe's idea of form and experience is thus re-read, as Cassirer notes, within the context of the argument for the logic of form Cassirer himself put forth in *Substance and Function*.[73]

In the modern era, Cassirer writes, there were essentially only two basic concepts of observation (*Betrachtung*) within logic for understanding the relation of the particular to the general. On the one hand, the general could be considered as an abstraction from the particular in the manner first developed by Aristotle. In this sense, it is a "schema and a species image," which allows us "to collect particular content on the basis of a common image."[74] This position is extensively critiqued at the beginning of *Substance and Function*.[75] On the other hand, the general can be considered a rule in which the particular is grasped, that allows us to link it with others – that is, the process of the function understood only as a serial order.[76] The first definition leads to a loss of definition (*Unbestimmung*), while the second leads to a "growing clarity" that comes at the price, however, of never being formed into a particular image even as it leads to ever greater connections of the relations between the intuited elements.[77]

Goethe's work is a means of combining these two basic approaches into a new hybrid form.[78] The role of functional "rules" is seen to enhance the immediacy and intensity of experience itself, since through the power of continuity we grasp the particular creatively through a relation to a pattern or idea.[79] It is this that Goethe described as "objective thinking" (*Gegenständliches Denken*) or "objective writing," and Cassirer considers it to fundamentally change the relation of "individuality and totality [...] Ideal and Real, the general and the particular."[80] Rather than simply being a "rule," the invariants of a form meet up as a "symbol" at the limit point of the present, with symbol here defined similarly to Leibniz as the determination of the particular by the plenum of the whole. "The whole is no longer grasped in the gradual progression of the particular," Cassirer writes, "but it appears and pulls itself together at the same moment in a point, in a concrete symbol."[81] This relation of rule to particular both changes the perception of the object while redefining the rule, so that in Goethe's startling description, "every new object, truly perceived, creates a new organ in [the perceiver]."[82] Goethe's philosophy in Cassirer's reading essentially provides an aesthetic definition of the relation of part and whole and the role of judgment, that is, a direct development of that which Cassirer had put forth in the reading of form and reality as they relate to perception in *Substance and Function*.

To explain Goethe's innovation, Cassirer traces its evolution in Goethe's natural philosophy, where the poet developed insights that allowed him to fundamentally change the dominant Kantian philosophy of the day. Although not limited to this version of natural science or to its technical application in Kantian philosophy, it is helpful to briefly sketch this starting point for Goethe's thought. The central concept of Goethe's theory of organic science was the definition of metamorphosis and of natural form, and its clearest application is found in Goethe's paradigm of perception in his initial pursuit of the problem of the *Urpflanze*, or primal/originary plant.[83] Just as Aristotle had developed his logic and theory of perception in close connection with his early focus on biology, notably the species/genus distinction, so here Cassirer will introduce through Goethe's biology a distinctly modern logic, first initiated by Leibniz but popularly transformed by the poet.[84]

During his journey to Italy, Goethe became newly interested in the *Urpflanze*, which might be the ancestor of all other plants. As Cassirer notes, although this concept was not focused on the causal mechanism of selection in the sense of Darwin, it was nonetheless – in conjunction with the Kant–Laplace theory – theoretically one of the key moments in the development of the theory of evolution. Walking in the public garden in Palermo, Goethe was struck with the vast assay of plants: "In the sight of so many forms both new and familiar, the old fancy occurred to me again: among this multitude could I not discover the original plant?"[85]

At first Goethe was interested in finding an *actual* "originary plant," but gradually he came to see the idea, or as he calls it "symbol," in a new manner defined by his broader philosophy of science. The "originary plant" would be an ideal form that yet enabled him to perceive the continuity underlying an entire developmental process. The relation of each plant and its morphology suggested a common theme, which in turn became visible through the ideal of the *Urpflanze* as its common ancestor. In Goethe's studies of plant morphology the relation of the individual takes on a new meaning, one that will be for Cassirer the model of the innovations Goethe makes on Kant's third critique.[86]

The Urpflanze is intuitively perceived in a manner similar to certain aspects of the Kantian schema, that is, as a framework for illuminating the relation of the sensory manifold and concepts to arrive at a particular, yet it is also more explicitly part of a dynamic process of relating phenomena in the world. In seeing each plant as part of a continuity with the *Urpflanze*, Goethe retains his emphasis on the particular plant, even as the meaning of this plant as well as its presence is transformed by the concept (in this case the concept of an Urpflanze) that illuminates it. The *temporal* element of plant development in world time, however, finds a crucial analogy in the basic phenomenon of perception itself: the presence of the particular plant is understood as in motion, so to speak, in relation to its structural connection with the other plants, diachronically and synchronically, as mediated by the promised Urpflanze.[87] In this manner, as Goethe understood it, he had taken the "stationary" idea of plant and life, and made it into something moving: "movement" on one hand of spatial and temporal comparison, on the other movement as a process of *becoming* in the present.[88]

This initial reconfiguration, in turn, will prove a cipher for changing the definition of all of experience, beginning with natural science. Goethe writes that he was allowed to develop "viewing [*Anschauung*] into a higher form [...] through the sensual [*sinnliche*] form of a transcendental [*übersinnliche*] originary plant."[89] The presumed actual temporal development of plant morphology serves as a model for understanding both theories of the natural world and perceptions of it as understood historically. Kant's original schematism of course depended on a temporal element through which imagination links concept and the sensuous manifold. Here, however, the structure of concept, perception and objective are all understood as temporal processes whose only mediation is found in the "sensual form" of the immediate present, which nonetheless is also placed in relation to an ideal structure. Cassirer's own theory of the necessary and constant evolution of structures of thought and reality are presented more distinctly in *Substance and Function*, but in Goethe he clearly thought he had found a classic source for this insight. In Cassirer's reading, it represented not only Goethe's greatest triumph but also that of German classicism, and it allowed Goethe to decisively transform the Kantian epistemology of his age.

The Problem of Aesthetic Freedom and Goethe's Transformation of Kant's Third Critique

For Cassirer, Kant's late philosophy in fact is only first clarified in "concrete form" (*konkreten Gestalt*) in its application in the work of Goethe. "The real affects of the 'critique of the teleological judgment,'" Cassirer writes, "first succeed through the work of Goethe's organic view of nature."[90] Cassirer understands Goethe's innovation as having a particular place in the Kantian system, namely in transforming Kant's original theory of the synthetic principles and schema in the first critique, which is principally focused on the epistemology of natural science, and providing a bridge from these to the problem of teleology in the third critique, which is principally focused on aesthetics. The teleological critique concerns the meaning of appearances towards their ends, a problem that draws together the interrelation of part and whole, a field that Kant considered in particular in relation to biology.[91] In transforming the Kantian schematism towards a dynamic model, Goethe is read by Cassirer as presenting key aspects of Cohen's Kant interpretation in an earlier form.

Where Kant looked for synthetic principles of judgment as static models, Cassirer writes in a later summary, Goethe sought "productive principles" at the heart of any particular perception.[92] In this way, Goethe redirects the entirety of the schematism and the principles to a new end that in Cassirer's reading goes beyond Kant.[93] The immediacy of the present is, in this reading, a point of immanent determination defined by multiple functions or meanings, but oriented tacitly towards a developmental ideal and experienced as a unitary living perception. "Kant," Cassirer writes, declares that nature is "'the existence of things, in so far as it is determined in accordance with universal laws.' Goethe cannot stop with such a nature, '*natura naturata*,' as artist and scientist he desires to penetrate into '*natura naturans*.'"[94] Nature is not to be taken as a given thing, a "natured nature" (in approximate translation), but as a productive developmental whole, a "naturing nature," of which humanity is a part, and which is continually producing new forms, new experiences and new knowledge. The emphasis is on the dialectic, or better dialogic, process through which, as Cassirer puts it, "the structure [*Gebilde*] of reality dissolves into process; but the process strives always towards the structure."[95] The play of structure and limit in *Substance and Function*, and with it the play of reality and the particular real, is revealed as having found its poetic and philosophical inspiration in Goethe.

For Kant, in Cassirer's reading, the central dictum is that "for the naïve world, we have things, in critical analysis, however, these are resolved into functions."[96] Functions are in turn grasped by Kant as *forms* of experience, such as in the first critique, where they are more often referred to as the categories and intuitions of time and space, as well as the forms of the schematism. Similarly, *form* is used to define the variegated forms of aesthetic experience in the third critique. In Goethe's definition of form the Kantian models are united into a broader definition relating any concept or idea to experience, and unifying structure and motion in a new manner.

Goethe's idea of form allows for the immanent perception of the core aspect of Leibniz's theodicy, that is, the promise of a unification of judgment and idea in the present. In its theological definition, the theodicy was the promise of "a necessary concordance of all directions of consciousness to a common outcome and goal."[97] But in its secular reception, this comes down to a premise – ultimately grounded on the development of life itself – of an ever greater meaning than what is simply given, and the possibility of perceiving this meaning. As Cassirer had defined it, an aesthetic "coordination between means of the observing and the means of judgment."[98] Goethe's definition of form, in Cassirer's reading, essentially completes the secularization of this idea. The fundamentally open-ended and exploratory mode of form links the immediacy of experience with the broader structure of "reality" as the promise of order and meaning in the world, a relation that Cassirer developed from Leibniz's philosophy through the problem of the "Real," or the ideal moment within the present that unifies these two aspects.

It would be difficult to overstate the importance of Cassirer's reading of Goethe in *Freedom and Form*. Cassirer describes Kant's placement in the book as secondary to Goethe's, and the same could be said of their relation to his future work.[99] When Cassirer writes in his introduction that his book provides a new philosophy of understanding the relation of "appearance and reality" as well as of understanding the relation of the "natural and cultural sciences," it is Goethe's theory of form he has foremost in mind. Yet the importance of this theory itself lies in how it allows Cassirer to recast core aspects of the Marburg project, notably the relation of the infinitesimal to judgment and perception, as well as his own theory, notably the theme of group theory, through a popular and living historical tradition. It is moreover through this broad definition of form he can do justice to the original question of the "being" of a people or other complex phenomena, as well as to the various definitions of legal, social and scientific form that are the topic of his essay.[100] It allows the element of "freedom" to be perceived in the critical appreciation of "form" as reality, even as it also insures that form is never taken as simply given either in the natural or social sciences.

If we move from the generalities of Goethe's discovery to its specific import, this centrality becomes clear. In the first place, the theory combines empiricism and rationalism in a new manner. Goethe's work, as Cassirer later writes, "forbids every 'abstraction,' every looking-away [*Absehen*] from the particular; it knows only an ever renewed and ever intensified looking closer [*Hinsehen*]."[101] The basis of Goethe's work in Cassirer's view is thus found in the concept that "pure seeing [*Schauen*] itself had become a form of ecstasy" – the process of exploration always occurs outside of a settled definition of subject or object, in a continual dehiscence of experience.[102] Goethe in this regard fundamentally transforms the notion of "subjectivity" by changing it from an "inner" reality to one that removes "all divisions between the world of reality and the world of the 'true appearance.'"[103] The radical immanence of Goethe's theory – and poetry – will in turn serve as the starting point for Cassirer's work.

As Goethe famously phrases it in a letter to Eckermann, his search can be defined as "imagination [*Phantasie*] for the truth of reality."[104] As in Cassirer's own Leibnizian reading of Kant, it is only in such an imagination that any scientific or other truth is possible. "Imagination is for him an organ that is not designed to build a new and 'otherworldly' world 'over' the reality, rather it grasps and clarifies the general form [*Gesamtgestalt*] of this reality itself."[105] In a later essay, Cassirer will highlight that despite Goethe's animosity to history as a discipline, his theory in fact also provides one of the best explanations of the relation of the instant of history or historical reconstruction to a general development.[106]

For Goethe as for Cassirer, the opposition of form and matter becomes completely codeterminative. In a manner only fully evident in the argument of *Substance and Function*, Cassirer sees the modern period as overcoming the traditional dichotomy of matter and form since both are codeterminative: "Matter *is* only with reference to form, while form, on the other hand, *is valid* only in relation to matter."[107] Form was initially defined in the Aristotelian tradition as one of four definition of cause, and placed in opposition to material; form is the *eidos* or idea of something, say as table is to wood, or the basis of its rule of change, as a tree is to wood. In the modern period, Leibniz introduces a new definition that will guide Cassirer's work, which eschews this opposition by claiming that all "transformations of determination" occur through "immanent form principles"[108] Form and matter are two perspectives on the same phenomena of determination of the particular, and different definitions of form lead to different definitions of matter. Kant continues this definition by focusing on the problem of "determination" (*Bestimmung*) as the intermediary term: matter is defined as "the determinable altogether" (*das Bestimmbare überhaupt*) while form is defined as "that which determines" (*dessen Bestimmung*).[109] It is this overall frame that Cassirer now calls "form" itself, whether it is covering Kant's original use of the term in the "form" of the categories and intuitions of time and space, or in more specific interests of the determination of any particular object. With Goethe's definition of form, Cassirer is able to broaden this definition of form out to encompass Kant's later definition of form from the third critique, as well as his own model of group theory from *Substance and Function*.

Goethe's definition of form changes not only the relation of appearances in the world and subjectivity, but the relation of any particular aspect of reality to the whole: "The structure and fabric [*Gebilde*] of reality, dissolves into process; but the process strives always towards the structure and fabric" – which is to say, the promise of the "Real" within experience.[110] This relation of structure and genesis will be the basis of Cassirer's own theory of reality and methodology of structuralism. For Goethe there was no "fixed world picture" (*fixiertes Weltbild*) at all, but rather only a world in which, through its "polarities" and "augmentation" (*Steigerung*), "is consistently renewed."[111] "The new 'objective truth,'" heralded by Goethe's philosophy, Cassirer writes, "is developed in the relation of all of the individual parts, not in the existence of the individuals themselves."[112]

The result of Goethe's work is a major transformation of the philosophy of science that in Cassirer's reading translates many of the key themes of the determination and the infinitesimal into modern terms that agree with "modern physics and the modern theory of knowledge."[113] By the former, Cassirer means principally the work of Planck and later Einstein; by the latter, we must assume that Cassirer meant the "modern theory of knowledge" he and Cohen developed. Yet Cassirer is also adamant about Goethe's originality and brilliance in this regard: "Goethe did not just give organic natural science a brilliant literary 'stimulus,'" Cassirer writes, but he "rather discovered their foundational reality and basic method."[114] Modern science allows us to perceive the "reality" of the world as ever more refracted through theory even as this theory proves, precisely through the succession of its forms, to be always provisional. Although Newton's first law of motion is "ideal" in the Kantian sense, for instance, we see the movement of objects in relation to it, even though it describes conditions – an object in motion without being affected externally – that strictly speaking could never exist for an observer. The key point for Cassirer is that "experience is only half experience" in that the definition of "reality" itself shifts in particular and local instances with broader changes in theory, and it does so precisely because the system of experience has no recourse to an "outside" standard. Within the present structure of knowledge, we can imagine new possibilities of relation that can then be experimented with, so that "imagination and nature here rival each other [*wetteifern*]," in Goethe's words on exploring organic phenomena, "as to which procedure *knows* more audaciously and rigorously."[115] This relation of imagination of knowledge is not an either/or decision, but a constant striving for unity on which experience is grounded.

Indeed, the deeper problem revealed by Goethe's definition of form as movement occurs as a stage of knowledge, and is revealed in the fact that, as Cassirer writes, "in principal one can say that no one asks a question of nature that could not be answered; for even in the very question lies already a type of answer, lies the feeling that one can think or anticipate something."[116] Our relation to nature is already a structure in the process of transformation, and this transformation occurs on the horizon of imagination and empirical investigation. For everywhere the assumptions of theory and appearance are interwoven, often in a manner that is difficult to discern in its very obviousness.

For the argument of *Freedom and Form* as a whole, the importance of Goethe's theory of form is that it provides the basis for understanding "objective spiritual [*geistige*] forms," such as those of "society and state, science and law," as both developing historical structures and as conveying an immediate and intuitive reality to those who participate in them.[117] It thus unifies the two forms of reality as process and intuition that were delineated in *Substance and Function*. The form of sociation within a literary salon, for instance, or of legal organization within a particular epoch would presumably define a particular local "reality" as functional form. This practical application in turn can explain how different forms create different modes of subjectivity and objectivity, and with them enable and constrain certain modes of freedom. *Freedom and Form* itself culminates in a more sophisticated reading of functional form in Hegel's concept of the "historical recognition of objective spirit," which Cassirer portrays as harmonizing the counterpoised concepts of Schelling's idea of nature and Fichte's idea of freedom.[118] The key concepts of Hegel's idea of objective spirit are for Cassirer, however, already outlined in Goethe's theory of form without the wider assumptions of Hegel's thought. Cassirer's own methodological use of functional form in the narrative of *Freedom and Form* is clearly meant to represent the culmination of Goethe and Hegel's development of this concept. In this manner the concept of functional form first developed in the study of science in *Substance and Function* is transposed by Cassirer into a new modality, and a more approachable style of thought, through the account of its "discovery" in German classicism.

Aesthetic Freedom and Basis Phenomena (*Urphänomene*)

One of the more important innovations in Cassirer's argument in *Freedom and Form* is his recognition of the primacy of Goethe's "basis phenomena" (*Urphänomene*): aspects of reality that are not capable of further reduction, but are also not "ungraspable metaphysical concepts."[119] Although the prefix *Ur-* could also be translated as "primal," the overtone of temporal relations in this term is confusing, so we will use "basis phenomena," as this will concern the future as much as the past. In his later "systematic" overview of *The Knowledge Problem, Objectives and Paths of Reality-Knowledge* (1937), as well as in a manuscript from 1940 entitled *Basis Phenomena*, Cassirer highlights the centrality of Goethe's basis phenomena as a key to his historical and theoretical projects, respectively, and this approach is arguably initiated with *Freedom and Form*.[120] Basis phenomena link Cassirer's early work on Leibniz with the treatment of Goethe in *Freedom and Form*, while also forming the foundation of Cassirer's later responses to the new philosophical movements of life-philosophy, existentialism and ontology.

Cassirer claimed in *Freedom and Form* that the concept of form developed by Goethe overcame most of the "basic oppositions" that ground modern philosophy, "the distinction expressed in such catchwords as rational-irrational; life-thought; intuition-concept; existence-value."[121] The starting point of the basis phenomenon is that there is no deeper basis to experience than its immanence itself. Rather than understanding philosophy or science as "copying" the truths of a single "outside" world of substantive things, the goal of criticism will be to see the world as an unfolding relational play based on initial relations of difference. Cassirer's description of the theme has a marked resonance with Cohen's philosophy of origin, particularly in its goal of seeking the most generative and "pure" concepts within an immanent field of experience. Cassirer summarizes Goethe's initial insight as follows:

> For explanations can only be valid in the sense that we follow a phenomenon back to another phenomenon, a complex appearance to a basis phenomenon [*Urphänomen*] and

can in turn derive it in reverse – but it is not possible from the combined matter [*Gesamtsache*] of these appearances themselves. Whoever does not grasp this, acts like a child, who when he has looked in a mirror, immediately turns it around, in order to see what is on the other side. Here we can see how the most "inborn" and crucial concept, the concept of causality and agency, brings us to catastrophe [*Verhängnis*] and to the source of innumerable ever repeated mistakes."[122]

Describing the relations of experience, Cassirer writes, "Only the direction, not the goal of this process is granted to us; but there is no outer transcendental goal, from which we could provide anything that would replace the abundance of its contents."[123] As Cassirer writes, the goal of life functions "just as in a game" in that "it is all about doing and not just about theorizing. Nature has given us the chess board, cut out the pieces, and now it is for us to play."[124] Unlike the example of chess, however, for Cassirer the structure of the game itself changes as we play.

In his later works, Cassirer emphasizes Goethe's delineation of three aspects to basis phenomena, but although presented in rich detail each can be seen to parallel Cassirer's readings in *Leibniz's System* and are consistent with the argument of *Freedom and Form*. The first, as he notes in *Objectives and Paths* is at its most basic level "life, the rotating movement of the monas in itself, which has neither rest nor quiet [...]. This characteristic nonetheless remains for us and for others a mystery."[125] The first basis phenomenon begins with the *natura naturans* of form giving life in itself, beyond subject and object, intention or fate, complete determination or freedom. Experience is grasped foremost in the sense of Leibniz's monads "as a process, as movement [...] unknown and revealed to all, the primary revelation itself."[126] Cassirer writes of Leibniz's philosophy, "The 'I' remains always the same mysterious open-basis phenomenon [*geheimnisvoll-offenbare Grundphänomen*]," which can be explained neither by ourselves nor by others, and certainly not by any form of "knowledge in the sense of mathematical-mechanical."[127] As the first basis phenomenon it culminates Cassirer's reading of the transcendental ego and transcendental unity of apperception as ungraspable in itself. Far from being "something static," it is best imagined as a "dynamic [...] exit and origin point for lines of force [*Ausgangs- und Ursprungspunkt von Kraftlinien*]."[128] This is revealed on the deepest level in experience through the concept of motion in the primal sense defined by Goethe's theory of form as at once a motion of empirical perception and the correlative "motion" of ideal structures relating to this perception.[129] This existential starting point of all history is epitomized by Goethe's pithy notion that "in the colorful refractions we have life," a nontrivial theme emphasizing the immanence of Cassirer's mirror simile suggested above, and developed throughout *Freedom and Form* on the basis of Goethe's theory of colors, and particularly his studies of the "entoptic" reflections of certain crystals.

The second basis phenomenon is the manner in which others, and for that matter ourselves, are only revealed through "the sense of doing – both action and reaction," which reveals "the inwardly boundless, the externally bounded."[130] It is essentially an emblematized version of Cassirer's focus on "doing" as the center of all meaning.[131] Finally, the third basis phenomenon is work. Cassirer summarizes this theme in *Objectives and Paths*: "Works no longer belong to us; they mark the first level of 'alienation.'"[132] It is through the sciences that work and action are structured in society on a level more primary than subject or object – a process Cassirer narratively develops from Luther's concept of work to Goethe's notion of activity and Hegel's concept of labor in *Freedom and Form*.

Cassirer frequently returns to Goethe's emblematic use of the concept of "entoptic colors" to summarize the process through which relations sustain our ability to see "real" phenomena in new ways, and he uses the model equally for both "subjective" and "objective" phenomena. Goethe develops the concept in his *Theory of Color*, and in Cassirer's view it summarizes his own theory of form and the methodology of *Freedom and Form*.[133] Entoptic colors arise from the resonance of reflections within certain crystals so as to lead to the appearance of new orders of light on their surface. Far from being "unreal" or epiphenomenal, they allow us to "see" the inner structure of the crystal as a single phenomenon, a fusion of its many interior angles and properties, when it is turned and held as a form of action combined from the whole.[134] Such entoptic colors epitomize for Cassirer Goethe's understanding of the fluid interrelation of the experience of reality and the structure from which it develops, of part and whole, ideal and real. Thus, describing his methodology in the first pages of *Freedom and Form*, Cassirer uses the example of Goethe's theory of color to epitomize a functionalist perspective: "We seek in vain, as Goethe has it in the introduction to his *Theory of Color*, to express the essence of a thing. We become aware of the effects of something, and a complete history of these effects comprehends at best the essence of a thing. It's foolish to try to sketch a person's character, one must place their interactions and actions together, and a picture of their character will appear for us."[135]

In this way Goethe can hold that "the true […] never permits itself to be known directly; we look upon it only in reflection, in example, symbol, in particular and related experiences."[136] Neither characters nor phenomena, neither subjects nor objects, are "things" at all, but they are nonetheless defined as having rules for their reality through the phenomenal unity of the resonance of all of their particulars. "What such an entity 'is' can not be articulated as an absolute, but can only be known in the manner its effects appear and are expressed."[137] Thus Cassirer claims that much as Goethe was fascinated with the "actions and sufferings" of light, so his text will address the "actions and sufferings" of history.[138] In this sense, Cassirer sees the relation of "appearance" and "being" newly defined by his study in *Freedom and Form*, but it is apparent that this is a form of unity in difference that he had long developed in his study of the symbol as it began with Leibniz.[139] Against both traditional idealism and the new ontologies, Cassirer will insist on a radically immanent philosophy that also critically situates the sciences.

It is telling of the importance of Goethe in this regard that in Cassirer's later Davos debate with Heidegger that he appeals to Goethe, not Kant, when he is asked the role "infinitude" in his philosophy – that is the theme that is usually taken as the core of his opposition to Heidegger's reading of finitude and ontology. Infinity appears in experience not through "privative determination but is instead a stranger sphere," Cassirer says, before – as we noted earlier but now are in a position to understand – somewhat cryptically citing Goethe: "'If you want to step into infinitude, just go in all directions into the finite.'

As finitude is fulfilled, i.e. as it goes in all directions, it steps out into infinitude. This is the opposite of privation; filling-out of finitude itself."[140]

The only feature of Cassirer's thought that appears to clarify this determination (*Bestimmung*) of finitude and the infinite, and with it a form of negation that is "the opposite of privation [...] the perfect filling-out of finitude itself," is limitative judgment as it develops from Cohen in Cassirer's early work. Such a reading is consonant with Cassirer's central claim in *Substance and Function* that "the particular 'representation' reaches out beyond itself, and all that is given *means* something that is not directly found in itself [so that] the particular appears as the differential, that is, not fully determined and intelligible without reference to its integral."[141] As we noted in earlier, this theme was further developed through Cassirer's central theme of symbolic prägnanz in his later work.

On this basis, one can understand why Cassirer would stress that a particular form of infinitude is the basis of experience, yet also why this infinitude is itself always grounded on the finite and particular. The argument intimated by Cassirer's statement is clearly far more nuanced than a frequent interpretation of his position at Davos, which holds simply that he is placing a humanist emphasis on the infinite creativity of the individual person. In grasping the appearance of the particular, of the infinite in the finite, such as in the entoptic reflections of the crystals, we grasp the infinite in the particular as immanent experience and what Goethe describe as "objective thinking." As in Cohen's work, the wider implication for Cassirer is, to put it in Heidegger's terms, that the ontic sciences structure an ontological horizon of meaning (a position Heidegger also holds), and for this reason must both be constantly critiqued and also only understood in their collective relation and delimitation to one another.

Unlike Heidegger, however, Cassirer does not believe that this ontological field can be delineated in itself beyond the very rudimentary sense of the basis phenomena, which are a mystery even as they are always only present in relation to particular local phenomena. Analogously to the entoptic play of light in crystals, experience is not merely ontic or finite, but also "reaches out" from this finitude to a deeper horizon of infinite meaning that also has an ontological presence, what Cassirer elsewhere called pragnänz, in itself. As we have seen, the parallel for Cassirer to the crystal metaphor in history is found in the experience of the present grasped through history and futurity, in which the present is revealed as "truly Symbolic, where in the terms of Goethe, the particular represents the general [...] as the living-instant revelation."[142] It is on this ground that Cassirer can later summarize his position against Heidegger as an alternate form of philosophical historicism, one that we have seen he develops from his very earliest works. "For us," Cassirer writes in summary of his position against Heidegger, "not only Dasein, but meaning – the idea – is primordially historical."[143]

Curiously, the starting point of both Heidegger's and Cassirer's positions in Davos are quite close, but they are essentially – as Cassirer claimed in his notes to the debate – incommensurable.[144] Both are arguably working out from Cohen's initial concept of *Ursprung*, with Heidegger claiming that an ontology of finitude can be established in itself as a deeper grounding of beings in Being, while Cassirer claims that such an ontology is always only defined in relation to the ontic or particular aspect of the forms in which it is developed. "The ontological cannot be separated from the ontic," Cassirer writes in a latter commentary, "nor the individual from the 'general' in the way that Heidegger tries to – rather the one is only within the other."[145] As Cassirer notes in a critique of the ontology of Nicolai Hartmann from 1927, to truly ground an ontological philosophy, one would have to explain its reciprocal relation with the language and forms in which it reveals itself. "The entire project of a philosophical ontology," Cassirer concludes, "is particularly susceptible to [a] form of 'language critique.'"[146] Such a critique (he has in mind the work of Fritz Mauthner) would simply be a particular application of Cassirer's general approach, however, for it would say that only in the particular structure (and indeed science or philology of language) can we grasp an ontological horizon of meaning, but for this reason we are always limited to the particularities of this language. As immensely valuable and even beautiful as Cassirer finds Heidegger's work to be, it cannot overcome this limitation. Cassirer, on the other hand, follows Cohen in holding that the ontological and even revelatory are woven through all ontic sciences, and that only their internal and collective critique can the wider grounding of ontology and potential of revelation emerge.

Humboldtian Liberalism and the Problem of Form

The central political importance of Goethe in *Freedom and Form* is found in a closely related but more prosaic study of the dialog of form as it defines a liberal political system. Humboldt's theory of the individual and the state are reinterpreted by Cassirer as having their middle term in a concept of form, which appeared in fields as varied as law, science and society. The meaning of "form" is itself, however, depicted as part of an evolution closely linked with Goethe's definition of the term. The same basic concept of form is then further developed as an interpretation of Hegel's "objective spirit," but now placed within a dialogic or agonistic theory of form rather than the development of the "idea" through dialectic. The goal of a liberal politics is not only to pursue individual liberty, Cassirer concludes, since individual liberty itself only exists and is defined through various forms of experience, but to change the forms of knowledge and practice through which individuality is itself defined.

The resolution of the meaning of the freedom of "aesthetic individuality," in Humboldt's era constituted for Cassirer the beginning of a new distinctly modern constitutional and social phase of this development. The "heteronomy" of the modern state provides new possibilities for developing autonomy in "empirical–historical life" that were not earlier available, yet the state itself also needs greater constraints than previously due to its increased power.[147] The present task for a liberal politics is to define the modern freedom of aesthetic "qualitative" experience and individuality championed by Humboldt himself. Even this stage of liberal individuality, however, is not an endpoint but only a path towards future, as yet unknown, forms of sociation.

The import of Humboldt's liberalism is now most widely known through Mill's classic statement on liberalism, *On Liberty*, which begins with a quote from Humboldt as promoting two requisites for individual liberty – "freedom, and variety of situations" – and announces its concept of liberty as derivative from his.[148] Despite Mill's enthusiasm for Humboldt's work, however, the basis of Humboldt's argument was in Cassirer's view largely lost in translation to the British world. The underpinning of Humboldt's idea of the individual's "harmonious development of powers" was not Locke but Leibniz, and the concept of the individual at play was not an insular individual derived from the state of nature or presumed by common sense, but the infinite monad that was always coextensive with its environment. In this context, what Mill understood as a commonsensical "variety of situations" needed for individual development was for Humboldt a problem that placed the clash of different forms of understanding, sociation and reality at the center of his philosophy.

The key to understanding why form, not pure individuality, becomes the key aspect of liberalism is that Humboldt's definition of individuality is itself always mediated. Humans are not social animals in the sense of other animals, for "defense, help and for rearing," Cassirer writes in summarizing Humboldt's argument, but "in the deeper sense, that it is only through others that one is raised to consciousness of oneself and an 'I' without a 'You' for its understanding and its perception is an impossibility [*ein Unding*]."[149] For this reason, Cassirer argues that "the concept of form and spiritual energy [*geistiger Energie*] is at the center of Humboldt's thought" – not individuality itself – since the forms or "organization" of sociation in their interaction determine both the individual and collective definitions of reality.[150]

Cassirer's reading of Humboldt is based on the insights already developed through the reading of Leibniz and Goethe, namely that 1) individuals are a world onto themselves; 2) these worlds are however unknowable to themselves other than through their mutual activities and works; 3) the greatest good is established through the maximal interaction of these worlds as individual entities, but this is accomplished through transformations of form, for instance in changes of constitution, law, art or mores.[151] Humboldt completes the political development of this concept by promoting "aesthetic individuality" as a radical defense of "particularity in itself," which steps beyond earlier legal or political protections of spiritual belief or freedom of thought in the Reformation and Enlightenment. "The grounding and foundational right of the individual," as Cassirer puts it, "which is to defend it against both [state and society], is no longer simply the right to freedom inside a specific walled off individual sphere, such as was the case in freedom of consciousness or freedom of thought. It is the right to peculiarity [*Eigentümlichkeit*] altogether."[152] The key element in Humboldt's argument is thus not simply the defense of individual rights in the sense of an "enclave" against state or other powers, as it arguably is in Mill's reception of Humboldt. It is rather to announce the role of individual particularity as a value in itself, and the central value of the free interaction of individuals both beyond the state and as a corrective to it. Placed in the context of Cassirer's wider description of form as it has evolved from Leibniz through Goethe, however, the means for this individuality to develop hinges not on simple self-actualization but on the problem of form in which this self-actualization occurs.

Humboldt's reflections on aesthetic individuality developed, Cassirer argues, in the same "circle of thought" as Goethe and the Schiller of the "Letters on Aesthetic Education." For Schiller and Humboldt, Cassirer argues, the aesthetic serves as a political alternative to the ideal of the isolated individual who understands himself or herself in the "reality" of a "state of emergency," such as in Hobbes's state of nature, and the individual defined by the "state of reason" as idealized in the Enlightenment commonwealth.[153] The aesthetic particularity of the individual cannot be reduced either to the necessities of nature nor to the delimitation of ideas, because it defines the sheer "given-ness" and contingency of individual experience and the creative development of this experience. In aesthetic experience, Cassirer writes, they found a form of experience that is "at once free and bound, which is dependent on nature even as it is free from it," and which as such can form a "new unity of real and ideal, of sense and spirit."[154] Aesthetic individuality reveals a form of freedom that finds its most serious application in attempting to change the "rule-forms" that first constrain human behavior, no matter whether they happen to appear as a product of nature or of humanity itself.

True individuality only appears in the deepest immersion in one's surroundings, which is how the individual learns to transform the "rule forms" that determine their own fate. Humboldt is arguing, Cassirer writes, that the individual is "symbolic in the sense of Goethe" in that she or he only "works through connections that spring out of the inner being of one and that make the fortunes of the one into that of others, and in this manner actualize and mediate the totality of humanity."[155] Just as Goethe holds the symbol to connect the particular perception to a larger totality in nature, so here the perception of particular human truth connects with, indeed *is*, the general truth of humanity as a whole, and this truth supersedes any later division of nature and culture. True individualism is not "egotistical" but rather based on relating the individual to the species ideal, and as such overcoming local pressures for conformity. In this manner, Cassirer writes, "The more individual the basis of actions are, the more comprehensive and general is the effect that derives from them," and this has to be held as true even if this effect is not seen by the mass of humanity at the time.[156]

Cassirer's understanding of "humanity" in Goethe and Humboldt is based not on presently existing humanity, but on an as-yet unknown human potential. It is this potential that guides the individual, and explains why placing one's "inner being ahead of all else" will be an eccentric or ecstatic phenomenon. Earlier in *Freedom and Form*, Cassirer highlights this argument as the foundation of the Kantian ethical ideal of treating people, including oneself, as "ends," which in turn is taken as Humboldt's starting point, somewhat paradoxically, of understanding oneself as working through the ideal of the human species. The pivotal concept is Kant's definition of the conceptual duality of the concept of "species" or "type" (*Gattung*). "The 'type,'" Kant writes of the species of humanity, "as historical idea of unity and ethical goal is in no way similar to the 'type' as a logical abstraction – whatever is raised against the latter confronts the former in no way."[157] Kant states the reason for this in relation to the concept of "humanity":

> Certainly whoever said: no single horse has horns, but the species of horses is horned, would be stating a clear absurdity. For in this sense "species" means nothing further than the details, in which all of the individuals among each other must share in common. When, however, the species of humans is considered the entirety of an infinite (indeterminable) series of descendants (as is the entirely normal use of the word), and it is taken, that this series approaches the line of its definition ever nearer,

so is it not a contradiction to say: that it is in each of its parts is asymptotic, and yet in its entirety it converges. In other words, that no part contains all of the evidence of humanity, but that only the species as a whole can fully reach this determination.[158]

This "indeterminable" series occurs through the historical development of humanity into the future, as well as in the plenum of human experiences in the present. In this case the "totality" is not merely based on its temporal continuity, much less on the totality of presently existing human beings, but would have to base itself on the "asymptotic" assumption of all different forms of knowledge, experience and individuality in the past, present and future.[159] This asymptotic relation, however, is experienced only in the open reality of the human present. In relation to the individual, this means that as she or he is striving to act towards their own sense of the harmonization of the whole, they act at once both towards what is "most human" and general, and out of the immediacy of their own perception. It is in this regard that Humboldt's definition of the individual is "symbolic in the sense of Goethe," since the individual is immersed in a present that at the same time connects with an infinite "form" of humanity in an act that is at once ethical and aesthetic. Just as in Goethe's theory of basis phenomena, the individual develops through the structure of forces of which they are part, even as the individual as unknowable *monas* or monad also serves as the register and spur to the development of society.

The concept of form developed in *Freedom and Form* serves as the middle point for this development of individual and humanity, part and whole. Individual aesthetic freedom exists within a historical world of forms that the individual harmonizes on a level that mediates between "free and bound," and between "ideal and real, sense and spirit."[160] True individuality in both the ethical and aesthetic sense derives from complete immersion in the immanent, sensual and demanding fabric of history, in the immediacy of individual experience this history meets up with the horizon of the unknown nature of humanity. The Kantian notion of "autonomy" for Cassirer allowed new permutations in the forms of objective spirit that first create and allow any form of empirical subjectivity. "Autonomy of will," Cassirer writes, "no longer relates to a unity that is given in the subject," but "to that which is sought and defined in the social and in knowledge itself."[161] As completed in the Goethean concept of form, this means that autonomy is principally based on transforming the "forms-rules" of society; it is, as Cohen had it, not an insular autonomy at all, but an *autotelie*, a paradoxical self-orientation defined by its context and operating as closely as possible out of a "self-less" absorption in this context. In the present, social form is the "iron cage" of Weber's studies of social forces.[162] Once understood in its temporal dimension, however, for Cassirer any definition of form is by definition open to some permutation.

The individual can fall back neither on a set definition of self nor on a ready-made identity in a group. They are in a dynamic and existential struggle to relate to the immediate balance of "ideal and real [...] free and bound" in experience. The difference of this form of monadic individuality and one that falls back on an "outside instance" of definition, Humboldt states, is the difference "between war games and actual war, between rented soldiers and actual soldiers."[163] The infinite power of the individual only unfolds through their interactions with others when their essence is at stake, and this only occurs when individual action is related to the wider concerns of the development or survival of humanity as a future possibility and ideal. This is an existential conflict both in the sense that it involves one's life as a microcosm of the life of the species, and in the sense that it continually casts the individual outside of the known or comfortable definition of social form.

In Humboldt's reading, the basis of individual worth is the "fine web" of interpersonal perceptions, which as a product of "spiritual energy and form" is beyond comprehension by any one actor and largely effaced once the "subject" is even defined in a particular manner. All activity for Humboldt has to spring from a reorganization of the individual's powers. "Every individual receives from others ever more, as he himself has to give from his content: the more individual the foundation of the relations are, ever more comprehensive and general is the affect that goes out from it. This inner wealth of giving and taking is immediately destroyed, however, in so far as it is placed under the tutelage of an outer instance."[164]

For Humboldt, the state is of course the principal form of this "outer" instance and, in a manner similar to Edmund Burke, Humboldt holds that the mechanisms of the state can only encumber this process of exchange. "It is enough," as Cassirer states in a summary of Humboldt's view, that "[the state] creates and defends the material foundations on which this spiritual relation can be developed as the free work of individuals."[165] In marked contrast to a figure such as Edmund Burke, however, Humboldt saw not only the state but any dominant forms of sociation and tradition as potentially an "outside instance" of power in relation to the individual, so that the forms of sociation had to always be redefined and revivified. The danger of the state, and indeed of any objectification of humans, is not merely to make people into a means rather than an end, but to not recognize the creative power that all people exercise simply in interacting, indeed simply in living, since the individual is always already socially defined. Thus not only is any outside instance of power liable to "misidentify humanity and make of humans machines," but it also tends to block a creative social power that reduces the infinitely greater power of human passions to those organized from the outside.

Importantly, however, neither individual nor any other entity, such as association or state, is ever known in itself in these relations, but only as a conjunction of forms and energies described as "character." It is only as an asymptotic limit that either individual or humanity is known in itself. It is in this sense that Cassirer understands Humboldt's call for the development of the "totality of character" as the basis of a form of liberalism that would resist the state. "Totality of character" is the critical interweaving of the forms of experience from one perspective.[166] It is not the province particularly of the individual person, but can occur through any individual social phenomena. The term "life" in the natural world means that a whole is governed by a particular "form of thought and law," as Cassirer writes summarizing Humboldt's concept of history, "so in the intellectual and moral world this is known as character. Humanity as a whole has one such character."[167] Yet each individual, association, organization and nation also has such a character, and only in their overlapping roles and ongoing competition can the whole be understood. Such an understanding of the multiplicity of competing forms can however only be understood in the historical development of the particular, and it is, as Humboldt writes, "for this reason that one must study nations and individuals." This understanding of the science of world history supersedes any definition that terminates in the nation state, "since this is itself drawn up into the spiritual energy and foundational life-power whose totality is grasped in world history."[168] Not only is the nation state composed of innumerable individual "characters" of different orders, but also between and above the nation state a similar emergent development is always occurring.

In this manner, the power of the state itself is held in check by the critique of knowledge and history, and the cultivation of a multiplicity of forms and characters, both within and outside of its sphere of influence. The active contest among all forms of experience, and the development of structures and ideals to facilitate this contest – notably through constitutions, mores and institutions – keeps a check on the state or other entities claiming a totalizing power.[169] Earlier freedoms both increased the power of the state, presumably through confessionalization and the development of a critical public, respectively, even as they also restricted the powers of the state. Thus, Cassirer writes, "the powers that were initially called up against the state gradually lead to the completion and deepening of the content of the state ideal."[170] The nature of this "ideal" itself is not as a total encompassing power, but rather as itself one of multiple forms of definition. The contest of forms at once refined the definition of the ineffability of the individual, while at the same time suggesting the impossibility of any totalizing form of concept or power. The concept of aesthetic individuality as applied to all levels of form completes this process. The ultimate goal is a relation of state to individual "that is not based on empty ideals, but on the forming of the state through the individual" by means of the cultivation of a multiplicity of interlocking forms of which the individual is part and through which the individual defines her or himself.[171] For Humboldt, however, the term "organic" defines the "fine web" of free relations between individuals in such a way that the state – and potentially any static objective form or traditional role – can never emulate. The state cares only "what people render [*leisten*], not what they are in themselves," and "thus makes human objects instead of individuals."[172] In this way, the state converts "pure subjectivity into pure objectivity by focusing on results rather than energy [*energia*]."[173] For Humboldt, Cassirer writes, "It is not the disturbance of this possession [of the individual sphere] that serves as the point of attack, but the fixing and approbation [*Bestätigung*] that the state gives the subject. Through this fixing, something is changed into a completed good that should only be the result of a free self-activity."[174] Simply creating an enclave for the individual is not enough. The basic method of critique has to be how the state or any other instance of power or tradition "fixes" the subject into a given role. The only means to prevent this is by cultivating the free play of formal agonism itself, and thus promoting a dynamic society that ultimately will lead to a dynamic, but delimited, state.

The metaphor of organicism thus comes to the defense in Humboldt's work, as in Burke, of the minimal liberal democratic state and not the autocratic state. But it also, as in Mill, encourages an antitraditional stance in order to break free of unreflexive definitions of form. The goal of the state for Humboldt is to approach the liberal ideal of minimizing its impact on citizens and their activities – understood foremost through developing forms of experience – while maximizing their freedoms, so that the organic nature of human interaction can arise. That these freedoms were highly likely to serve as a faster catalyst for social change is implicit throughout Cassirer's discussion.

Cassirer's Humboldtian Liberalism in Light of Cohen's Concept of Individual and Association

To understand Cassirer's double reading of Humboldt's concept of freedom – as a freedom of individuals and as a reciprocal transformation of "rule-forms" of society – it is helpful to note the analogies found in this reading to Cohen's ethical and political philosophy. In Cohen's work, multiple forms of association overlap to determine the individual, even as the individual is paradoxically the point of resistance to any passive imposition of definition. The tension of these two movements creates the basis for a dynamic society and guarantees the ineffability of the individual. Indeed, Cohen claims that the principle that individuality is determined by overlapping forms of association is nothing less than the "guiding principle of [my] ethics."[175] He introduces the problem at the beginning of his *Ethics of Pure Will* (1904) in the following statement: "How about the human person, is he an individual? By no means is he this alone; he also stands as rank and file within a plurality, or better, within many pluralities. And he is not only this; rather in this totality he first completes the circle of his being. And this totality also has many degrees and levels until its completion is a true unity, namely, in humanity, which is at the same time an eternal new beginning."[176]

The individual is defined foremost as the refraction of the many forms of association that converge in them. Just as in Cassirer's Goethean analogy of the entoptic crystal at the beginning of *Freedom and Form*, so the single phenomenon of the person only exists through the multiple aspects that give rise to it. There is not first an individual, who is then multiplied to create society, but rather society principally forms and defines the individual. For Cohen as for Humboldt, however, the individual is also the point of resistance to all forms of collective. "Every society has the purpose of creating the true individual, for this individual can arise only from a real, healthy and developed community. [...] Ethics has as its task both problems, the community and the individual, and these concepts are to be seen not as standing beside each other or as one after the other, but as requiring and determining each other, as reciprocal concepts."[177]

To guarantee an awareness and respect for the ineffability of the person, Cohen held that the individual of law and political theory had to be defined foremost as the legal person. Such a legal person may in turn define an actual physical person, but could define many other forms of legal unity such as the association, the cooperative or a political entity such as a municipality or state. Through this device, Cohen creates a means of retaining the ineffable ethical person on the one hand, while on the other linking this person to the politics of associations and state. The legal person represented an "ideal, representative, ideal will," understood not as a fiction but as a *hypothesis*, as the premise or promise of an actually existing but unknowable individual.[178] The individual person exists in relation to the science of law or of ethics, then, in a manner parallel to the way in which the individual object exists in relation to any natural science or to aesthetics. In relation to the human person, this use of the legal person guaranteed that the system would not describe or prescribe what is at heart an ineffable phenomenon. "[The person] is the hypothesis of ethical self-consciousness, of the ethical subject, that which is realized in the juridical person of association."[179] Again, a basic premise of Cohen's thought is that, as the philosopher Harry van der Linden puts it, "structural political change is a precondition for human emancipation rather than the other way around."[180] In Lange and Cohen's ethical socialism, the culmination of this political philosophy is a socialist society that first allows us to see others as ends in themselves, and indeed for this reason capitalism is seen as inherently opposed to this goal. The categorical imperative, Cohen writes, "contains the moral program of a new era and the entire future world history. [...] The idea of the priority of humanity as an end becomes the idea of socialism, which defines each human as an end in itself, as purpose."[181]

Cassirer uses the form of Cohen's argument in *Freedom and Form*, but to notably different ends as the basis of a new form of liberalism. The Kantian ethical ideal is most clearly held up in his reading of Humboldt's program of preventing any "fixing" of the individual by "outside instances," and truly takes humanity as an end in itself. Perhaps due to his grim experience of state power and misinformation during the war, the emphasis in Cassirer's reading is, however, liberty defined through recognizing the ineffable difference of individuals, with equality defined in Leibnizian terms solely as this difference. On the problem of how this balance of liberty and equality could be sustained in the face of material inequality, Cassirer is notably silent.

For Cassirer, Kant's categorical imperative is socially recast through form and humanity: in the developing forms of society – such as those of freedoms, conscience, opinion and "aesthetic individuality" – we become progressively more able to see other people as ends in themselves. The imperative to treat other people as ends rather than means is, for Cassirer, by definition always oriented towards the future, and indeed not only the future of this particular person but of humanity in general. In Cassirer's later reflections on law, particularly when placed in dialog with their reception by Hans Kelsen, as we'll see in the next chapter, we see a particularly dramatic application of science, in this case the science of law, to this future concept of humanity.

In place of the Romantic holism that dominated wartime discussion, Cassirer developed a reading of Goethe that demanded that all holistic form be explained, and be open for critique, through functions. Similarly, the oppositions of a putatively organic German state against a mechanical British state, as wartime propaganda had it, were inverted. Only in a Humboldtian liberalism based on law, which insists on the defense of the individual, does the organic nature of society become central. Finally, against the frequent argument for the qualitative mode of representation found in Germany as opposed to the quantitative mode found in England, Cassirer reversed the terms using Humboldt. Only a state with strong forms of functional defense of the individual on all levels could develop a qualitative understanding of the person, whether it be the individual person, the association or even the "personality" of the state itself.

Notes

1. Mannheim, Károly. "Rezension von Ernst Cassirer, Freiheit und Form. Studien zur deutschen Geistesgeschichte," *Atheneum* 3 (1917): 409–13 (German translation in part C (III)). My thanks to Prof. David Kettler, Bard College, for a copy of this review.
2. FF 392.
3. FF 159.
4. Cohen, Hermann. *Kants Begründung der Ethik*, 237ff.; Cassirer, "Kant," in *Encyclopedia of the Social Sciences* (1932), 539b.
5. FF 166.
6. FF 392.
7. Cassirer, Ernst. "Philosophy and Politics" (1944) in SMC 232.
8. Cassirer saw Spengler's "system of historical fatalism" as the opposite of Hegel's "progress in the consciousness of freedom," which he advances, on different grounds, in *Freedom and Form* (SMC 227).
9. FF 389.
10. EK1 x.
11. Ibid.
12. Ibid.
13. Ibid.
14. EK1 7–8.
15. EK1 x.
16. It is an example that Cassirer develops later, but is useful for its clarity: in special relativity, the drastic shift is of course from invariants of absolute time and space to the invariant of the absolute speed of light. "The invariants shift 'from place to place,' such as in the theory of relativity" (PSF4 120). *The Knowledge Problem* presents the historical role of the sciences, for instance, in gradually overcoming the "dogmatism of the usual view [*gewöhlichen Ansicht*]" of relations, such as that of "subjective sense to object," but then demonstrates how the sciences in turn create their own false substantives in the modern period, and indeed how these are popularly taken as "real," in rendering ideas such as "material" or "atom" to be existent invariant relations. Only in following the further development of chemistry or physics in the concept of force, for instance, can a relational view of these phenomena be intuited, but even here there is a danger of reification (EK1 x).
17. EK1 5.
18. EK1 8.
19. EK1 9.
20. Ibid.
21. EK1 8.
22. EK1 7.
23. EK1 6.
24. Ibid.
25. EK1 8. Panofsky was not technically a student, but took the unusual approach of sitting in on Cassirer's lectures as a fellow faculty member. Holly, Michael Ann. *Panofsky and the Foundations of Art History* (Ithaca, NY: Cornell University Library, 1985), 122.
26. EK1 9.
27. EK1 14.
28. EK1 5.
29. EK1 xi.
30. Ibid.
31. EK1 13.
32. EK1 12.
33. EK1 5.

34 EK1 17; EK4 217ff.
35 EK1 8.
36 EK1 45–142. "Cusanus becomes, in a sense, the exponent of that whole intellectual circle to which Leonardo belongs [...]. They dealt with concrete, technical and artistic problems, for which they sought a 'theory'" (IC 50ff.).
37 EK2 397ff.
38 EK4 21–118.
39 ZWW.
40 L 428.
41 ZWW 174.
42 PSF4 202–5. In this regard, Cassirer's description here is an extension of his basic concept of an "ineluctable duality of determination" between the particular and the infinite plenum which supports it and is indicated in part through it – in this case of history itself (ZWW 4ff.).
43 PSF4 85.
44 PSF4 110.
45 PSF4 203.
46 FF 1.
47 FF 388; FF 387.
48 FF 20.
49 Ibid.
50 EK2 669; ECW3 559.
51 FF 21.
52 FF 13ff.; 327ff.
53 FF 25.
54 FF 392.
55 FF 43.
56 FF 381.
57 FF 392.
58 FF 390.
59 FF 327.
60 FF 331; Cassirer also describes this connection to Blackstone in his speech on the republican constitution (ECW17 301).
61 FF 343.
62 FF 359.
63 FF 381ff.
64 FF 384.
65 Ibid.
66 FF 386.
67 FF 347.
68 SL 409.
69 IC 36, FF 306.
70 EK1 61, 142; IC 35ff.
71 FF 392.
72 FF 20.
73 SF 4ff.
74 FF 226.
75 SF 4ff.
76 FF 226.
77 Ibid.
78 FF 226n121.

79 FF 229.
80 FF 225, 227.
81 FF 212.
82 FF 227.
83 FF 220.
84 FF 26; SF 4ff., 220, 329.
85 RKG 75, citing Goethe *Italienische Reise* (17 April 1787).
86 FF 180.
87 FF 229.
88 FF 230.
89 FF 235.
90 FF 180.
91 Kant, Immanuel. *Critique of Judgment*, trans. Werner Pluhar (Indianapolis: Hackett Publishing, 1987), 301–324, sections 79–84.
92 RKG 93ff.
93 Ibid.
94 Quoting *Prolegomena*, par 14, Werke iv, 44. The original form of this distinction is probably Spinoza's distinction of "natura naturans" and "natura naturata," which Kenneth Burke translates as "naturing nature" and "natured nature" in *A Grammar of Motives* (Berkeley: University of California Press, 1969), 140.
95 FF 257.
96 FF 175.
97 LS 400, FF 297.
98 L 422.
99 FF 392. The following study of Cassirer's reading of Goethe is originally largely in linking this text to the early development of Cohen's ideas and further development of it by Cassirer within the context of Humboldtian liberalism. More complete readings of Cassirer's Goethe interpretation can be found in Naumann, Barbara. *Philosophie und Poetik des Symbols: Cassirer und Goethe* (Munich: Fink Verlag, 1998) and Naumann, Barbara. "The Genesis of Symbolic Forms: Basis Phenomena in Ernst Cassirer's Works," *Science in Context* 12, 4 (1999): 575–84. Also see Naumann, Barbara and Birgit Recki, *Cassirer und Goethe* (Berlin: Akademie-Verlag, 2002).
100 FF 390.
101 ECW24 287, from "Thomas Mann's Goethe-Bild, Eine Studie über 'Lotte in Weimar'" (1945).
102 FF 223.
103 FF 392.
104 RKG 94; FF 182.
105 FF 182.
106 Cassirer, Ernst. *Goethe und die Geschichtliche Welt* (1932). ECW 18, 353ff.
107 SF 311.
108 L 368; EK2 163ff.
109 "Logicians formerly gave the name 'matter' to the universal, the name 'form' to the specific difference" (ibid., *Critique of Pure Reason*, A267).
110 FF 257.
111 Ibid.
112 FF 239.
113 FF 251.
114 FF 260.
115 FF 236.
116 FF 237.
117 FF 20.

118 FF 381.
119 FF 234; ZWW 8.
120 ZWW 1, 8ff. Cassirer, Ernst. "Vorwort der Herausgeber," in *Nachgelassene Manuskripte und Texte*, vol. 2: *Ziele und Wege der Wirklichkeitserkenntnis*, ed. Klaus Christian Köhnke and John Michael Krois (Hamburg: Felix Meiner Verlag, 1999), 1. To make any product of society comprehensible, "we must reawaken the functions, out of which it originally developed," a project that is principally historical (11). In the text, Cassirer argues that this is defined by three primary, and analogous, forms of "knowledge of reality" – those of mathematics, perception and human science – which together negatively define a certain domain of freedom as, in Goethe's terms, primary phenomena or *Urphänomene* (ZWW, 8ff.).
121 FF 241.
122 FF 270. Quoting Goethe, "Über Naturwissenschaft im Allgemeinen, Naturwiss Schriften"; and letter to Eckermann, 18 February 1829.
123 FF 257.
124 FF 219.
125 ZWW 9.
126 ZWW 8–9; PSF4 128.
127 FF 39.
128 ZWW 10.
129 FF 234.
130 ZWW 9.
131 ZWW 10; FF 294. Cassirer develops this argument most clearly through his functionalist reading of Fichte's process of revelation in the "I and the not-I," a reading indebted to the limit concept and its purported influence by Maimon (FF 291ff.).
132 ZWW 10ff.; PSF4 130.
133 von Goethe, G. W. "Color Theory – Didactic Section," *Goethe's Collected Works*, vol. 12, ed. and trans. Douglas Miller (Princeton: Princeton University Press, 1988), 57.
134 FF 131, 134, 141, 210.
135 FF xii; ECW7 390.
136 RKG 78.
137 Cassirer, Ernst. "Erkenntnistheorie nebst en Grenzfragen der Logik und Denkpsychologie" (1927). ECW 17, 75.
138 FF 390.
139 Ibid.
140 KPM 201.
141 SF 300.
142 ZWW 174.
143 PSF4 204.
144 PSF4 205.
145 PSF4 203.
146 Specifically, Cassirer wrote that "the entire project of a philosophical ontology is particularly susceptible to the form of 'language critique' of the type popularized by Fritz Mauthner" (EGD 76). Mauthner's *Philosophical Dictionary*, a wildly popular skeptical turn-of-the-century text that resurfaced later in the century through the works of, among others, the Argentine writer Jorge Borges, had argued that all philosophical definitions rest on other definitions and language, and thus could not be seen as self-sufficient (Mauthner, F. *Wörterbuch der Philosophie* (Vienna: Böhlau, 1997), vii). Cassirer noted that he does not fully agree with Mauthner's position, which he considered overstated, but that "a case has never seemed clearer to me in which a critique of language is necessary as the 'prologue to all future metaphysics'" than in Hartmann's text (EGD 76). The relevance of this approach as a response to ontological philosophy is more fully brought out in volume three of the *Symbolic Form* series, subtitled

"The Phenomenology of Knowledge." In the essays collected as the fourth volume, "The Metaphysics of Symbolic Forms," Cassirer more fully clarified his phenomenological method in relation to Heidegger.

147 FF 327; ECW7 343.
148 Mill, J. S. *On Liberty and the Subjection of Women* (New York: Penguin, 2007), 5, further developed at 66ff. and 82ff.; referring to Humboldt (1791), 11–13.
149 FF 352.
150 FF 353.
151 FF 330.
152 FF 343.
153 FF 344.
154 Ibid.
155 FF 346.
156 Ibid.
157 FF 339.
158 FF 342.
159 ETR 454–5.
160 FF 342.
161 FF 160.
162 See Scaff, Lawrence. *Fleeing the Iron Cage: Culture, Politics, and Modernity in the Thought of Max Weber* (Berkeley: University of California Press, 1991).
163 FF 345–6.
164 FF 346.
165 Ibid.
166 FF 345.
167 FF 352.
168 FF 353.
169 FF 20.
170 FF 343.
171 FF 351.
172 FF 345.
173 FF 346.
174 Ibid.
175 Cohen, Hermann. *System der Philosophie, Zweiter Teil: Ethik des reinen Willens* (Berlin: Bruno Cassirer, 1904), 49 (8 of original). I am indebted to the translation and analysis from Harry van der Linden, *Kantian Ethics and Socialism* (Indianapolis, IN: Hackett Publishing Company, 1988), 206.
176 Ibid.
177 SP2 207. Translation from Van der Linden, 207.
178 EPW 231–2.
179 Ibid.
180 Van der Linden, 216.
181 EPW 320ff. Translation by Van der Linden, 224.

Chapter Seven

LAW AS SCIENCE AND THE "COMING-INTO-BEING" OF NATURAL RIGHT IN COHEN, CASSIRER AND KELSEN

In an overview of Ernst Cassirer's Weimar career, the philosopher Jürgen Habermas commented on a central contradiction. "In the realm of the German Mandarins, [Cassirer was] one of the few courageous exceptions who defended the Weimar Republic against its despisers among the intellectuals," yet it is "all the more astonishing" that nowhere in Cassirer's key writings on symbolic forms from the Weimar Period does the concept of right and morality, and with it the realm of politics, find a clearly defined place.[1] In this, Habermas voices a disappointment common to many readers of Cassirer: even as Cassirer was clearly aware of the perils of his era, his Weimar philosophy appears to have retained a "Mandarin" distance from politics. Habermas responds to this lacuna by noting the continuity and importance of Cassirer's concept of symbolic form with a "theory of civilizational processes," an argument that has been developed from a different perspective in Drucilla Cornell's recent work on Cassirer.[2] As we have already suggested from Cassirer's work in the Wilhelmine period, however, he had earlier presented the outlines of a surprisingly robust theory of law and the state that greatly strengthens Habermas's reading of Cassirer. Indeed, Cassirer's theory of law, ethics and his liberal theory of something resembling "civilizational processes" prove integrally related to his later work, and only by placing these aspects of his project together can we understand the political meaning of his philosophy.

Cassirer notably sought to defend the "idea of a republican constitution" in his 1928 *Rektorsrede* as rector of the Hamburg University on Constitutional Day, saying that the republican constitution is "in no regard a foreign concept, but rather [grown] from our own ground and through its fundamental power, and nourished through the power of idealistic philosophy."[3] In light of his early work, Cassirer did not consider this as a secondary defense of the embattled republic, but as part of a new understanding of state and law, most clearly developed in *Leibniz's System* and *Freedom and Form*, one that would work in combination with a continual critique of form across the sciences.[4] Far from leaving his earlier work behind after the First World War, Cassirer's thought was everywhere dependent upon it. Although the issues Cassirer confronted in the Weimar period were of paramount importance and shaped the focus of his public narrative, the critique of science from the first half of his career remained important both for the content of his work and the completion of key statements within it.

In the reception of Cassirer's theory within the science of law (*Rechtswissenschaft*), we can see two major aspects of his early work both consequentially developed and fundamental sundered from each other. A study of these two aspects will allow us to retrospectively assess Cassirer's earlier political and legal philosophy. The "functionalism" of Cassirer's early work in relation to one science was perhaps most influentially applied in Hans Kelsen's (1881–1973) highly abstract "pure theory of law." Kelsen was a central figure, perhaps the central figure, in developing both the modern Austrian constitution following the First World War and the United Nations charter following the Second World War, so his reception of Cassirer's work is particularly consequential. Yet at the same time, from a nearly diametrically opposed set of assumptions, Cassirer was presumed to present a "common sense" defense of the ideals of democratic liberalism in the Weimar Republic on the basis of his presumed "liberal humanism." The necessary fusion of these two forms of Cassirer's philosophy can only be understood by returning to its earliest foundations, which in turn can cast light on the meaning of this philosophy as a whole.[5]

Law can be considered a capstone to Cassirer's early project, acting as a form that both epitomized his functional reform of the sciences and acted as the basis of a liberal politics that encompasses different forms of truth within the horizon of a common human reality. In *Leibniz's System*, law provides the principle framework through which the conflict and dialog of forms of sociation and knowledge occur in a progressively ordered fashion, and thus forms the basic means for facilitating the questioning of state and other forms of power.[6] Law functions, as Cassirer noted in *Leibniz's System*, as "the link between general ethical rules and the reality of history [...] in the developmental forms of empirical communal life," and this empirical communal life is the basis in Cassirer's reading of a liberal state.[7] Law thus acts as the most practical and empirical of the four modes of social and human sciences Cassirer outlined in the text – that is, history, aesthetics, theodicy and law – but it is also capable of radical transformation in itself. Indeed, as *Freedom and Form* argues, the history of law is one of the clearest demonstrations of the structure and development of society, particularly in the changing definitions of individual freedom.

As we have seen, Cassirer was not advancing a "traditional" form of Western liberalism in Germany, as has been assumed by David Lipton and other commentators.[8] He was rather advocating a generalized Humboldtian liberal reform of Germany that entailed a continuous critique of society in the sciences in conjunction with a form of the Leibnizian definition of right and law. Cassirer distinguished Leibniz's critique from Western philosophy by claiming, as we have seen, that it was "more sharply argued and more consequently executed than by Locke, Montesquieu or Rousseau," and Leibniz's philosophy clearly forms the cornerstone of Cassirer's own position.[9] The necessary corollary of this project was a functionalist development of each of the sciences, which would then form the basis of a critique of society. For legal science, this approach was important enough, as noted in the introduction, that when Siegfried Marck wished to write an overview of the state of German legal philosophy in 1925, he summarized the basic issues of dispute in his title by referring to Cassirer's earlier work, *The Substance Idea and Function Idea in the Philosophy of Law*.[10]

Despite the increasingly wide reception of Cassirer's functionalism in a number of fields, his own work developed in a strikingly different direction following the establishment of the Weimar Republic's constitution in 1919. Cassirer largely underplayed his subtle early theory of liberalism in the Weimar period due not only to the tenuous success of the fragile republic – and its consequent need for clear rhetorical support – but also because Cassirer's argument depended upon a definition of science and knowledge that he found the Weimar generation had little patience for. Nonetheless, Cassirer never abandoned his early position. He clearly understood it as the basis for a more durable form of liberalism and for a future of permanent epistemic transformation of the sort outlined in *Substance and Function*. He returned to it frequently in his later works, notably in *The Philosophy of the Enlightenment* as well as works such as his study of the Swedish philosopher Axel Hägerström.[11]

It is perhaps telling for Cassirer's sense of the role of law that his decision to emigrate was closely related to the breakdown of German legal order. Soon after the Nazi's Enabling Acts (*Ermächtigungsgesetz*), the laws that kept German jurisprudence in place but granted plenary powers to Adolf Hitler and his cabinet, Cassirer read aloud to his wife the sentence, "Law is that which serves the Führer."[12] He immediately declared that if all German jurists did not stand up the following day, Germany was lost. To the shock of their friends, the Cassirers left Germany on 12 March 1933 – 11 days before the law was enacted. While many took the enabling acts as either a passing phenomenon or a technical transformation of law, for Cassirer, the structure and implementation of law was the structure of the state and its civilization. Just as the establishment of the republican constitution in 1919 was for Cassirer a monumental event that formed the basis for a new society, so its legal negation in 1933 effectively and immediately destroyed this society.

The Context of Cassirer's Functional Definition of Law and State

Cassirer's philosophy of the state culminated in his reading of German classicism, specifically the relation of Fichte and Hegel, but his understanding of law developed largely in relation to specifically modern circumstances. Although Cassirer principally wrote about law within his general treatment of political theory in *Freedom and Form*, and did not apparently consider himself an expert in it, it is clear that for him law is perhaps the most politically important form of knowledge and social form. His basic approach in this regard again appears to follow Leibniz, where law on the one hand "refers back to the principal questions [*Prinzipienfragen*] in its foundations while on the other it brings to representation [*zur Darstellung bringt*] in the manifold of its development the transformation of individual historical determinations."[13] Law is the most important fulcrum for transforming "empirical communal life" and is a means for shaping the form of liberal polity, and as such is a particularly vivid example of Cassirer's understanding of the political import of science and form, as well as of the application of his definition of truth and reality to one field.

Cassirer's response to the crisis of German sciences and society was intended to redefine two largely opposed extremes of legal and political debate at the turn of the century. On the one hand, law and politics were increasingly approached through positivistic, technocratic and systemic means. This development began with Christian Wolff's popularization and systemization of Leibniz's work, and culminated in the German Civil Code of 1900. On the other hand, often under the banner of the "personality of the state," law and politics were approached as affective, physiognomic and symbolic. Leibniz provided Cassirer with the initial template for understanding how to criticize this concept of "personality" by demonstrating how all "physiognomic wholes" could be defined through functions, and how a critical system could develop into a critique of reality. Cassirer sees Leibniz's work as initiating a movement that culminates in Kant and Cohen, and will attack all areas in which "one dresses certainty in the intuition of objects" as a dangerous carry-over of medieval mysticism and its objective corollary in the concept of substance as *in se et per se esse* (in and of itself).[14]

Freedom and Form has as its central goal the functional redescription of the concept of substantial personality or the essence of a people.[15] As such it demands a critique of the forms of life, particularly the institutions of "objective spirit" in Hegel's sense, so as to "place us in the middle point of the active and productive forces that will affect the future formulation [*Gestaltung*] of our being [*Dasein*]."[16] Within this argument, Cassirer's understanding of law occupies the key position. The basic principle, as we have seen developed from Cohen in Cassirer's reading of Humboldt, is that the individual has to be treated as infinitely complex. The focal point of liberal theory is not the ineffable "person" per se, but the forms of objective spirit and law that negatively allow individuality and sociation to fully develop. Before formulating this position more clearly, we can look at the importance of Cassirer's splintered reception in the 1920s, which will then allow us to trace the broader unified problem of his work in his pre-Weimar writings.

The Two Legal Receptions of Cassirer: Abstract "Functionalist" and Liberal Humanist

A. The pure theory of law and Cassirer's functionalism

If we look at Cassirer's reception in the 1920s in regard to the problem of a distinct theory of rights and the state, we find a surprising split characteristic of the dichotomy in the era's legal theory. On the one hand, Cassirer's thought was taken to be the focus of a thorough rationalization of legal science, particularly in the development of a "pure" theory of law by Hans Kelsen and functional theories of law by others. Here the goal of removing false "substantializations" in law, whether of objects or of personality, appears to be radically executed: law is treated as a purely formal play of relations, and defined solely within its own ambit of meaning. On the other hand, Cassirer himself in a 1931 lecture to the Hamburg Legal society appeared to argue for precisely the opposite result in suggesting the primacy of what he called "the being and becoming of natural law."[17] It was an argument that existed outside of the purely immanent rules of the legal profession and seemed to be deeply grounded in humanist principles. Here Cassirer seems to fit squarely with his later North American reception as a latter-day exponent of the Enlightenment and a humanist. In this line of reception we find a reading amenable to Habermas's attempt to use symbolic forms as a basis of civilizational thinking, although one anchored in Cassirer's earlier definition of form in itself. A true understanding of this position, however, depends on seeing how it fits into a dialectical relationship with Cassirer's scientific conception of law and his philosophy of difference. Indeed, on closer reading we can see that Cassirer's concept of "natural law" – and with it his concept of rights – and his "humanism" are far from common-sense definitions of either term.

That Leibniz's work is both the foundation of modern German legal jurisprudence and a modern technocratic definition of law itself has recently been persuasively argued in Roger Berkowitz's recent *The Gift of Science: Leibniz and the Modern Legal Tradition*. As Berkowitz summarizes it, "Leibniz presided over the birth of positive law" in Germany leading up to the *Bürgerliches Gesetzbuch* (1900), and Leibniz's understanding of law as a science was dominant throughout the modern period.[18] Cassirer similarly saw Leibniz's concepts as the beginning of a distinctly modern form of law based on human will understood through a science that "had no recourse to a law giver, whether human or godly" and that was thus for the first time, along with Grotius, "removed in principle from the question of the existence of God."[19]

Law for Leibniz is principally a means of structuring community (*empirisches Gesellschaftsleben*), and with it of connecting "general ethical rules" with the form of "empirical life."[20] Law is based exclusively on a system of norms, of "should" (*Sollen*) relations and is not to be confused with relations of being (*Sein*) even as it frequently conditioned them. Law is thus an organized system of statements; if *x* occurs then *y* should result, both in the relation of expectations between legal actors and in the relation of crime to punishment. In this regard, Leibniz saw Hobbes's philosophy of politics and the state as fundamentally incorrect. Although it was well founded in trying to develop an interconnected system of rules for human society, it replaced the actual form of law as based on *Sollen* relations with a reading, indeed a political physics, of *being* relations. "The leveling of law and power is indeed," Cassirer notes of Leibniz's critique, "generally expressed, a symptom of a fundamental misrecognition of the relation of being [*Sein*] and should [*Sollen*]."[21] The relation of *Sollen* statements is not based on what is given, but rather functions regulatively and guides the creation of positive law. For Leibniz, however, the overall relation of *Sollen* statements itself formed a unified system, and thus a science.

In this regard, the purist modern exemplar of Leibniz's system at first appears to be the enormously influential jurist Kelsen, who attempted to develop a "pure theory of law" purged of any "substantial" definitions of the legal subject or the state. Kelsen's goal was to define law solely as a play of functional relations or "norms." In theory, these relations could define any state, and one need not enter into any sociological or empirical study of society to understand its system of law. Kelsen describes his work in relation to Cassirer as follows: "As with every personification, so that of the state is accomplished in order to make a multiplicity of relations simple and concretely intuitive for thought [...] this is simply a particular case of the 'general tendency of thought to transform the pure means of knowledge into ever so many objects of knowledge' (Cassirer)."[22] As Samuel Paulsen and others have argued, although Kelsen was certainly influenced by Hermann Cohen, Cassirer's functionalism is also an important influence at the root of his thinking.[23]

Kelsen proposed a radically nominalistic legal theory, in which all "fictional" aspects of the law were debunked and replaced with a pure study of relations. "The idea of the state has the same place as the idea of 'force' in physics, of the idea of the 'soul' in psychology and generally of the idea of 'substance' in natural science."[24] All of these are "fictional" concepts that lend intuitive immediacy to what on analysis proves to be pure systems of relations. As long as one can assume one "grounding norm" of a system of law, the entirety of its functioning can be considered purely relational. Kelsen was part of a general movement that emphasized relations over substance and that contemporaries often linked with Cassirer's "functional" definition of science. Marck's *The Substance Idea and Function Idea in the Philosophy of Law* could thus describe Kelsen as a central figure in a far broader movement developing the "function idea" in law.[25]

B. Humanism and the coming-into-being of natural law

At odds with Cassirer's reception in the functional definition of law was his widespread reception as the quintessential liberal humanist. To understand the bridge between these two themes, we begin with Cassirer's own defense of the concept of natural law and its relation to norms in society in an essay to the Hamburg jurists in 1932 entitled "On the Nature and Coming-into-Being of Natural Law." Despite the late date of the lecture, Cassirer's views were deeply interwoven with his earliest work, and indeed as with his other Weimar statement on politics, only fully comprehensible in reference to it.

Cassirer begins his essay by noting the extreme disfavor into which the concept of natural law has fallen in the modern world. Indeed, from Kelsen at one extreme and Carl Schmitt at the other, natural law had by the late Weimar period largely ceased to be a topic in German jurisprudence. Cassirer's lecture suggests that the concept of natural law has to be understood as a developmental concept, focused prospectively on future human development. By natural law, Cassirer here means the concept that certain rights and privileges are inherent to human beings independent of their place in civil society. Cassirer's argument hinges on the definition of nature in his philosophy of science, which held that the "natural," like the "objective," is simply that which reveals a greater level of invariance, as epitomized by the "universal invariant" structures such as number. As such, there is no essential or initial difference for the early Cassirer between what is typically called "nature" and "culture." Again, Leibniz is the original source of this definition, since his work forms the basis for Kant's conclusion that community understood ethically "is no second nature, but the means of a new viewpoint of judgments, over which we can relate to this [one] nature."[26] In many ways, forms of sociation construct and reveal invariants in the world in a manner effectively just as free from – and for – human volition as do forms of knowledge about the natural world.

Although Cassirer would later put great effort into distinguishing the particularities of the field of "culture science" from that of natural science, he never failed to insist on the importance of tracing both back to a single origin, in Cohen's sense of origin as *Ursprung/ arché*, and along the lines first outlined by Leibniz.[27] Natural law in this sense is an ideal limit case that guides positive law to define those structures that allow human beings gradually to perceive each other as protean, rational and infinitely complex creatures.

In short, as ends rather than means. Much as depicted in the early Renaissance studies by Jacob Burckhardt that Cassirer frequently referred to, this fluidity emphasized that humans were capable for better or worse of changing the forms in which they perceive themselves and others, and nowhere was this more true than in law.[28]

In his lecture, Cassirer begins with the foundation of modern natural law by Hugo Grotius, who attempted to develop a "rational" idea of law based on mathematics. At first this appears to suggest law as a form of positive science in the manner that will come to dominate the German legal code. But Cassirer's argument about Grotius ultimately concerns the necessary interrelation of law with broader questions of ethics and other forms of objective spirit. Just as Cassirer would end each of his studies of natural science by suggesting its necessary interrelation with fields such as history and art, so here the science of law is dependent on these external relations.[29] Cassirer notes the initial apparent absurdity of Grotius's rational model of law, but he is also careful to say that the form of rationality Grotius had in mind was not necessarily the same as that of mathematics. Rationality is not an abstract or transcendent principle, but as an ideal shapes the tacit reality of our sense of other people. In revealing features of humanity that are relatively invariant, but were themselves often latent until ordered by law, the rationality of law can be called natural.

Cassirer appears to take Grotius as arguing that through use certain patterns or rules in society reveal larger forms of order and more complete definitions of what "humans" are. For Grotius, as for Leibniz, this does not lead to empiricism, but to a quest for the rules that first allow order to exist. Leibniz's basic approach is that "the science of right and not-right does not have its foundation in experience and details, just as little as does that of logic, metaphysics, arithmetic, geometry or dynamics. Rather it serves to make details themselves calculable [*von den Tatsachen selbst Rechenschaft zu geben*] and to regulate them in advance."[30] In the context of the argument in *Freedom and Form*, we can see Cassirer's broader point, since the "rational" principles of law are what first allow us to "see" the rights inherent to a person – indeed to even understand what a person is – in the first place. The battles over religious freedom, political opinion and the inviolability of individual perception were plays of forces that in time revealed broader ideals of law culminating in the "open" or "empty" definition of the individual, the individual as "hypothesis" as Cohen put it, in the modern state.[31] Civic freedoms first allowed for the possibility of imagining a political space in which others are considered equal, which in turn becomes a "rational" principle for the founding of a constitution and system of law based on "naturally" identical civic legal persons.

In a curious parallel to Kelsen's concept of a "fundamental norm," Cassirer writes that there is an "inescapable basic moment" (*unentbehrlicher Grundmoment*) of the ideal structure of natural law that exists as the horizon of any community.[32] Whereas for Kelsen the "founding" norm is an ideal or judgment, for Cassirer it is clearly something like the "asymptotic" point of humanity suggested by Kant but developed solely, as in Kant's "cosmopolitan ideal of history," through the progressive unfolding of new structures of society. Being in any communicative assembly at all presupposes, as the basis of communication itself, a minimal definition of the "coming-into-being" of natural law that provides the underlying structure of community. The horizon of human activities always suggests some means of triangulating with a new, more foundational definition of humanity, and thus of nature. In this regard the science of law, just like any other science for Cassirer, never formulates its objects solely internally, but rather depends as well on the multiple other sciences and fields that exist outside of it. In this case, these fields together form the horizon of human conduct, expectation and civility that in turn inflects the functional, and minimal, definition of the person.

Cassirer developed this reading at greater length in other treatments of Grotius, whose work he had already described in *Leibniz's System* as both of decisive importance for Leibniz and most fully developed by him.[33] Thus, as he writes in the *Philosophy of the Enlightenment* (1932) concerning Grotius, a year after his discourse to the Hamburg jurists:

> Law is not simply the sum total of that which has been decreed and enacted; it is that which originally arranges things. It is "ordering order" (*ordo ordinans*), not "ordered order" (*ordo ordinatus*). The perfect concept of law presupposes without doubt a commandment affecting individual wills. But this commandment does not create the idea of law and justice; it is subject to this idea; it puts the idea into execution, though the execution must not be confused with the justification of the idea of law as such. [...] The assertion that there can exist and must be a law even without the assumption of divine existence, is therefore not understood as a thesis but a hypothesis.[34]

Far from being restrained by a concept of "nature" in the sense of natural science, divine nature or even a stable anthropology, natural law in Cassirer's sense is similarly a "hypothesis" based on the ideal of a radically heterogeneous individuality. As law, however, all forms of rights operate to amplify something like Georg Simmel's "social *a priori*," as conditions of the possibility of social action. Natural law is always "coming-into-being," rather than simply given, since this process is always developmental, and can for Cassirer be destroyed by "outside instances of power" – such as he later defined occurring through the political myths of the Nazis.[35]

In this regard, law is just one form of objective spirit among others, and is defined in part as a reflection or concrescence of all the other struggles of society. The background of Cassirer's understanding of this problem, and the realistic context of his placement of it, only comes out in the last paragraph of the essay to the Hamburg jurists. Here Cassirer suggests a "personal reflection" on the problem of natural law that occurred to him and that contains a "typical meaning and a symbolic value."[36] Cassirer notes an observation he made during the trial of Alfred Dreyfus, a Jewish French captain, which acted as a lightning rod for anti-Semitism in its time.[37] A jurist in the trial asked a long line of questions pertinent to the case, to which the defense replied "*cette question ne sera pas posée*" – since the question stood outside of what could legally be asked it was not admissible. Cassirer says he is sure that at the time the lawyer in charge believed that these questions were technically inadmissible, and that they were operating within the proper ambit of the law. These same questions – which presumably were those having to do with the justice of Dreyfus case, the fact that he was a Jew being tried for military espionage, etc. – were however asked everywhere outside the court, Cassirer notes, and ultimately led to the case's ongoing trial and retrial.

Natural law develops in conjunction with those "unwritten laws" that structure how people imagine themselves and their society.[38] These "unwritten laws" develop through all of the forms of objective spirit in society, which is why Cassirer's philosophy culminated in a historical critique of forms of objective spirit in society and nature. Law, however, provides an axis for the interrelation of these forms based on the maximal definition of individual difference and human future potential, a definition that could only be developed functionally. The very idea of the constitution epitomizes this goal, since a constitution is for Cassirer only ideally valid when it expresses the will of the people.[39] In Cassirer's view the best route to this goal was to define civilization as a complex and heterogeneous process, and law as a minimal defense of "natural" human rights and an open structure for debate.

The basis of Cassirer's reading is here Cohen's "concept of origin," which suggests that behind the relations of any two or more given statements or propositions, a certain "next lower" level of logic must exist that supports their coherent connection. Thus, in a manner anticipating Habermas's later theory of communicative action, any cultural form that allows for communication suggests a certain logic that can be considered reflective of a broader rule. In human interaction this rule reflects on some level the nature of "humanity" and thus gradually comes to define human or "natural" rights. Any particular law or communicative act in the present is similarly based on an ideal of difference that always resolves into an infinite horizon of past acts and towards an infinite future horizon of ethics that determines what "human" might be.

For Cohen, a guarantee of this infinity was God, whereas for Cassirer, following Leibniz's turn towards an immanent and non-transcendent definition of God, it is the horizon of all possible means of ordering the world and the hope for its gradual coordination on ever higher levels. As Cohen put it in a 1907 essay on "Religion and Morality": "The irreplaceable, very deep ground of the ethics of humanity can no longer lie in the prophetic hope of the future, but, in accordance with its method, in the reality of the future. This reality is an ethical idea: the idea where the idealism of ethics separates from the utopia of a transcendent bucolic world."[40] In this manner, the necessary context of law for Cassirer is not only the variegated forms of history and aesthetics, but the teleological vision of a future humanity in the sense he earlier developed in Leibniz's theodicy and Kant's definition of the human species.[41]

Cassirer's Synthetic Position between Technocracy and the Personality of the State

The foundation of Cassirer's synthetic position lies with Leibniz, and it is by returning to Leibniz's philosophy that we can grasp the distance of Cassirer's ultimate position from Kelsen's purely functionalist understanding of law. The broadest feature of Leibniz's work is again its contrast of discursive and intuitive knowledge, which is to say knowledge gained through signs or symbols as opposed to that given immediately. Similarly, as we saw, this split suggests the antinomy of freedom at the basis of Leibniz's project – for God, the interaction of monads is fully determined, whereas for humans and other conscious monads it is pure spontaneity. It is for this reason that Cassirer's work consistently portrays Leibniz as beginning a modern form of knowledge that defeats the "immediate intuition" of mystics, whether in its religious form or, through its reception of Kant against Swedenborg, or in its Romantic form.[42] Cassirer's argument is not only against the "quietism" of mysticism, however, but the form of dogmatism it shares with all forms of "substantialist" thinking from materialism to empirical psychology.

In relation to law, this dichotomy in Leibniz's philosophy has profound implications, particularly once it was received and transformed by Kant. As we have seen, Leibniz's understanding of the monad was in Cassirer's reading refined through the work of Kant to form the basis of the distinction of the transcendental and empirical ego. This tradition in Cassirer's reading suggested that there is no stable definition of the ego that is accessible to the subject or between subjects. The "transcendental ego" represents simply the mysterious *factum* of connection in the world (and is thus equivalent to Leibniz's monad as a unity of apperception), while the "empirical ego" is always shaped by contingent forces and accessible only within the mediation of these forces. "The self thus always comprehends and constitutes itself," as Cassirer summarized it, "only in some form of activity."[43] The transcendental ego cannot be known intuitively, since even as it is directly given, it correlates with no object or delimitation; the empirical ego, on the other hand, can be known only as a functional determination of signs.

The relations between individuals can also not be perceived in their true individual form, but only through signs, that is, through form. For Leibniz, following medieval definitions of reality and virtuality, other individuals cannot be directly perceived as "true" individuals other than intuitively – that is, by God or, in the medieval definition, after death in heaven by "beatific vision," where the infinite difference of individuals can be seen. In human knowledge, one can only perceive relative individuality through comparison and form.[44] The culmination of Leibniz's basic insight thus occurs for Cassirer with Wilhelm von Humboldt. Here the more it is assumed that the "fine web" of individual communication exceeds what is directly perceived, the more it must be protected from outside simplification and false placement "under the tutelage of an outer instance."[45] As we have seen, a liberal society in Cassirer's view is based foremost on the historical critique of the forms that allows this immediacy to be cultivated and placed in ever richer architectures.

It is on this basis that we begin to see the outlines of Leibniz's theory of natural right, which will in turn be the basis of Cassirer's understanding of the "Coming-into-Being of Natural Right." For Leibniz, the basis of the "scientific knowledge of the natural law" is an infinite horizon that is in itself rationally unknowable to humans. In its complexity and in the infinite terms at its basis, it is not accessible to discursive thought – it is only perceivable through God's intuitive thought, which could not in any case be communicated as such. Within discursive thought, however, the movement towards natural law structures the "experiential knowledge of positive law," which we have access to.[46] It is for this reason that Leibniz's definition of both right and the individual will be radically different than under Locke. Leibniz "could not have had the inventory of general and unchangeable rights develop from the original 'equality' of all subjects: because for him this 'equality' is a purely abstract concept that does not have any reality in the world."[47]

Ironically, it is the quantitative definition of the individual that was most immediately developed from Leibniz, and indeed became in Cassirer's reading the centerpiece of the Western definition of the individual. The basis of this development was the popularization of Leibniz's work by Christian Wolff, who immediately systematized Leibniz's reading of the individual as one in which there are inherent "equal" rights. Equivalence is defined in the sense of identity. Cassirer is unambiguous about the gravity of this mistake: "Whereas Wolff takes the original equality of right to develop ontologically from the similar being of the individual [thing], for Leibniz it constructs the universal society [*Gemeinschaft*] of ethical organization."[48] The two are contradictory: if one assumes rights inherent to the similar being, then there can be no ethics in Leibniz's sense. For Leibniz it is through ethics and human activity that this equivalence can first be perceived as a limit case. Wolff's systemization of Leibniz's ideas was popularized by Blackstone's *Commentaries on the Laws of England*. These in turn played a key role, as argued by Georg Jellinek (1851– 1911), in both the development of the Virginia constitution and, through this, the French and American declaration of rights.[49]

It is here that Leibniz's broader theory of truth is pivotal, since it will demonstrate that law is not a copying of one reality, or an identification of substantive similarities between people, but a process for discovering invariables that come to define these subjects themselves. Leibniz's break with scholastic logic was epitomized by his premise that truth is defined simply by a relation of subject and predicate in which "the difference between the two becomes ever smaller."[50] Similarly, natural law is always being approximated and revealed in positive law, since positive law allows us to see the complexity and difference of others moving towards a more complete definition of human individuality.

Knowledge in Cassirer's reading further allows that invariants can be reinscribed in different permutations. Law is a process of finding ever more elegant invariants, forming an ordered and harmonious system, and being continuously oriented towards possible future permutations. Just as in projective geometry as a form of metageometry point and line were revealed as definitionally equal in some cases, so in law as a form of metapolitics surprising links might lead to the postulate of equality being developed in ways never originally contemplated. An example of the practical implications of this approach can provide a general sense of this project and its Leibnizian origins. We noted the axiomatic importance Cassirer ascribes to Leibniz's statement that "the science of right and not-right […] serves to make details themselves calculable and to regulate them in advance."[51] The law in this case, like any other science, has a generative power to reveal new aspects of legal subjects and connections between them, and indeed new ways of perceiving relations of these subjects or objects in the world by uncovering different layers of the logic that sustains them. The same would be equally true in Cassirer's view for understanding "objective" relations in the successive physics of Aristotle, Newton or Einstein.

To avoid the historical details of German law that might be unfamiliar to many anglophone readers, we can use an example from North America to demonstrate Leibniz's claim, although it has analogies in many contemporary countries. Although the case is contemporary, the argument is similar to that described by Cassirer in *Leibniz's System* and *Freedom and Form*, and suggests how a field of knowledge or science progressively "reveals" different subjective and objective realities. Earlier definitions of particular liberties pertaining to guilds or estates, Cassirer had argued in *Freedom and Form*, were recast by the French and American Revolutions' idea of a negative determination of "rights" applied to any neutral subject. The concept of "equality" that developed from this basis was further elaborated in the United States' Civil Rights Act of 1964 to explicitly include equality of access to facilities regardless of race, color, creed or gender. This redefinition of equality in turn went through another permutation with the Americans with Disabilities Act of 1990, in which equality is redefined as literal "equality of access" for those with physical disabilities, particularly to structures accessible to others. In each case, a characteristic transformation uses a negation to broaden the definition of who will be defined as "equal" and in what circumstances, but then also broadens what equality itself means in relation to a new phenomenon, say, in the last example architecture and infrastructure. Despite what Cassirer considers its misleading reception by Blackstone and others, equality is not simply given in Leibniz's theory, but itself develops differently and implies different forms of subjectivity.

In Cassirer's reading, law can in this case effectively form rules of invariance through the complex relations among individuals. Law as science can be allowed to develop with maximum subtlety so that individuals can come to see each other more clearly, to place it in Kantian terms, as ends in themselves rather than means. Indeed, the emphasis in Cassirer's reading of Leibniz is not the actual equality of identical subjects, which Leibniz held to be impossible even in theory, but the unfolding of a greater awareness among incommensurably different rational beings.

The key issue is the relation of the form of the science of law to the appearance of subjects and objects, here found in the different forms of legally and culturally defined subjects as well as the objects of their lived environments. Each of the other fields of the "human sciences" (*Geisteswissenschaften*) outlined in *Leibniz's System* – those of history, aesthetics and a secularized theodicy – as well as their modern variants must, in Cassirers view, similarly be understood as not "copying" a given state of affairs, but first creating a continuum of possibilities in which new realities come to light. The same process is equally applicable for fields linked with "objectivity" and "nature" as it is for the legal example in terms of "subjectivity" and "culture." Indeed, for Cassirer, as for Leibniz, there is no initial underlying difference between the fields of culture and nature, subject and object, only historical changes in the categories of knowledge.

In Cassirer's reading, law effectively forms rules of invariance through which individuals can come to see each other as infinite, as ends in themselves, and in which the complex relations among individuals can be allowed to develop with maximum complexity. In this context we can return to Cassirer's central statement on law, noted earlier in his statement on Fichte, that "the content of law does not develop from the being [*Dasein*] of various different subjects, but rather the recognition of the necessary validity of a norm of law is itself the *condition* [*Bedingung*] for the [possibility] that the 'I' recognizes outside of itself not only, for instance, empirical things, but free, self-determining [*sich selbst bestimmende*] reasonable beings."[52] In Cassirer's reading of Leibniz, taking oneself as a "reasonable being" implies not similarity with other reasonable beings, but infinite difference from them mediated by invariants of law – invariants that allow one to see them as "reasonable" and "equal" in this difference.

At first glance it may seem that the development of positive law as a science that culminated with Kelsen fulfilled Leibniz's initial insights. In Cassirer's reading, however, such definitions of "pure law" could represent a fundamental effacement of Leibniz's initial premises. The distinction Leibniz drew between discursive and intuitive knowledge, between an invariant and "copy" theory of truth, meant that law continually has to acknowledge its inability to grasp its object as a totality and acknowledge its incapacity of making final definitions of its objects. Similarly, the provisional science of law has to always be complemented by ethics and aesthetics, which will outline humanity's relation to both other forms of society and to the infinite plenum of human relations themselves.

It is precisely this issue that Cassirer raises at the end of his "coming-into-being of natural law" essay, and it is indeed this aspect that fundamentally separates Cassirer's interest in the "inescapable basic moments" that ethically structure a community from the "grounding norms" that form the structural pivot of Kelsen's pure law. The Wolffian reception of Leibniz that dominated German jurisprudence through the modern period created a technocratic law that, in Roger Berkowitz's words, "retreats behind reasons and grounds [...], loses its natural connection to any ideas of truth and justice except those that are given as its justification."[53] Berkowitz's own study of the affect of this "technocratic" dominance of German law offers recourse to a Heideggerian understanding of an ethics beyond all instrumentality. For Cassirer, however, it was precisely from within Leibniz's own philosophy, particularly his understanding of the infinite as it was played out in his aesthetics and ethics, that this resolution could best be accomplished.

Once we place Leibniz's basic insight into the architectonics of Leibniz and Kant's philosophy, this becomes evident. For here the dichotomy of intuitive and discursive understanding suggests a radical disjunction between the sort of calculation that can be done by the system of law and the actual moment of decision contained in ethical action. The first is by definition based on the asymptotic limit suggested by the interactions of all forms of objective spirit, codified and organized within the particular form of "positive law" so as to best approximate the "natural law" that reveals human society in its greatest complexity. The second, however, is a moment beyond discursive knowledge, even as it can be informed by it: here a choice cannot be simply "calculated" because it actually occurs in an infinity of relations.

This infinity of relations, however, would for Cassirer pose an "absolutely insoluble labyrinth" if it were taken, in the manner of Schopenhauer, as "the willing subject" giving itself "in its determinate essence once and for all in a primitive act underlying its empirical existence."[54] Rather, this infinity can only connect the "in itself" of actual infinite complexity with our discursive knowledge through its orientation to the future, and indeed the future not of the subject but of humanity as an unknowable whole. "One and the same act stands on the one hand under the compulsion of causes that are past and gone," as Cassirer writes of Kant, while on the other hand it is seen from "the point of view of future ends and their system unity."[55] It is in this sense that Kant's second critique asks that the individual consider the categorical imperative as bidding "the will to act as if the maxim of its action were through it to become a law of nature."[56] As Cassirer emphasizes, however, the nature Kant has in mind "is not the sensuous existence of objects, but the systematic interrelation of individual ends and their harmonious composition in a 'final end.'"⁵⁷

Even as positive law contains its own organizational structure and norms, it is always informed by its relation to present and future forms that could define "humanity." The immediate objective of "humanity" as a topic of law is, as Leibniz and Kant insisted, never immediately given even in theory. It is defined in relation to both a future and an infinitely complex present in a manner that obliges positive law to always have reference to other aspects of human experience, and explicitly demands the transcendental imagination as the means for matching schema to intuition.⁵⁸ It is on this ground that we can understand why for Cassirer the basis of politics and law is not the individual, the state, or even the constitution, but the concept of form itself, which is our best means of grasping the various and interrelating formation processes of "humanity."

The hermeneutics of understanding how we are shaped by, and how we shape, forms of objective spirit express the principal mode of freedom available to humanity. Autonomy does not principally relate to the subject, but the social phenomenon of form, and through this balance we see from a new direction how it "overcomes the contradictions between freedom and form [...]. True freedom is directed at the generation [*Erzeugung*] of rules of form [*Gesetzesformen*.]"⁵⁹

Cassirer summarized his project at the end of *Freedom and Form* by writing in relation to Fichte, as we noted in our introduction, that the "the constitutional moment is not the historical past of a country [...] but its future."⁶⁰ The goal of law and the constitution is to allow human beings to perceive ever more clearly their mutual complexity and, as in Humboldt, a far greater level of subtlety of communication and sociation than could be perceived from any "outside instance." The necessary corollary for this project, however, is the historical critique of all forms of society and the constant questioning of the form of "human" itself. It is this project that is at the center of Cassirer and Cohen's reading of law, as it is of any other science. Kelsen's pure law developed only one component of the scientific aspect of Cassirer's theory of law in functionalism, whereas the historicist and humanist reception of Cassirer's project tended conversely to overlook its integral critique of law and science. Cassirer's own notion of a "coming-into-being of natural law" suggests the relation of the system of law with the "foundational norms" that generally structure society. The state is never a totality but rather a plenum whose unlimited horizons can only be grasped in ethics, not science, even as it is only through science that ethics can first locate its modes of activity. In modern society, however, this relation is further distorted by the misreading of natural science, which leads to a tacit definition of "reality" that needs to be actively recognized as always "hypothesis." It is this project that grounded both Cassirer and Cohen's mutual critique of society and science, and which actively underlay the evolution of Cassirer's work.

Notes

1 Habermas, Jürgen. "Die befreiende Kraft der symbolischen Formgebung: Ernst Cassirers humanistisches Erbe und die Bibliothek Warburg," in *Vom sinnlichen Eindruck zum symbolischen Ausdruck* (Frankfurt: Suhrkamp, 1997), 9.

2 Cornell, Drucilla. "Symbolic Form as Other: Ethical Humanism and the Vivifying Power of Language," in *Moral Images of Freedom: A Future for Critical Theory* (London: Rowman & Littlefield, 2007), 75–103.
3 Cassirer, Ernst. "Die Idee der Republikanischen Verfassung: Rede zur Verfassungs Feier am 11. August 1928" (public address in Hamburg, 1929). ECW17 292.
4 ECW17 295.
5 We could argue in this regard that Cassirer's reception not only illuminates original aspects of his argument, but fundamentally changes them. Cassirer would argue for his own work, as he does elsewhere for that of Leibniz and Kant, that "the decomposition [*Zerfall*] of the critical system into its individual varied elements means at the same time the condition and the beginning of a new understanding of its conceptual architecture" (GK 201). By looking at Cassirer's reception and development of his earlier ideas in the Weimar period, then, we can thus return more fundamentally to the meaning of his original works.
6 L 401–87.
7 L 403.
8 Lipton, David R. *Ernst Cassirer: The Dilemma of a Liberal Intellectual in Germany, 1914–1933* (Toronto: University of Toronto Press, 1978); Krois, John Michael. "David R. Lipton, Ernst Cassirer: The Dilemma of a Liberal Intellectual in Germany, 1914–1933," *Journal of the History of Philosophy* 20, 2 (1982): 209–12.
9 FF 315.
10 Marck, Siegfried. Foreword to *Substanz- und Funktionsbegriff in der Rechtsphilosophie* (Tübingen: J. C. B. Mohr, 1925); on legal personification, 23–5, 83–147; on von Gierke, 92–104.
11 Cassirer, Ernst. "Axel Hägerström: Eine Studie ur schwedischen Philosophie der Gegenwart," *Göteborgs Högskolas Årsskrift* 45, 1 (1939); PE.
12 Cassirer, Toni. *Mein Leben mit Ernst Cassirer* (Hamburg: Felix Meiner Verlag, 2003), 195.
13 L 403.
14 FF 163.
15 FF 390.
16 FF 388.
17 WW 1.
18 Berkowitz, Roger. *The Gift of Science: Leibniz and the Modern Legal Tradition* (Cambridge, MA: Harvard University Press, 2005), 7.
19 L 403.
20 Ibid.
21 L 405.
22 Kelsen, Hans. "Das Verhältnis von Staat und Recht im Lichte der Erkenntniskritik," in *Die Wiener Rechstheoretische Schule*, ed. R. M. Hans Klecatsky, Herbert Schambeck (Vienna: Europe Verlag Wien, 1968), 101.
23 Paulson, Stanley L. Introduction to Kelsen, Hans. *Introduction to the Problems of Legal Theory*, trans. Hans Kelsen, Bonnie Litschewski Paulson (Oxford: Clarendon Press, 1996), xxxix–xl; on the general setting of neo-Kantianism and Kelsen's theories, see Paulson, Stanley L. "The Neo-Kantian Dimension of Kelsen's Pure Theory of Law," *Oxford Journal of Legal Studies* 12 (3) (Autumn 1992): 311–32; "Kelsen and the Marburg School: Reconstructive and Historical Perspectives," in *Prescriptive Formality and Normative Rationality in Modern Legal Systems: Festschrift for Robert S. Summers on his 60th Birthday*, ed. Werner Krawietz et al. (Berlin: Duncker & Humblot, 1994).
24 Kelsen, "Das Verhältnis," 101.
25 Marck, *Substanz- und Funktionsbegriff*, 1, 35ff.
26 FF 163.
27 SP1 79 and LC.
28 EK1 74; I 3ff.
29 See, for instance, the ending to section one of SF 233; ETR 455–6; DI 211–13.

30 L 405.
31 FF 333.
32 WW 227.
33 L 403.
34 PE 240.
35 M 273ff.
36 WW 227; R 26ff.
37 Toni Cassirer notes the trial as her first memory of a political event (*Mein Leben*, 9).
38 WW 227.
39 WW 223–4.
40 Cohen, Hermann. "Religion und Sittlichkeit. Eine Betrachtung zur Grundlegung der Religionsphilosophie," in *Jüdische Schriften*, vol. 3 (Berlin: Schwetschke, 1924), 148.
41 L 424ff.; FF 375ff.
42 FF 56.
43 SF 342.
44 The concept of differing knowledge in this regard, as not only intuitive (*Realiter*) and discursive (*Idealiter*), but also virtual (*Virtualiter*) – that is knowledge that is not accessible intuitively by the human mind but can be known as real through the help of divine signs – is usually traced back to Duns Scotus, particularly the fifth section of his *De Cognitone Humana* ("Concerning Human Understanding"): "The fact then that the divine intellect, the true Uncreated Light, has a twofold causality (viz. that it produces objects in intelligible being and that it is also that in virtue of which the secondary objects produced actually move the intellect), this fact can supply as it were a third type or mode of interpretation as to how we can be said to see truly in the Eternal Light" (*Duns Scotus: Philosophical Writings*, ed. Allan Wolter (London: Thomas Nelson, 1962), 125). Leibniz was, of course, deeply familiar with nearly all of scholasticism, and was probably familiar with this term.
45 FF 346.
46 Berkowitz, *The Gift of Science*, 49, citing Leibniz, letter to Conring, 13/23 January 1690 in *Leibniz: Philosophischer Briefwechsel*, vol. 1, 29.
47 FF 333.
48 FF 334.
49 Blackstone's text was in fact perhaps the most important legal text affect the American Revolution, although the monarchism of the book was as much the object of derision as its concept of rights was of emulation. Jellinek notes that the "bill of rights" was derived from Blackstone (Jellinek, *Die Erklärung der Menschen und Bürgerrechte: Ein Beitrag zur modernen Vergassungsgeschichte*, cited in FF 315).
50 EK2 149.
51 L 405.
52 FF 359.
53 Berkowitz, *The Gift of Science*, 52.
54 EK3 431; ECW4 415ff.
55 K 257.
56 Kant, Immanuel. *Grounding of the Metaphysics of Morals*, trans. James W. Ellington (Cambridge, MA: Hackett Publishing, 1993), 20.
57 K 259.
58 FF 341; L 134ff. and 148ff.
59 FF 159.
60 FF 342.

Conclusion

CRITICAL SCIENCE, THE FUTURE OF HUMANITY AND THE RIDDLE OF *AN ESSAY ON MAN* (1944)

Cassirer's work before the First World War set out the conditions for a new form of liberalism specific to central Europe and the development of the modern sciences, one based on a historical critique of knowledge revealing different definitions of subjectivity, objectivity and meaning for different eras. This form of liberalism was expressly different than that of other Western European traditions since it was rooted in Leibniz's philosophy, which had developed a political theory of rights, liberalism and democracy that Cassirer considered, remarkably, superior to that of Locke, Montesquieu or Rousseau.[1] Rather than simply defending a liberal vision of the individual as enclave, Cassirer's liberalism aimed to develop a continuous critique of all forms of knowledge, and with them the development of all forms of sociation and reality. From this critical project, new forms of institutions and ideals could be developed that were more durable, more open to permutation and creative development, and less presumptuous of an essential human nature – and thus open to new definitions of human freedom – than their predecessors.

Despite his role as perhaps the most famous German liberal of the Weimar period, the original content of Cassirer's liberalism and its relation to the sciences as it developed in the imperial period, from his first studies of Leibniz through books such as *Freedom and Form* (1916), has remained largely undeveloped by political and social theorists. In the practice of specific disciplines, however, Cassirer's model of critical science has been widely influential, largely in a manner removed from his original liberal project and its Wilhelmine context. Particularly when placed in relation to the early definition of functionalism presented in *Substance and Function*, Cassirer's work proved tremendously influential in fields focused on the progressive definition of objects and different "realities," notably in the history of natural science (Koyré, Kuhn) and art history (Panofsky), as well as those focused on the history of subjectivity and forms of sociation (Bourdieu, Elias, Foucault), and finally on the role of systems of knowledge and experience in framing these relations (most recently Blumenberg and Luhmann).[2] In addition to shedding light on Cassirer and Cohen's original program of Wilhelmine reform, the present study has sought to provide the basis for a dialog of Cassirer's Wilhelmine liberalism with the present practices of the social, natural and human sciences, as well as a dialog of his version of the late Marburg modernist philosophy of difference with the postmodern or poststructuralist theories of the last half century.

Cassirer's early work is anchored in his reception of the critical science of Leibniz, whose texts he edited and translated for a widely circulated German edition in 1904 and who was the subject of his first book in 1902. From Leibniz and his own teacher Hermann Cohen, Cassirer developed a sophisticated process philosophy that can be read as foundational for much of his work in the Wilhelmine and Weimar periods. Similar in some regards to the philosophy of his contemporary Alfred North Whitehead, Cassirer's early process philosophy, developed in texts such as *Substance and Function* (1910), presents a systematic and measured form of antiessentialism as the basis of a program of Wilhelmine reform.

Unlike many other authors working on a process or functional philosophy, however, Cassirer's writing style and mode of presentation were themselves democratic: they were designed to be accessible to any dedicated reader from a number of different starting points, even as they also guided this reader into an ever more complex, encompassing and challenging worldview. Cassirer's frequent claims that an author needs to be read on the basis of the entirety of his work are in this regard central as a clue to his own style. His writing found multiple voicings for different audiences that could in turn be read against each other or traced back to earlier core ideas in his history and philosophy of science, history of ideas and philosophical functionalism.

Cassirer knew that his early work was both demanding and politically unpopular on both the right and left of the political spectrum. The language of science in which he developed many of his key early ideas was increasingly misunderstood by the educated public, and the form of liberalism Cassirer championed suggested a thoroughgoing critique of all of society, yet eschewed any single essential or even commonsensical base – economic, psychological or political – from which this critique might win ready adherents. His liberalism also appeared to ultimately constitute a break from the innovative socialism of his predecessors in the Marburg school, such as F. A. Lange and Cohen, as well as from that of friends such as Dimitry Gawronsky, in its relative silence on this issue. Yet from *Leibniz's System* to *Freedom and Form*, Cassirer clearly saw an application of functional philosophy and model of critical science as pivotal to a modern, democratic and liberal definition of society. Perhaps due to the challenges of this context, he gradually developed an extremely clear but often indirect style for promoting his thought and its practical consequences, particularly through his histories, while pointing interested readers to its more challenging foundational ideas through reference to his earliest works and selective development of their themes.

Cassirer summarized the philosophical foundation of this process philosophy in terms of his invariant theory in 1940: "The final 'invariants' are not given, they must be *searched* out, and 'established.' Yet this 'establishing' is never something 'absolute,' but rather depends on the continuing course of science."[3] Invariants can – following the model of group theory – be radically redefined by their context so that, Cassirer tellingly continues, they "'shift from place to place' [...]."[4] Frames of invariants reveal different subjects and objects, forms of time and space, and relations of matter and form. Human freedom is defined by the ability to find permutations in these forms, so that, as he wrote in *Freedom and Form*, the "freedom problem and the *a priori* problem are simply different expressions of one and the same fundamental claim."[5]

It was the project that Cassirer understood as the overlap of philosophy and science but which culminated in a critique of the underlying principles that drove "concrete historical" practices found in "objective spiritual forms" in areas such "society and state, science and law."[6] Such changes of "framework," he continues in his 1940 notes, occur not only in the sciences, but further apply through "analogous" relations to "expressive function," such as in myth or the physiognomic perception of our sense of a landscape, a city or so forth. Humanity's freedom is grounded in the constant "searching out" of reality and new constitutions of generative form that occurs through a middle voice that is neither merely active nor passive, internal nor external, and is the foundation of what Cassirer defined as a "productive" use of history and science for the future.[7] In his late assessment of Albert Schweitzer from 1944, Cassirer argued that the key problem of philosophy since the mid-nineteenth century was that it "philosophized about everything except civilization" and that it had thus "in spite of all its learning, become a stranger to the world."[8] It is difficult not to assume that Cassirer included his own work in this indictment, but it also seems that his early work provided a path to this greater specificity and worldliness, one born out less in Cassirer's own application of it than by its reception in the specific disciplines.

Perhaps nowhere in Cassirer's works is the demand for a unified reading of his own texts, and thus a return to his earliest works, more clearly suggested than in his late *Essay on Man*, published in the same year as his appraisal of Schweitzer. By the time the book was published in the depths of the Second World War, Cassirer presented the value of his functionalism as primarily that of a perspectivalism and "self-liberation" through different forms, and he appears to take this as the broadest liberal value he could readily communicate in popular form.[9] As such, his position was readily mistaken as a basic and somewhat naïve idealistic view of the creativity of the human spirit. Cassirer's earliest work reveals a more sophisticated approach: the focus of his functional philosophy of difference is the political imperative to find the optimum architecture of the sciences, institutions and mores of society, that is, of civilization. Through a study of form and invariants grounded in science in the widest sense, the forces of society could be both illuminated and creatively opened to permutation. Works such as *Leibniz's System* and *Freedom and Form* had, moreover, sketched out some possibilities of how this could occur, such as in their respective definitions of rights and Humboldtian liberalism.

Following the First World War Cassirer had largely focused on new problems in the philosophy of symbolic forms, notably for our purposes the political problem of the abuse of expressive meaning and myth that culminated in his *Myth of the State*, the text he completed right before his death in April 1945.[10] In light of the development of the Weimar Republic, Cassirer's subtle Leibnizian arguments for the nature of right, liberalism and democracy seemed largely unnecessary, particularly in a context in which their "scientific" foundation was increasingly against the temper of the era. In emigrating to Sweden, England and then the United States, he found thriving liberal democracies that developed in large part from separate traditions. Given the context of the rise of fascism, and Cassirer's painfully detailed knowledge of the disaster for the European Jewish community (his daughter worked at the American Jewish Committee), it is not surprising that his earlier nuanced reading of the sciences became a relatively secondary feature of his work.[11] Nor of course was he any longer in a position to redefine or even transform the nature of academic endeavour in these foreign lands, as the original Marburg project had boldly sought to do, much less to believe that such a project could change society in its new setting.

Despite this break from his early project, in *Essay on Man* the necessity of connecting Cassirer's earlier and later works for understanding his philosophy as a whole, particularly in its relation to ethics and politics, becomes evident. Cassirer begins the book by attacking any essential definition, any "Procrustean bed," as he puts it, from which to develop the definition of *human*, whether through substantial definitions of nature or culture, or through more complex foundational narratives such as those put forth by Freud or Marx.[12] In a manner the text itself only indirectly addresses, but which can be seen as indebted to Cohen, Cassirer argues that "man has no 'nature' – no simple or homogenous being. He is a strange mixture of being and non-being."[13] Indeed, the *Essay* argues that the modern "essence of humanity" cannot be defined by any "substantial unity" or by any "homogeneity" of human nature at all.[14] "Man is no longer considered as a simple substance which exists in itself and is to be known by itself," Cassirer concludes. "His unity is conceived as a functional unity."[15]

For society at large, this implies that there is no "general orientation, frame of reference, to which all individual differences might be referred."[16] Far from being just a given state of affairs, as it might be in postmodern thought, Cassirer holds that this lack of an "intellectual center" for understanding humanity constitutes a "grave theoretical problem" and nothing less than an "imminent threat to the whole extent of our ethical and cultural life."[17] Amazingly, though, the *Essay* never directly explains how this "immanent threat" is to be avoided, and never clearly describes the "functional unity" that was to oppose traditional "substantial unities" in philosophy and could provide what he describes as a "clue of Ariadne to lead us out of this labyrinth."[18] With this, the *Essay* poses more of a riddle to the question of humanity than an answer. This appears to be entirely intentional.

The *Essay on Man* functions as a sort of practical prolegomena to Cassirer's philosophy for a primarily North American audience, and as such, he writes in the introduction, avoids "taxing [the reader's] attention [...] with a difficult and abstract subject" in the form contained in the *Philosophy of Symbolic Forms*, much less the earlier *Substance and Function* to which the text also directs readers.[19] Cassirer defines his approach in the *Essay on Man* largely in terms of his later philosophy of culture, but the manner in which a philosophy of culture is itself to avoid substantialism or essentialism – the key problem defined in the introduction – is never spelled out. Symbols in the *Essay* are defined as "having only functional value," but what "function" is in this context, or how it relates to a "synthetic unity" of judgments that are both "particular and universal" is also not clearly explained, even as it is a problem that reaches back to one of the earliest leitmotifs of Cassirer's work.[20] In this context, Cassirer's ideas of culture and symbolic forms can be, and often were, mistaken as a lens through which we view the underlying reality of the world (a mistake that Cassirer tellingly saw as characteristic of the earliest reception of Kant and Fichte), and the book as a whole an anodyne plea for diversity of perspectives lacking any real basis of common agreement – a reading that indeed would shape Cassirer's North American reception as an good humanist but not a particularly trenchant philosopher.[21]

The provisional answer to the central question of the meaning of humanity in the *Essay on Man* is to send the reader back to a philosophy of symbolic forms, and with it a definition of *homo symbolicus*, but this postpones the "imminent threat" rather than resolves it, since – as Habermas and others have argued – relatively little of a positive ethics or politics are presented in Cassirer's philosophy of symbolic forms.[22] Indeed, Habermas's claim for the "astonishingly" limited lack of discussion of morality and politics in Cassirer's later work is vividly illustrated by the text.[23] For not only are these themes barely covered – in the case of politics not mentioned at all – the lack of a common substantial foundation for humanity appears at first, and by Cassirer's own admission, to undercut any possible basis for coherent political action or an ethics, yet very little of a functional replacement is suggested in its place.[24]

If we take seriously Cassirer's claim that the *Essay on Man* is suggesting an alternate "functional" definition of humanity, however, we find a strong resolution of the riddle of the text not in the book itself or even in the *Symbolic Forms* project to which it refers, but in Cassirer's early works. Here we find a functional theory that from the first defines individuals through difference, not similarity, and which indeed uses this definition as something resembling what Habermas had called a "theory of civilizational processes."[25] The keystone of such a theory is the negative or, more precisely stated, differential definition of the individual in relation to forms of knowledge and practice, and the critical reflection on these forms by different modes of science. This process was clearly outlined as early as *Leibniz's System* and described by Cassirer as the most "enduring" aspect of Leibniz's work into the modern period. Through an imminent critique, Cassirer sought to describe how this definition progressively developed in the social, human and natural sciences through the modern era, while also suggesting how it might act as normative basis for future development. If the functional definition of humanity was to act as the "clue of Ariadne" to lead humanity of its present impasse, these early works provide us with an important sense of both how this functionalism was defined and how, if combined with its application in other fields and with the later philosophy of symbolic forms, it might serve as a guide to "lead us out of this labyrinth."[26]

That Cassirer occasionally understood many aspects of his later work as a popularization and necessary simplification of his philosophy as a whole is suggested both by his initial recognition of the difficulty of his early functional project, as for instance in his claims in the *Problem of Knowledge* that its direct explication is "as difficult as it is necessary," and also by his claim that the *Essay* itself is "more an [...] explanation and illustration than a demonstration of my theory."[27] Cassirer expected serious readers of the text to return to the foundational arguments of his philosophy not only in the *Symbolic Forms* project, but also in texts such as *Substance and Function*.[28] Cassirer's own readings of authors such as Nicolas of Cusa, Leibniz and Kant consistently promoted this "genetic approach" to reading an author, one developed from Leibniz's philosophy not as a single linear development of thought but based on permutations of forms of understanding through time, a model based on permutations of groups outlined in *Substance and Function*.[29]

A telling clue that Cassirer intentionally left key parts of his argument out of the *Essay* is found in the penultimate chapter of on science, a topic that Cassirer describes as nothing less than "the most important subject of a philosophy of man" and "the highest and most characteristic attainment of human culture."[30] Despite these claims, however, the chapter pointedly avoids a survey of the philosophy of science altogether, somewhat oddly footnoting the title "Science" with a note that the chapter "does not of course claim to given an outline of a philosophy of science or a phenomenology of knowledge."[31] The reader is instead directed back to the earlier *Substance and Function* for the former and the third volume of *Symbolic Forms* for the latter, with the further implicit claim being a request to place both texts in dialog. The *Essay*, Cassirer continues, can only "indicate briefly the general function of science and determine its place in the system of symbolic forms."[32] In this way, Cassirer appears to signal both the necessary foundation of his early work for these problems – the topic of science is, after all, nothing less than the "most important subject" to a "philosophy of man," which is to say, the topic of the *Essay on Man* – and his deliberate rhetorical strategy of avoiding any in-depth description of it.

Without Cassirer's definition of science, however, it is in fact very difficult to understand the basic features of his philosophy, most importantly the problem of reality as it relates to symbolic forms and the sciences. For most readers, natural science of this era was assumed to entail precisely a copy of a pre-given "reality" in the world. As such there is an inherent dualism of subject and object, inner and outer, form and matter that Cassirer's philosophy of science, particularly in *Substance and Function*, argues against at length. The focus of Cohen and Cassirer's early reading of natural science was precisely to contest this "copy" theory of knowledge and the culturally dominant understanding of experience that it anchored. In Cohen's terminology, they suggested instead that any particular "Real" moment was determined in time through form, even as it rested on an always already present, but always partially inchoate, existential field of "reality."

It was this model that we have argued persisted in Cassirer's work with remarkable continuity from *Leibniz's System* to the central phenomenological concept of symbolic prägnanz in the *Philosophy of Symbolic Form*. It claims to describe how "perception as a sensory experience contains at the same time a certain nonintuitive meaning which it immediately and concretely represents," and it does this through the "interwovenness of the singular perceptive phenomena given here and now, to a characteristic total meaning [so that] every perception embraces a definite 'character of direction' by which it points beyond the here and now. As a mere perceptive *differential*, it nevertheless contains within itself the *integral* of experience."[33] Far from being limited to functions of signification (for instance, the appearance of "evidence" in a particular moment of scientific experiment), or representation (the practical navigation of space in everyday life), this key moment of a "phenomenology of knowledge" also claimed to encompass with equal validity the reality of expressive functions (the physiognomy of a city or the appearance of a god). Without the critical foundation of Cassirer's philosophy of science in *Substance and Function*, this later breadth of the reality of symbolic form becomes difficult to understand.

Cassirer's conclusion to the *Essay* nonetheless gives the reader a poetic, but enigmatic, further guide to the functional argument on which his text as a whole rests. Cassirer begins and ends the conclusion with a quote from Heraclitus suggesting that humanity can ultimately only be understood functionally as a "dialectical unity, a coexistence of contraries."³⁴ "Men do not understand," Cassirer quotes the pre-Socratic philosopher, "how that which is torn in different directions comes into accord with itself – harmony in contrariety, as in the case of the bow and the lyre."³⁵ Yet, characteristically, the *Essay* provides very little definition of the meaning of this quote, or what such a functional or dialectical unity is, how it can also be a "coexistence of contraries," much less how such ensembles fit with the problem of symbolic forms and the problem of a unified direction of human endeavor that is the purported theme of the text.³⁶ Human culture "may be described as the process of man's progressive self-liberation," Cassirer writes, in which each "all these functions, [while revealing] strong contrasts and deep conflicts," nonetheless "complete and complement one another," so that "each one opens a new horizon and shows us a new aspect of humanity."³⁷ They reveal a "unity" that is "not simplicity."³⁸ Lacking evidence, though, the reader is inclined to feel that while Cassirer may hope for such unity, its basis has not been presented in the text. It is thus a bit puzzling when the last sentence of the book returns, using the same Heraclitean citation, to claim that humanity is not substantial but is based on functions "that cannot be reduced to a common denominator […] in which nonetheless there is harmony in contrariety, as in the case of the bow and the lyre."³⁹

A unitary system of functional difference with no "common denominator" is thus the culmination of Cassirer's system. Cassirer had in fact earlier presented a longer exegesis of this Heraclitian epigram in a 1925 essay on pre-Socratic philosophy that suggests why he would consider it an apt summary of his own process philosophy. He claims here that all of Heraclitus's philosophy could be considered a commentary on this "foundational thought," which underlies "his physics, his mathematics, his ethics" in a manner, we could add, that is similar to its place in Cassirer's own project.⁴⁰ His explanation of it is as follows:

> The "thing" as well as the "property" are nothing in themselves, rather they are only an incision [*Einschnitt*] in the continuum of being; they present phases of becoming in which everything only has a sense and a content in that it is related to other phases that precede or succeed it. The form of this relation determines the kind of thing or property [it is]. That every individual point does not posses any self-sufficient or self-contained content, but is rather only to be thought of as starting point [*durchgangspunkt*] and junction [*Knotenpunkt*] in the process of becoming [*Prozess des Geschehens*], so in turn does our way of determining and naming it depend fundamentally on the approach we take to it.⁴¹

Within this definition, Cassirer continues, Heraclitus held that every "being" existed "out of its opposites" and was under a "two-fold sign" – presumably of the past and future in which it is an "incision" – that not only does not undermine its unity but first creates it.⁴² For every being is "only constructed out of contraries […] and the inner tensions that are continually constructed and continually resolved."⁴³ In this manner, as Cassirer's more evocative German translation of Heraclitus has it, "that which strives against itself comes together [*Auseinanderstrebede ineinandergeht*] […] like the unifying tension [*gegenstrebige*

Vereinigung] of bow and lyre."[44] Through this approach, Cassirer writes, Heraclitus was able to avoid the perspective of both "naïve experience" and the newly developing "sciences of nature" (presumably with Leucippus) of understanding the world as made out of "individual substances."[45]

Cassirer wrote in the introduction of *Substance and Function* that his era witnessed the demise of the long reign of Aristotelian logic as well as the more recent substantial definition of the modern sciences, in which "the work of centuries in the formulation of fundamental doctrines seems more and more to crumble away."[46] It is in this regard not surprising that in place of a substantial definition of logic and science his own work would emphasize this Heraclitian vision of a "process of becoming." Indeed, the overtones in the passage above of Dedekind's "cut" – the theme he used to modernize Cohen's work in his debate with Renouvier – to the Heraclitian "incision" perhaps suggest his awareness of this link.[47] In his dialog with Meyerson on the nature of science, Cassirer argued we are not left with "ultimate substantial things," and with them a stark contrast of science and reality, but a contrast of "functional orders and correlations" in science and nature that are "determined only in and by [...] the element of diversity and change."[48] The progressive historical development of these orders, and their experience in time, Cassirer wrote, "re-establishes the harmony of being"[49] as the process of science, even as this harmony is always a state of difference and fluid process towards wider unities of knowledge.

Cassirer's concluding use of Heraclitus's theme of the bow and lyre is in fact quite intelligible in terms of his development of Leibniz's and Cohen's philosophies. As *Substance and Function* had argued, any individual moment of being or consciousness – like Leibniz's individuals or monads – is a moment of determination between differences, with the emphasis on the determination (*Bestimmung*) of this consciousness as a differential by its place within form. It cannot be broken into elements, faculties or even discreet moments of attention. The modern model of calculus used by Leibniz and Cohen, and further developed by Cassirer's studies of "series and limit" in *Substance and Function*, provides a more sophisticated basis for Heraclitus's "foundational thought," but it is closely analogous. With Leibniz, the negative moment of determination suggested by Heraclitus can for the first time be explained as a positive expression of a derivative or force. As Cassirer explained for the application of the model of calculus to biological life in Leibniz, but has applications throughout his system, "the particular moments of being are directed to a context in the whole [that is] thought of as a preceding and determining ground of the part."[50]

A "clue of Ariadne" out of the dilemma of modern humanity could thus be the model of process, and with it the absolute difference of individuals, that had its earliest example in Heraclitus and was then initiated in the modern period by Leibniz's monadology.[51] This allowed for recognition – and critique – of the particular structures of difference, histories and forms of knowledge that ground a developmental architecture for humanity.[52] It is this basic position that *Substance and Function* defined as a model for the sciences as they seek invariants within different structures of knowledge, and which *Freedom and Form* developed into a model Goethean perception and Humboldtian liberalism.

The unity of experience or of humanity for Cassirer cannot in this regard be the closed unity of substance, or even the unity of "one"; it has no common denominator.

It is rather the unity of a plenum of possibility realized in the knotted form of each moment. It was this theme that was logically grounded for Cohen in the problem of infinite judgment, and which Cassirer wove through his own philosophy as the basis of an "open unity" and form of determination through which the various sciences could be constituted.[53] This philosophical resolution, however, is coextensive with an ethical and political resolution. For it culminates in the idea, as Cassirer puts it at the end of *Leibniz's System*, that "Being [...] is never closed, but it rather develops in the historical collective-process out of the work and interaction of individuals," so that Leibniz's philosophy – like Cassirer's – culminates in the "historical experience of humanity."[54] Within this definition of unity even the most violent oppositions create some similarity and complementarity, some form of more foundational norm, and can thus be put in the ethical framework of a striving towards greater cohesion in the sense found in Kant's cosmopolitan history. In this regard, Cassirer writes, in place of a "harmonization of interests," Leibniz's philosophy demands "a critical determination of limits or boundaries [*kritische Grenzbestimmung*]" – which is to say, a new development of the sciences in all of their forms.[55]

As abstract as Cassirer's reading of science, law and politics could be, it underlay his claim in his 1928 *Rektorsrede* that the "idea of a republican constitution" had "[grown] from our own ground" – by which he presumably meant particularly the philosophies of Leibniz, Humboldt, Fichte and Hegel as he had outlined them 12 years earlier.[56] Following Leibniz, in this system a true ethical and legal theory of equality and rights is not based on the identity of individuals, but on their radical difference, and not on their substantial presence but on their functional relation to one another. Politics and political science is not the demand for a universal basis of agreement, but the ability to develop different constitutions of partial agreement and constructive agonism. In the broader sense, a critical science is also a "metapolitics," as Cohen had it, that seeks to constantly critique the basis of common assumptions of a given society and seeks to find productive new permutations for them. If the functional definition of humanity was to be a clue to the riddle of the *Essay on Man*, Cassirer thus provided at least some minimal earlier guides to "lead us out of this labyrinth," guides that can be combined with the development of functional science in particular disciplines and with his own later work on what Drucilla Cornell and Kenneth Michael Panfilio nicely summarize as "symbolic forms for a new humanity."[57]

Notes

1 FF 331.
2 The link between Kuhn and Cassirer occurs particularly through Koyré. See in particular Friedman's discussion of its complexity in "Ernst Cassirer and Thomas Kuhn: The Neo-Kantian Tradition in the History and Philosophy of Science," in *Neo-Kantianism in Contemporary Philosophy*, ed. Rudolf A. Makkreel and Sebastian Luft (Bloomington: University of Indiana Press, 2010), 180ff. and "Kuhn and Logical Empiricism," in *Thomas Kuhn (Contemporary Philosophy in Focus)*, ed. Thomas Nickles (New York: Cambridge University Press, 2003), 19–44.
3 PSF4 120.

4 Ibid. The breadth of Cassirer's model here is suggested by his use of the example of the theory of relativity as a model for shifting invariants combined with his conclusion that "analogous considerations [apply] to the *expressive function*."
5 FF 308. Cassirer is discussing Kant's philosophy here, but on this topic it is coextensive with his own development of the theme in *Freedom and Form*.
6 FF 18–19.
7 FF 386.
8 SMC 232. Cassirer's move towards an ever more popular presentation of his philosophy in the late years of his life is in part explained by this review of Schweitzer. Whereas, he continues, philosophy in the eighteenth and early nineteenth century "produced in a perfectly natural way a living popular philosophy" this was lost in the late nineteenth century, when philosophy "did not realize that the power of ideas about civilization had become a doubtful quantity." Although we now have an acute sense of how closely European attitudes towards civilization were linked with the colonial process, it is crucial to remember that Cassirer understood civilization as principally the process of self-critique of objective and intellectual forms of development through science. It is likely that Cassirer chose Schweitzer as his principal example in large part on the basis of his indictment of the historical basis of Christianity and its possible ethical redefinition in his *The Quest of the Historical Jesus: A Critical Study of its Progress from Reimarus to Wrede* (New York: Macmillan, 1966).
9 E 228.
10 MS vii.
11 Cassirer, Toni. *Mein Leben mit Ernst Cassirer* (Hamburg: Felix Meiner Verlag, 2003), 289. Of course, Cassirer continued to develop a sophisticated reading of the sciences throughout his later work, notably in the reading of the natural sciences presented in the third volume of the *Philosophy of Symbolic Forms* and in the important study of the cultural sciences (*Kulturwissenschaften*) in his *The Logic of the Cultural Sciences*.
12 E 21. Cassirer also included Nietzsche in this list, but this seems misplaced given Nietzsche's clear antiessentialism, and appears to be derived more from the anti-Nietzschean sentiment of the wartime years than an accurate reading of his project. Cassirer does present a favorable use of Nietzsche in his reading of Burckhardt in EK4 269–70, 280.
13 E 11.
14 E 21, 222.
15 E 222; PSF4 211.
16 Ibid.
17 EM 22.
18 Ibid.
19 E vii.
20 EM 32, 186.
21 For Cassirer's argument against the "copy theory" of knowledge, see particularly PSF1 105ff. and SF 283ff. as well as the introduction to PSF3. The prospect that the *Essay* is presenting an intentional set of rhetorical simplifications is strongly suggested if we look at Cassirer's readings of the initial reception of Kantian philosophy. Here Cassirer outlines how a similar popular dualism, noted as completely false, was developed in the reception of Kant by Heinrich von Kleist's famous description of our forms of knowledge as "green glasses" that irrevocably "distort" our view of a "pre-given" world. Cassirer argues that in fact Kleist was referring to Fichte in this description of the "Kantian" philosophy, but the widespread reception of Kleist as in fact talking about Kant is emblematic of this popular misunderstanding of Kant. See "Heinrich von Kleist und die Kantische Philosophie," in IG 401ff. (Cassirer, Toni. *Mein Leben*, 296).
22 The major exceptions will be Cassirer's discussions of politics and ethics in *The Myth of the State* (1945, published in 1946) and *Axel Hägerström: Eine Studie zur schwedischen Philosophie der Gegenwart* (Göteborg: Göteborgs Högskolas Årsskrift, 1939).

23 Jürgen Habermas, "Die befreiende Kraft der symbolischen Formgebung: Ernst Cassirers humanistisches Erbe und die Bibliothek Warburg," in *Vom sinnlichen Eindruck zum symbolischen Ausdruck* (Frankfurt: Suhrkamp, 1997), 9.
24 Cassirer discusses ethics in passing in relation to Kant and Bergson (EM 60, 88–9).
25 Habermas, "Die befreiende Kraft," 9. For a further development of this reading of symbolic forms, see Cornell, Drucilla. "Symbolic Form as Other: Ethical Humanism and the Vivifying Power of Language," in *Moral Images of Freedom: A Future for Critical Theory* (London: Rowman & Littlefield, 2009), 75–103.
26 E 22.
27 E viii.
28 A characteristic anecdote is his reaction to a radically re-edited version of his *Myth of the State* that appeared in Fortune Magazine in June 1944. At first he noted to his wife that it "was very good, but I did not write a word of it," but then on realizing it had been formed by recoupling aspects of his original article, he laughed and decided he was entirely content with it (Toni Cassirer, *Mein Leben*, 320).
29 On Cassirer's "genetic approach" to knowledge see SF 341; it is further developed in FF through a reading of Schelling and Fichte (246ff.), and appears frequently as both a method and a theme in Cassirer's work.
30 E 206.
31 E 207.
32 Ibid.
33 PSF3 203.
34 E 228.
35 E 223. As Barry Sandywell has noted, Heraclitus's quote is "an ideal image for a world structured as a dynamic system of forces" (Sandywell, Barry. *Presocratic Reflexivity: The Construction of Philosophical Discourse 650–400 B.C.*, Logological Investigations, vol. 3 (London: Routledge, 1996), 262).
36 EM 222.
37 E 228.
38 Ibid.
39 E 223, 228.
40 Cassirer, Ernst. "Die Philosophie der Griechen von Anfängen bis Platon" (1925). ECW16.
41 Ibid.
42 Ibid.
43 Ibid.
44 Ibid.
45 Ibid.
46 SF 3ff.
47 Dedekind had argued, in Cassirer's words, that "number now functions in its originary and most general sense no longer as an answer to the question of 'how many' parts are contained in a definite whole, but rather function as an exact denotation of the positional relation [*Stellenverhältnis*] of any ordered series altogether" (U 110). But this definition was suggestive for Cassirer not only for mathematics, but any phenomena at all. "Removed from the relational forms of experience there could be no phenomenal 'content' [or] empirical objects" (U 115).
48 SF 324–5.
49 SF 325.
50 L 367.
51 As we saw earlier, however, Cassirer's reading of the "monadology" strips it of its more fantastic components to reveal its inner logic. "The outer form of the monadology must be broken, in order to bring to reality ever more clearly and freely the conceptual motives which help to build the system" (L 431).
52 Ibid.

53 It is this "open unity" of infinite judgment that I take Cassirer to be referring to in describing the form of determination on which his work is based as a "stranger sphere," in which: "In infinitude, it is not just an opposition to finitude which is constitute but, in a certain sense, it is just the totality, the fulfillment of finitude itself. But this fulfillment of finitude constitutes exactly infinitude [...]. It is the opposite of privation, it is the perfect filling-out of finitude itself" (DV 201). It is this position that defines Cassirer's "double form" of determination as both enabling particular sciences of the particular yet also grounding them in the infinite plenum of experience, which is experience in the ineluctable infinitude of the present. Particularly given Jakob Gordin's extensive description of this theme of infinite judgment with Cassirer's consultation during the same period, it is seems to me very likely that this is what Cassirer is referring to, which I take as the basis of his position at Davos.
54 L 432.
55 L 432. Cassirer had earlier written that the "Goal of the new theory of knowledge seeks ever more a critical determination of the limits between nature and habits, to a line between the methods of natural and historical knowledge," but that such determinations can only start from a common ground, which is what Leibniz provided (L 380).
56 Ernst Cassirer, "Die Idee der Republikanischen Verfassung: Rede zur Verfassungs Feier am 11. August 1928" (R); ECW17 292.
57 Cornell, Drucilla and Panfilio, Kenneth Michael. *Symbolic Forms for a New Humanity: Cultural and Racial Reconfigurations of Critical Theory* (New York: Fordham University Press, 2010).

INDEX

Aesthetics of Pure Feeling (Cohen) 18, 64
Analyst, or a Discourse Addressed to an Infidel Mathematician, The (Berkeley) 51
antinomies 53–4, 81n160, 145–6
anti-Semitism xviii, 37–9, 200
Aristotelian logic 125–6, 128–9, 133, 174, 216

Baeumler, Alfred 119n159
Bahr, Hermann: on philosophical importance of Cassirer's work xvii–xxii, 20; political affiliations xviii; interest in Cassirer's work xl–xlin13
Bambach, Charles xlvn67
basis phenomenon xlviin89, 177–9, 183
Bauch, Bruno 37–9, 43n58
Bergson, Henri 4, 17, 219n24
Berkeley, Bishop George 51
Berkowitz, Roger ix, 197, 205
Bernstein, Eduard 3, 29
Blackstone, William liin132, 168, 189n60, 203–4, 208n49
Blumenberg, Hans xxvii, xliiin41, ln121, 123, 143, 209
Bourdieu, Pierre xxvii, xlivnn57–8, 141, 148, 209
Burckhardt, Jacob 169, 199, 218n12
Burke, Edmund 184–5

Capeillères, Fabien xxxvii, 22n10
Cassirer, Ernst: Bahr's reading of xvii–xxiii, xl–xlin13, xlin17; works as reform project and role in Marburg school xx, 3–21, xlin15; during Weimar Period xxii–xxiii, 193–5, 207n5; historicism xxiii–xxvi, xlvn67; on relation of "subjectivization and objectivization" xxv–xxvi; idea of functionalist philosophy xxvii–xxviii; idea of process philosophy xxix; on humanity xxxiii; invariant theory of truth xxxiii, 132–4, 116n91; neo-Kantianism xxiv–xxvii, ln121, linn124–6; *Hermann Cohen and the Rejuvenation of Kantian Philosophy* 9–12; relationship with Cohen 33–5; response to Bauch 38–40; schema of knowledge and reality 56–8; definition of symbol 69; definition of "ultimate principle of knowledge" 70; epistemology 68–72; form 72; symbolic prägnanz xxxix, xlviin89, 72–5, 81n160 153n60, 179, 214; on Leibniz's worldview 88–90, 114n28, 118n145; on Leibniz's theodicy (*see* Leibniz); structural realism 103; on Leibniz's concept of law 108–10; definition of form 128; logic (*see Substance and Function* [Cassirer]); series and mathematics 128–9; mathematical group theory 131–4, 150–52; concept of the atom (*see* chemistry in *Substance and Function* [Cassirer]); truth and reality 139–40; problem of the individual 143–4; liberalism 169, 209–11; reading of Goethe's work 172–83, 190n99; three aspects of Goethe's basis phenomena 177–8; interpretation of Humboldt's theory of the individual and state 180–87; theory of law 194–206; law and functionalism, reception of 196–8; reception as a humanist 198–201; synthetic position 201–6; positive law as science 205–6; the problem of humanity (*see Essay on Man* [Cassirer]) 209–17, 218nn8–12, 218n21
Cassirer, Paul xln13, 37, 43n54
Cassirer, Toni 34, 36–7, 39, 42n31, 208n37
calculus: as a model for Cassirer and Cohen xxx–xxxii; xlviiin101, 6–8, 22n16, 25n81; in Cohen's *Infinitesimal Method* 47–58, 76n28, 79n108; in Leibniz's definition of reality 103, 119n159; in relation to set theory 153
chess game analogy, Saussure's 141
Chickering, Roger 37, 43n55
Cohen, Hermann xix; "third way" of Marburg school xxii; philosophy of origin xxix–xxx; philosophy 6–21; neo-Kantianism 9; critique of knowledge (*see Hermann Cohen and the Rejuvenation of*

Kantian Philosophy [Cassirer]); Marburg program and reform 12–16; God-idea and atheism 16; psychology 17–18; comedy and irony 18–21; aesthetics 19–21; Lange's critique 28–30; Cassirer as a student 33–5; *Infinitesimal Method* 47 (see also *Infinitesimal Method* [Cohen]); nature of experience 54–5; anticipations of perception 58–62; problem of continuity 62; theory of reality 62–8, 76n29, 79n97, 79n100, 144–5; necessary preconditions 64–5; Cassirer's analogies to, in reading Humboldt 186–7

Commentaries on the Laws of England (Blackstone) 203

Copernican system 104, 142

copy theory of knowledge 138–9, 146, 214, 218n21

Cornell, Drucilla xxxvii, 193, 217

Das Erkenntnisproblem in der Philosophie und Wissenschaft der neueren Zeit: see *Knowledge Problem* series

Davos debate xviii, xxxvi, lin126, 4, 178–9, 220n53

Descartes: in Western thought 90; ego and contrast with Leibniz 90–94; ideas of extension and space 94–7; role of God 97–102

Descartes' Critique of Mathematical and Natural Scientific Knowledge (Cassirer) 90

Determinism and Indeterminism in Modern Physics (Cassirer) 123, 136, 144, 155n134

Dilthey, Wilhelm lin126, 48, 68

Dreyfus, Alfred 200

Einstein, Albert 12, 87, 116n91, 142, 175, 203

Einstein's Theory of Relativity (Cassirer) xlvn68, 87, 144

Elective Affinities (Goethe) 19–20

Elements of Psychophysics (Fechner) 48

Elias, Norbert xxvii–xxviii, xlivn59

Enabling Acts xlviiin98, 35, 195

entoptic colors 177–9

epistemology: see *Substance and Function* (Cassirer)

Erdmann, Benno xxxv, 126, 131–2

Erlanger Program (Klein) 116, 132

Essay on Man (Cassirer) definition of symbol 212–13; definition of science 213–14; concluding functional argument 214–17

expressive motion 18

Fechner, Gustav: philosophy in natural and social sciences 48, 81n142; psychophysics 48–50; deism and cosmology 49–50; dualism 58, 66–8; Fechner's Log Law 76n15

Ferrari, Massimo xxxvii, xxxix, lin124, liin128

Fichte, Johann Gottlieb xxxiv, 167–8, 191n131, 204, 206, 218n21, 219n29

Foucault, Michel xxiii–xxvii; xxxvii, xliiin41, xliiin43, xliiin47, xlivn54, 149, 210

Freedom and Form (Cassirer): reception xliin27; critique of science 22n10; World War I 36–40; on Leibniz 112–13; Baeumler's response 119n159; continuation of earlier works 150–52; outline of liberal civilization 159–1; critical history 165–9; definition of form 166–7; Cassirer's reading of Goethe 173–80; analogies to Cohen's philosophy 186–7; Humboldtian liberalism 180–87; definitions of law and state 195–206

Frege, Gottlob xlviiin10, xlixn104, xlixnn106–107, 3

Freiheit und Form: Studien zur deutschen Geistesgeschichte: see *Freedom and Form* (Cassirer)

Funkenstein, Amos x, xx, 64, 96

functionalism: Cassirer's early philosophy xxvii–xxxi, xlivn56; in *Substance and function* 121–4; in historical analysis 161–3; political and legal philosophy of Cassirer 194–206

Gawronsky, Dimitry xlviiin99, 63, 156n186, 210

Germany: during World War I 36–40; anti-Semitism 37–8; problem of the "essence" of the German people 165–6, 169; Cassirer leaves 195

Geschichte der Materialismus: see *History of Materialism* (Lange)

Gift of Science: Leibniz and the Modern Legal Tradition, The (Berkowitz) 197

Goethe, Johann Wolfgang von: theory of form xlin15, 167–72, 170–77 experience of history 165–6; natural philosophy of 171–2; transforming Kant's schematism 172–5; basis phenomena (*Urphänomene*) xlviiin89, 176, 191n120; theory of color 177–9; in theory of form in Cassirer's interpretation of Humboldt 180–85

INDEX 223

Gordon, Peter E. xxxvi, xliin19, 25n68, 80n116, 155n133
Grotius, Hugo 109, 167, 197, 199–200
group theory: Cassirer's use of as model xxvi, xxx–xxxi, xliiin41, xlvin80, xlixn103, 68; invariants in a group xxxiii–xxxiv, 116n91, 132–4; in illustrating the idea of series 131–2; model for historical transformation 140–45; in *Substance and Function* 165–7

Habermas, Jürgen 193, 196, 201, 213,
Heidegger, Martin: Davos debate xviii, xxxvi, lin124, lin126, 4, 165; historicism xlvn67; Cassirer's critique 140, 155n133; ontology 178–80, 191n146
Heidelberger, Michael 50
Helmholtz, Heinrich 79n103
Helmholtz, Hermann von 10
Heraclitus of Ephesus 214–16, 219n35
Herder, Johann Gottfried 20–21, 26n97, 26n99, 65, 88, 100, 167
Hermann Cohen and the Rejuvenation of Kantian Philosophy (Cassirer): *see* Cassirer
History of Materialism (Lange) 4, 12, 27–8, 79n103
hodegetics 18
Holzhey, Helmut xiiin128, 42n41, 79n97, 79n101
Human Sciences (*Die Geisteswissenschaften*) (Buek and Heere) 32
Humboldt, Wilhelm von: development of Goethe's form xlin15, 167, 190; concept of the individual 169, 186–7; liberalism and form 180–85, 190n99; definition of individual 181–2; power of the state (*see* Cassirer, interpretation of Humboldt's theory of the individual and state)
Hunter, Ian 8
Husserl, Edmund 75, 82n160, 129, 131

infinitesimal method 4–5, 10, 81n142, 85
Infinitesimal Method (Cohen): content and context 45–7; dichotomy of calculus 50–52; process of determination and experience 52–5; and judgement of reality 52–5, 62–8; universal schema 55 8; principle of anticipation and problem of continuity 58–62; (*see also* Cohen; Cassirer)
invariants: *see* Cassirer, invariant theory of truth

Judaism 31, 34

Kant, Immanuel: *a priori* truths xxxiii–xxxiv, 28, 133–2, 160; in development of Cassirer's thought and the Marburg school xxxviii–xxxix, xliiin20, xlivn50, xlivn54, xlivn56, xlvin81, xlvin88, xlviin89, xlixn106, ln115 ln121, linn124–6, liin128 liin132; conflict of faculties 5–8; Cohen's studies of 9–12, 45–7, 78n69; aesthetics 18–20; in development of Marburg critique 27–30; antinomies 53–5; axioms of intuition 58–9; reality 63, 79n100; pre-established harmony 101; reception of Leibniz 92, 98; as an intermediary between Leibniz and Goethe 167, 182–3
Kant and Marx: A Contribution to the Philosophy of Socialism (Vorländer) 29
Kant-Studien xxxi, 9, 38, 106
Kant und Marx: Ein Beitrag zur Philosophie der Sozialismus: see *Kant and Marx: A Contribution to the Philosophy of Socialism* (Vorländer)
Kants Theorie der Erfahrung: see *Kant's Theory of Experience* (Cohen)
Kant's Theory of Experience (Cohen) 3, 210, 51, 67, 89
Kelsen, Hans xxi, 29, 194, 196–9, 205–6, 207n23
Klein, Felix xxxiii, 53, 116n91, 132, 150
Knowledge Problem series (Cassirer): content and context of xxiv–xxv, xxvii; calculus 52; compared to *Leibniz's System* 85, 113n3; approach towards historical functionalism 161–5
Köhnke, Klaus Christian xxxvii, ln118, ln121
Krois, John xxxviii, xlviin89, xlixn107, 73

Lange, Friederich Albert: as a founder of the Marburg school xx, 10, 12; Marburg critique 27–30, 41n14; reading of Fechner 49–50
Leibniz, Gottfried Wilhelm von: philosophical method xxxii-xxxiii; applied and understood by Cassirer xxxix, xlvin88, xlviin89, xlviin92, 22n16, 23n20, 25n82, 86–90, 114n28, 210, 216–7, 272; as Kant's "teacher" 67, 92; infinitesimals 80n124, 81n114; monadology 89–94, 102–3, 118n145; Molyneux Problem 117n120; critique of Descartes 90–102; space, time, and motion 95–7, 134; theodicy 98–102, 111–13; definitions

of truth and pre-established harmony 99–101; theory of reality 102–3; concept of symbol 104–6; reality of the individual 106; role of history 106–8; concept of law 108–10; aesthetics 110–11; theory of natural right, Cassirer's synthetic position 202–6, 207n5, 208n44

Leibniz: Key Works for the Foundation of Philosophy (Cassirer) 86

Leibniz's System and its Scientific Foundations (Cassirer): as Cassirer's first work xxx–xxxi, xlviii92, li–lii; as a presentation of Cohen's philosophy 23, 85–6; continuities with *Substance and Function* 121–2, 138–9; law 194; continuities with *Freedom and Form* 203–4, 210–11; *see also* Leibniz

Levy, Heinrich xlviiinn94–5, 114n21
Linden, Harry van der 40n6, 187
Lipton, David 194
Locke, John 86, 90, 109, 133, 202
Logical Investigations (Husserl) 129
Lotze, Rudolf Hermann 61, 126–7, 130
Luhmann, Niklas xxviii, 61, 148, 209
Lukács, Georg xix

Maimon, Salomon xxxii, xlvin88, 56–8, 65, 67, 77n59
Maimonides, Moses xxxii, xxxix, 16, 25n68, 60
"Mandarin" culture (Ringer) xxi, xxvii, 31–2, 193
Mannheim, Karl 159
Marburg school: foundation of xvii–xxi; in terms of Cohen's and Cassirer's philosophies xxviii–xxxix, xlinn13–7, xliin19–20, xlvn67, ln121; reform and philosophy of science 3–9; in Wilhelmine society and developing a metapolitics 12–15; critique of science 27–30; limitations affecting reception 30–31; summarized in *Leibniz's System* 85–6
Marck, Siegfried xxviii, 124, 194, 198
Metahistory (White) 20
metapolitics: of Cassirer and the Marburg school xxxi–xxxii, xlvin78; developed by Cohen 12–6, 24n56, 106, 108, 159, 163; developed by August von Schlözer 14–15
Meyer, Thomas xln8, 6, 85, 113n3
Meyerson, Émile xxvii, xliv, 145–6, 216
Michelangelo 19–20
Mirandola, Pico della xxxii–xxxiii

Molyneux problem 103, 117n120
monad: *see* Leibniz
Müller, Johannes 64n79

Natorp, Paul xx, xliin20, xliinn23–4, 33, 35, 119n175, 113n3
Naumann, Barbara 26n96, 190n99
negative theology xxxii, 16, 25n68, 87, 98, 164
neo-Kantianism xxxvii–xxxviii, ln121, lin124–6, 9, 22n6, 207
Newton, Isaac: calculus and Cohen's *Infinitesimal Method* 51–2, 153n66; Cassirer's examination alongside Leibniz 86–7, Cartesian natural philosophy 95; definition of space 133–5; mechanics as model of "functional form" 141–2
Nietzsche, Friedrich xvii, 27, 65, 79n107, 218n12
Nipperdey, Thomas 20, 31, 83

On Liberty (Mill) 180–81
Order of Things, The (Foucault) xxv, 149

Patton, Lydia xlivn50, xlviiin101
Paetzold, Heinz xlviin89
paradigm shift (Kuhn) 66, 142
Parsons, Talcott xxviii
philosophy of origin: *see* Cohen
Philosophy of the Enlightenment, The (Cassirer) xxiii–xxvii, 118n144, 200
Poma, Andrea xxxii, xxxvii, 26n87
positive law 108–9, 197–8, 202–3, 205–6
Preconditions of Socialism and the Mission of Social Democracy, The (Bernstein) 29
Principle of the Infinitesimal Method and Its History: A Chapter on the Foundations of the Critique of Knowledge, The (Cohen) 3, 45
Principle of the Conservation of Energy, The (Planck) 4

Renouvier, Charles 55–6, 59, 216
Renz, Ursula xxxviii, xlviin89
Repp, Kevin xxi, 32–3
Rickert, Heinrich xxxviii, xlvn67, li–liin126, 146
Ringer, Fritz xxi, 31
Romantic movement: Cohen's reading of Kant 18–19; Fechner's themes 48; reception of Leibniz 86–7; 107, 111
rule-forms 151, 159–60, 166–7
Rudolf, Enno xxxvii, xxxix, lin124, liin128
Russell, Bertrand xxxv, xlixn103–106, xlviiin101, 80n124, 114n28, 126
Ryckman, Thomas xxxvi–xxxvii

Schmitz-Rigal, Christiane xxxvi–xxxvii, xlixn110, ln115, lin124, liin126, liin128, 87, 114n21
Schweitzer, Albert 160, 211, 218n8
Schlözer, August von 14–15
Schopenhauer, Arthur 10, 19, 27, 205
Skidelsky, Edward xxxv–xxxvi, xlin16, xlviiin99, xlixn106, 42n38
Sieg, Ulrich xxxvii, 85
Star Maker (Stapledon) 137
Substance and Function (Cassirer): influence of, and Cassirer's philosophy xxv–xxviii, xlvn68, xlixn107, lin124; Cohen's philosophy of origin in xxx; addressing Cohen's philosophy xxxv–xxxvi, 22n10; process philosophy 210; group theory xlvin80; Cohen's opinion and response 34, 149–50; epistemology 68; Cassirer's basic argument 121–4; Cassirer's critique of knowledge 125; logic 124–32; chemistry 134–7; "the subject and object of knowledge" 138–40; structuralism 140–42; theory of reality 145–50, 161; perception 150–52; history of form 161–4; Cassirer's reading of Goethe's form 170–72
Substanzbegriff und Funktionsbegriff: Untersuchungen über die Grundfragen der Erkenntniskritik: see *Substance and Function*
symbolic forms 72

symbolic prägnanz: *see* Cassirer
system of experience 17
systems theory 132

theory of relativity 12, 116n91, 142, 188n16, 218n4
Third International Congress for Psychology in Munich (1896) 13
Tolstoy, Leo 48
transcendental invariants 11–12

Urpflanze: *see* Goethe, natural philosophy of
Urphänomen: *see* basis phenomenon

Verene, Donald xlvn68, xlviiin94
Vorländer, Kurt xxi, 29

Weber, Max 37, 162, 183,
Weber-Fechner Law 48–9, 76
Weimar Republic xxxiv, xxxix, xviii–xxii, 35, 193–5, 209–11
Western philosophical tradition 86, 90, 100
White, Hayden 20, 26n97, 26n99
Wilhelmine Germany: science (*Wissenschaft*) and second industrial revolution in xxi–xxii, xxxii–xxxvi, 30–33, 47; technocracy 108
Wolff, Christian 87, 110, 168, 196, 203

Zeno of Elea, paradoxes 52, 65, 93

www.ingramcontent.com/pod-product-compliance
Lightning Source LLC
Chambersburg PA
CBHW021821300426
44114CB00009BA/274